ADVANCED MATERIALS FOR NEXT-GENERATION TECHNOLOGIES

Challenges and New Prospects

ADVANCED MATERIALS FOR NEXT-GENERATION TECHNOLOGIES

Challenges and New Prospects

Edited by

Shrikaant Kulkarni, PhD

Vipul Srivastava, PhD

Rabah Khenata, PhD

Yarub Al-Douri, PhD

AAP | APPLE ACADEMIC PRESS

First edition published 2025

Apple Academic Press Inc.
1265 Goldenrod Circle, NE,
Palm Bay, FL 32905 USA

760 Laurentian Drive, Unit 19,
Burlington, ON L7N 0A4, CANADA

CRC Press
2385 NW Executive Center Drive,
Suite 320, Boca Raton FL 33431

4 Park Square, Milton Park,
Abingdon, Oxon, OX14 4RN UK

Library and Archives Canada Cataloguing in Publication

Title: Advanced materials for next-generation technologies : challenges and new prospects / edited by Shrikaant Kulkarni, PhD, Vipul Srivastava, PhD, Rabah Khenata, PhD, Yarub Al-Douri, PhD.
Names: Kulkarni, Shrikaant, editor | Srivastava, Vipul, 1973- editor. | Khenata, Rabah, editor. | Al-Douri, Yarub, editor.
Description: First edition. | Includes bibliographical references and index.
Identifiers: Canadiana (print) 20240465237 | Canadiana (ebook) 20240465296 | ISBN 9781774917800 (hardcover) | ISBN 9781774917817 (softcover) | ISBN 9781003558712 (ebook)
Subjects: LCSH: Materials—Technological innovations.
Classification: LCC TA403.6 .A38 2025 | DDC 620.11—dc23

Library of Congress Cataloging-in-Publication Data

CIP data on file with US Library of Congress

ISBN: 978-1-77491-780-0 (hbk)
ISBN: 978-1-77491-781-7 (pbk)
ISBN: 978-1-00355-871-2 (ebk)

About the Editors

Shrikaant Kulkarni, PhD

Adjunct Professor, Faculty of Business, Victorian Institute of Technology, Melbourne, Australia; Adjunct Professor, Centre of Research Outcome and Impact, Chitkara University, Punjab, India

Shrikaant Kulkarni, PhD, is currently an Adjunct Professor in the Faculty of Business at the Victorian Institute of Technology, Melbourne, Australia, as well as an Adjunct Professor at the Centre of Research Outcome and Impact, Chitkara University, Punjab, India. Dr. Kulkarni has been a senior academician and researcher for over four decades. He has delivered invited lectures and conducted sessions at national and international conferences and faculty development programs. He has guided many major and minor projects in engineering chemistry, green chemistry, nanotechnology, analytical chemistry, catalysis, chemical engineering materials, industrial organization, management, and AI. He has published over 100 research papers in national and international journals and conferences of repute. He has authored more than 50 book chapters. He has edited 25 books published by international publishers. Another 10 books are in the offing. He authored four textbooks in engineering chemistry. Dr. Kulkarni possesses MSc, MPhil, and PhD degrees in Chemistry apart from master's degrees in Economics, Business Management, and Political Science. He has expertise in materials science, green chemistry and engineering, analytical chemistry, green nanoscience, nanotechnology, AI, and computational intelligence. He has expertise in wastewater treatment, green and analytical chemistry, and advanced areas in chemical engineering and materials science. He has published two patents and applied for two more.

Vipul Srivastava, PhD

Professor of Physics, Department of Physics, School of Chemical Engineering and Physical Sciences, Lovely Professional University, Phagwara, Punjab, India

Vipul Srivastava, PhD (Physics), is a Professor of Physics in the Department of Physics, School of Chemical Engineering and Physical Sciences at Lovely

Professional University, Phagwara, Punjab, India, where he is involved in both teaching and research. His current areas of interest are materials science, condensed matter physics, electronic structure calculations, density functional theory, optoelectronics, thermoelectrics, Heusler alloys, and perovskites materials. He has a total of 21 years of research and teaching experience. He has handled three research projects funded by the Government of India. He has published over 100 research articles. His h-index is 20. He has attended more than 50 conferences in India and abroad and has organized many workshops, seminars, and conferences. He is an active reviewer for more than 30 indexed journals in Physics, such as *International Journal of Energy Research, Journal of Magnetism and Magnetic Materials, Journal of Physics and Chemistry of Solids, Energy Storage, Chemical Physics Letters, Ceramics International,* etc. In 2020, he was awarded a research excellence award and in 2021, a research appreciation award by his university. He received a Young Scientist Award from the Government of India, Department of Science and Technology, New Delhi in 2005. He earned his BSc (Physics), MSc (App. Physics), and PhD (Physics).

Rabah Khenata, PhD
Professor, University of Mustapha Stambouli, Mascara, Algeria

Rabah Khenata, PhD, is a Professor at the University of Mustapha Stambouli, Mascara, Algeria. In addition, he is Head of the Laboratory of Quantum Physics of Matter and Mathematical Modelling (LPQ3M). He was nominated in 2020 as the vice rector of the external relations at Mascara University. He was elected as a founding member of the Algerian Academy of Sciences and Technologies. He has directed or co-directed 15 master's and 16 doctorate theses with more than 20 research projects completed. He has published over 570 peer-reviewed scientific publications in international journals (with over 12500 citations and H-index of 57 on Google Scholar and 54 on ResearchGate). To date, he has participated in more than 40 national and international conferences. He has edited a research book titled *Recent Advances in Novel Materials for Future Spintronics* (2019). He was recognized by the President of the Algerian Republic (Lagouat-2011) for his research excellence and received the Scopus Prize (Algiers, 2013) for the same. His main scientific work is focused on the study of mechanical, magnetic, optoelectronic, thermodynamic, and thermoelectric properties of the crystalline materials with potential for applications in diverse sectors. He received his engineering degree in Electronics and his master's degree in Physics (Materials Science) from the University of Sidi Bel-Abbès

Djilali Liabès, Algeria. He was awarded a PhD degree at the Faculty of Sciences from the same university and was promoted to the rank of Professor.

Yarub Al-Douri, PhD

Professor, Al-Bayan University, Baghdad, Iraq; Piri Reis University, Istanbul, Turkey; University of Malaya, Kuala Lumpur, Malaysia

Yarub Al-Douri, PhD, is a Professor at Al-Bayan University, Iraq & Piri Reis University, Turkey. Dr. Al-Douri is a fellow of the European Academy of Sciences. He is one of the most renowned scientists of the Middle East, Northern Africa, and Southeast Asia, known for his contributions in renewable energy and nanotechnology. Dr. Al-Douri has a Doctorat D'etat in Materials Science. He has initiated a Nanotechnology Engineering MSc Program and Nano Computing Laboratory, the first in Malaysia and Southeast Asia, in addition to founding the Applied Materials Laboratory in Algeria. Dr. Al-Douri was a Head of the Department of Nanomaterials and Department of International Networking and Collaboration, in addition to Secretary of Department of Physics. He has received over 78 national and international prizes and awards, including the the prestigious Khalifa International Award for Date Palm and Agricultural Innovation 2024, IAAM Scientist Award from the International Association of Advanced Materials, Sweden 2022; winner of the World's Top 2% Scientist Career-Long Citation Impact by Stanford University, USA 2020; named as one of the World's Top 2% Scientists by Stanford University, USA 2024, 2023, 2022, 2021 & 2020; OeAD Award, Austria 2020; Japan Society for the Promotion of Science (JSPS) Award 2019; Asian Universities Alliance (AUA) Award 2019; and others. He has more than 32 years of experience of research, teaching, administration, editorial board management, organizing events, and handling research grants and consultations. He has more than 900 publications with renowned international publishers such as Elsevier, Springer, AIP, IOP, Taylor & Francis, and Wiley, and has accrued more than US $5.1M in research grants. Dr. Al-Douri's work has been cited many times. He has supervised many postgraduate students. He is Associate Editor of *Nano-Micro Letters* (Springer) and Editor-in-Chief of *Experimental and Theoretical NANOTECHNOLOGY* (Scopus-indexed). His research field focuses on nanotechnology, nanoelectronics, nanomaterials, renewable energy, modelling and simulation, semiconductors, and optical studies. Finally, Dr. Al-Douri is a public figure at international media in the UK, Singapore, Malaysia, Qatar, and UAE.

Contents

Contributors

Archana
Department of Physics, School of Chemical Engineering and Physical Sciences,
Lovely Professional University, Phagwara Punjab, India

L. Beldi
Laboratoire de Modélisation et Simulation en Sciences des Matériaux, Université DjillaliLiabès de Sidi Bel-Abbès, Sidi Bel-Abbès, Algérie

B. Bouhafs
Laboratoire de Modélisation et Simulation en Sciences des Matériaux, Université DjillaliLiabès de Sidi Bel-Abbès, Sidi Bel-Abbès, Algérie

Parvesh K. Deendyal
School of Physical Sciences, Jawaharlal Nehru University, New Delhi, India
Government Polytechnic for Women, Faridabad, Haryana, India

K. G. Deepa
Laboratory for Energy and Environmental Devices, Department of Physics, University of Kerala, Thiruvananthapuram, Kerala, India

Shweta Dhakla
School of Physical Sciences, Jawaharlal Nehru University, New Delhi, India

Y. Al-Douri
College of Health and Medical Techniques, Al-Bayan University, Baghdad, Iraq; Department of Mechanical Engineering, Faculty of Engineering, Piri Reis University, EflatunSk, No: 8, 34940 Tuzla, Istanbul, Turkey; Nanotechnology and Catalysis Research Centre, University of Malaya, 50603 Kuala Lumpur, Malaysia

Farzana Farooq
Department of Physics, School of Chemical Engineering and Physical Sciences,
Lovely Professional University, Punjab, India

S. Heera
Laboratory for Energy and Environmental Devices, Department of Physics, University of Kerala, Thiruvananthapuram, Kerala, India

Manish K. Kashyap
School of Physical Sciences, Jawaharlal Nehru University, New Delhi, India

Jasmeen Kaur
Department of Physics, School of Chemical Engineering and Physical Sciences,
Lovely Professional University, Punjab, India

Prabhpreet Kaur
Department of Physics, School of Chemical Engineering and Physical Sciences,
Lovely Professional University, Punjab, India

R. Khenata
Laboratoire de Physique Quantique et de Modélisation Mathématique, Université de Mascara, Algeria

Mukesh Kumar
Department of Physics, Lovely Professional University, Phagwara, India

Sarvesh Kumar
Inter-University Accelerator Centre, Aruna Asaf Ali Marg, New Delhi, India

Agnibha Das Majumdar
Department of Physics, School of Chemical Engineering and Physical Sciences, Lovely Professional University, Phagwara, Punjab
Department of Physics, Madanapalle Institute of Technology and Science, Madanapalle, Andhra Pradesh, India

Om Mishra
Department of Physics, School of Chemical Engineering and Physical Sciences, Lovely Professional University, Phagwara, Punjab

Arya Mohan
Laboratory for Energy and Environmental Devices, Department of Physics, University of Kerala, Thiruvananthapuram, Kerala, India

Neha Munjal
Department of Physics, School of Chemical Engineering and Physical Sciences, Lovely Professional University, Phagwara, Punjab

V. J. Anjana Nair
Laboratory for Energy and Environmental Devices, Department of Physics, University of Kerala, Thiruvananthapuram, Kerala, India

Prabhjot
School of Chemical Engineering and Physical Sciences, Lovely Professional University, Phagwara, Punjab, India

Sangeeta Prasher
Department of Physics, Kanya Maha Vidhayalya, Jalandar, Punjab, India

Suman Rani
Department of Physics, School of Chemical Engineering and Physical Sciences, Lovely Professional University, Punjab, India

Muzafar Salam
Department of Physics, Lovely Professional University, Phagwara, Punjab, India
Kawaljeet Singh Samra
Department of Physics, Lovely Professional University, Phagwara, Punjab, India

Riyadh A. Al-Samarai
Electromechanical Engineering Department, Engineering College, University of Samarra, Samarra, Iraq

F. Semari
Laboratoire de Physique Quantique et de Modélisation Mathématique, Université de Mascara, Algeria

Amit Kumar Sharma
Department of Physics, Lovely Professional University, Phagwara, India
Department of Physics, Maharana Pratap Govt. College Amb, Himachal Pradesh, India

Sahendra P. Sharma
School of Chemical Engineering and Physical Sciences, Lovely Professional University, Phagwara, Punjab, India
Division of Research and development, Lovely Professional University, Phagwara, Punjab, India

Vipul Srivastava
School of Chemical Engineering and Physical Sciences, Lovely Professional University Phagwara, Punjab, India
Department of Physics, School of Chemical Engineering and Physical Sciences, Lovely Professional University, Phagwara Punjab, India

Ankur Taya
Department of Physics, Mukand Lal National College, Yamunanagar, Haryana, India

Abbreviations

0D	zero-dimensional
10 H	10 layers hexagonal
1D	one-dimensional
2D	two-dimensional
3D	three-dimensional
AE	active electrode
AFM	atomic force microscopy
AFM	anti-ferromagnetic
AIHP	all-inorganic hybrid perovskite
ARPES	angle-resolved photoemission spectroscopy
ASE	amplified spontaneous emission
BQD	bismuth-based perovskite quantum dot
CASP	convolution approximation for swift particles
CB	conduction band
CBD	chemical bath deposition
CBM	conduction band minimum
CCN	charge collection narrowing
CCT	correlated color temperature
CF	conducting filament
CIGS	copper indium gallium diselenide
CIS	copper indium selenide
CMR	colossal magneto resistance
CNTs	carbon nanotubes
CRI	color rendering index
CVD	chemical vapor deposition
CW	continuous wave
DBR	distributed Bragg reflector
DFB	distributed feedback
DFT	density functional theory
DJ	Dion–Jacobson
DSPE PEG	1,2-distearoyl-sn-glycero-3-phosphoethanolamine-poly (ethylene glycol)
DSSC	dye-sensitized solar cell
ECD	electrochemical deposition

EDLC	electric double-layer capacitor
Eg	energy band gap
EIL	electron-injecting layer
EL	electroluminescence
EQE	external quantum efficiency
ERD	elastic-recoil-detection
ETL	electron transport layer
FCC	face-centered cubic
FF	fill factor
FL	fluorescence spectroscopy
FP-LAPW	full-potential linearized augmented plane wave
FTIR	Fourier-transform infrared spectroscopy
FWHM	Full-width half-maximum
GGA	generalized gradient approximation
HAXPES	hard X-ray photoelectron spectroscopy
HCP	hexagonal close-packed
HDP	halide double perovskites
HepG2	hepatoblastoma cell line
HEX	simple hexagonal structures
HHP	hybrid halide perovskite
HIL	hole-injecting layer
HMT	hexamethylene tetramine
HP	halide perovskite
HTL	hole transport layer
HTM	hole-transporting material
IHP	inorganic halide perovskite
ITO	indium tin oxide
IUAC	Inter-University Accelerator Center
KeV	kilo electron volt
LASER	light amplification by stimulated emission of radiation
LCAO	linear combination atomic orbital
LDA	local density approximations
LED	light emitting diode
LMTO	linear muffin tin orbital
LO	longitudinal optical
LSDA	local spin density approximation
LSS	Lindhard, Scharff, and Schiott
MemPV	memristive photovoltaic
MeV	mega electron volt

MIM	metal–insulator–metal
MOCVD	metal organic chemical vapor deposition
MOF	metal organic framework
MR	magnetoresistance
MT	muffin tin
NAND	not and
NBE	near-band edge
NC	nanocrystal
NCs	nanocrystals
NIR	near infrared
NMTO	nth-order muffin-tin orbital
NP	nanoparticle
NS	Northcliffe and Schilling
NW	nanowire
OER	oxygen evaluation reaction
OLED	organic light emitting diode
ORR	oxygen reduction reaction
PAR	photosynthetically active radiation
PAW	projected augmented waves
PCE	power conversion efficiency
PEC	photoelectrochemical
PECVD	plasma-enhanced chemical vapor deposition
PEG	polyethylene glycol
PeLED	perovskite light emitting diode
PeMOF	perovskite metal organic framework
PGEC	phonon glass electron crystals
PL	photoluminescence
PLD	pulsed laser deposition
PLQE	photoluminescence quantum efficiency
PLQY	photoluminescence quantum yield
PM	photomultiplier
P-MBE	plasma-assisted molecular beam epitaxy
PMMA	poly(methyl methacrylate)
PMN-PT	pb(mg1/3nb2/3)o3-pbtio3
PQD	perovskite quantum dot
PRM	perovskite-related material
PSC	perovskite solar cell
PV	photovoltaic
PVDF	polyvinylidene fluoride

QD	quantum dot
RAM	random access memory
ReRAM	resistive random-access memory
RF	radio frequency
ROM	read only memory
RP	Ruddlesden–Popper
RS	resistive switching
RT	room temperature
SAED	selected area electron diffraction
SEM	scanning electron microscopy
SOC	spin orbital coupling
SRIM	stopping and range of ions in matter
STEM	scanning transmission electron microscopy
TDPAD	time differential perturbed angular distribution
TE	thermoelectric
TEM	transmission electron microscopy
TEOS	tetraethyl orthosilicate
TGA	thermogravimetric analysis
TMAH	tetramethylammonium hydroxide
TMN	transition metal nitrides
TO	transverse optical
UC	upconversion
UV	ultraviolet
UWOC	underwater wireless optical communication
VB	valence band
VBM	valence band maximum
VCA	virtual crystal approximation
VCSEL	vertical cavity surface emitting lasers
VEC	valence electron concentration
VLS	vapor–liquid–solid
VOC	open-circuit voltage
VOC	volatile organic compound
WGM	whispering gallery mode
XRD	X-ray diffraction
ZEH	zinc 2-ethylhexanoate

Preface

The book aims to present the fundamentals as well as give deep insight into the advances in materials with special reference to structural, physical behavior, and applications, keeping in perspective the generation of new technologies for improving upon the quality of life of people and for environmental sustainability. The book explores the structure–property–utility relationship exhibited by the materials. The behavior of materials in response to stimuli is attributed to their property profile while property profile is a reflection of structure in terms of basic building blocks constituting the materials, the nature and magnitude of forces holding together the basic building units in the materials.

This book sheds light on the organizational and orientational order of the atoms responsible for getting characteristic and distinct architectures in materials. These variations in the architectures reflect upon a plethora of properties that further reflect upon the purpose for which they can be utilized.

This book also discusses how the materials can be maneuvered for attaining optimality in the structure and, in turn, provide the property profile in tune with the applications. Further advances in the novel materials for upcoming technologies attributed to their interesting and exciting properties, such as electronic band structure, band gap, half-metallicity, multi-ferroic behavior, piezoelectricity, thermoelectricity, thermodynamics, and optoelectronic behavior, are presented in the book. These properties are responsible for material applications in spintronic devices, semiconducting devices, thermoelectric generators, and sustainable sources of energy.

This book focuses on the advanced novel materials for their technological applications. It discusses protocols for the preparation of novel materials with respect to crystal structural analysis, electronic band structure, half-metallicity for spintronics, thermoelectrics for generators, optoelectronics for optical devices, sustainable energy devices, etc. It also discusses density functional theory and different empirical tools to analyze the performances of the materials apart from presenting options for enduring objectives in energy.

This book is intended for researchers, academicians and scientists in the fields of materials science and technology, computational physics, and industrial technology. etc.

The book is divided into three parts. Part I is devoted to novel materials for applications in frontier areas and presents four chapters. Chapter 1 discusses perovskites as promising materials for future technology. Chapter 2 deliberates on the multifunctionality of perovskites in creating wonders in different application areas. Chapter 3 gives an overview of the role of perovskites in upcoming technologies. Chapter 4 throws light on photovoltaic applications of perovskites.

Part II is dedicated to the synthesis protocols for the design and development of some novel materials. Chapter 5 sheds light on spinel material synthesis and its structural analysis. Chapter 6 discusses the methods of green synthesis of metal oxides for application as grow light. Chapter 7 dwells on materials processing of metals, ceramics and polymers for industrial products and their applications.

Part III is meant for exploring property profile of the materials for behavioral change. Chapter 8 gives an idea about morphological analysis and optical studies of ZnO nanostructures using different analytical tools. Chapter 9 discusses the first-principles of LDA+U prediction of electronic, magnetic, and structural characteristics of the ternary iron arsenide $CaFe_2As_2$. Chapter 10 gives a comparative view of opto-electronic behavior of some Cs-based halide double perovskites for solar cell applications. Chapter 11 gives a first principle study to determine the structural and elastic properties of metal nitride (XN; X = Ta, W) while Chapter 12 provides a comparative study of electronic energy loss and ranges of different ions in germanium.

This book is comprehensive in both theoretical understanding as well as experimental findings integrated with theory for furthering research in materials science and engineering.

PART I
Introduction of Materials for Technologies

CHAPTER 1

Perovskite Pathways: A New Way to a Sustainable Technological Future

PRABHJOT[1] and SAHENDRA P. SHARMA[1,2]

[1]*School of Chemical Engineering and Physical Sciences, Lovely Professional University, Phagwara, Punjab, India*

[2]*Division of Research and Development, Lovely Professional University, Phagwara, Punjab, India*

ABSTRACT

For a variety of optoelectronic and photonic device applications, perovskite materials have emerged as the most effective and promising low-cost energy materials. Perovskites have the potential to play a significant role in Solar Energy, Photodetection, Display, Lighting, 5G, and Wireless Communication technology. This article presents a systematic review of structural analysis, development and applications of perovskites. Particular emphasis is put-on carbon-based perovskite solar cell, their fabrication process and the opto-electronic properties of perovskite materials are described. Perovskite materials are potential for usage in solar cells because of their high photovoltaic efficiency, large absorption coefficient, and short carrier diffusion length. Perovskite materials have a high absorption coefficient that enables them to absorb a large amount of light and a high photovoltaic efficiency that allows them to convert the absorbed into electricity. Perovskite materials also feature a narrow bandgap, a high radiative recombination rate, and a high exciton binding energy, which makes them appropriate for usage in LEDs. On the other hand, perovskite-based solar cells have great potential particularly for achieving sustainable development goals by providing affordable and clean

Advanced Materials for Next-Generation Technologies: Challenges and New Prospects. Shrikaant Kulkarni, Vipul Srivastava, Rabah Khenata, & Yarub Al-Douri (Eds.)

energy. However, there is still much work to be done in this direction. A multidisciplinary approach is necessary to overcome the significant problem of lowering the toxicity of perovskite materials. To lessen the toxicity of these materials and make them more appropriate for commercial uses, various techniques are being employed, including the use of lead-free perovskites, alternative perovskite materials, encapsulation, and recycling.

1.1 INTRODUCTION

Perovskite material is the most well-known and extensively studied inorganic compound calcium titanium oxide ($CaTiO_3$). In recent years, calcium titanium oxide perovskite has also been explored as a potential material for photovoltaic applications due to its unique properties, such as its high absorption coefficient and tuneable bandgap. This material was named after the Russian mineralogist Lev Perovski, in 1839, Perovski discovered a new mineral in the Ural Mountains of Russia, which he initially named calcium titanate.[1] This mineral was later identified as a new mineral species and was named perovskite in honor of Perovski's contributions to the field of mineralogy. Perovskite materials can be made from a variety of elements, including: Lead (Pb), Tin (Sn), Cesium (Cs), Sodium (Na), Magnesium (Mg), Aluminum (Al), Calcium (Ca), Strontium (Sr), and Barium (Ba). However, in recent years, the term "perovskite" has been more commonly used to refer to a family of hybrid organic–inorganic metal halide materials with the general formula $APbX_3$, where A is an organic cation, Pb is lead, and X is a halide ion such as chloride, bromide, or iodide.[2,3] The transition from an insulator to a metal, ionic conduction behavior, dielectric properties, various solid-state phenomena, and metallic or superconducting features are only a few of the traits that perovskite materials display.[4,5] These characteristics make them very beneficial in a wide range of physics and chemistry applications.

There has been an increasing demand for high-performance piezoelectric materials that can operate at cryogenic temperatures, particularly in applications such as actuators for adaptive optics. Adaptive optics systems are used in astronomical telescopes and laser communication systems to correct for atmospheric turbulence and other optical distortions, which can severely degrade the quality of images or signals.[6,7]

At cryogenic temperatures, some perovskite piezoelectric materials such as $Pb(Mg_{1/3}Nb_{2/3})O_3$-$PbTiO_3$ (PMN-PT) have been shown to exhibit significantly improved piezoelectric properties, such as higher electromechanical coupling coefficients and larger piezoelectric coefficients. These materials are promising candidates for the development of high-performance piezoelectric actuators and other devices that can operate at low temperatures.[8] Furthermore, perovskite materials have a high exciton binding energy, which means that they require less energy to form excitons (electron–hole pairs) compared to other materials. This property is beneficial for the efficient generation of light in LEDs. The narrow bandgap of perovskite materials makes them suitable for use in LEDs that emit light in the visible region of the electromagnetic spectrum. In addition, the high radiative recombination rate of perovskites allows for efficient light emission, which is essential for the development of high-performance LEDs.

Some of the commonly studied perovskite halides include methylammonium lead halides ($MAPbX_3$, where X is a halogen), which have been widely used in solar cells due to their high-power conversion efficiency.[9] Other halides that have been investigated include cesium lead halides ($CsPbX_3$) and mixed-halide perovskites, which have shown promise in optoelectronic applications such as light-emitting diodes and photodetectors.[10,11]

The toxicity of perovskite materials is a major challenge that needs to be addressed in order to make these materials more suitable for commercial applications. There are several strategies being used to reduce the toxicity of perovskite materials. Researchers are using techniques such as molecular beam epitaxy and solution processing to produce high-quality perovskite crystals with reduced defects and improved crystal homogeneity.

1.2 STRUCTURE OF PEROVSKITE

Perovskite materials can exist in different forms or structures, depending on the composition and processing conditions. The cubic unit cell of the perovskite structure, a particular type of crystal structure, has the general formula ABX_3, where A and B are cations and X is an anion, usually oxygen. An octahedral coordination of anions surrounds the B atoms in the perovskite structure's center, which is where the A atoms are placed in the corners of the cubic unit cell. Around the B atom, the oxygen atoms create another octahedral coordination by occupying the cube's faces. The A cation is in 12-fold cub-octahedral coordination, which means that in the perovskite structure, it is coordinated to both oxygen and anions.[12] The perovskite structure has a distinctive arrangement of atoms that makes it intriguing for a variety of applications. However, the perovskite structure is also discovered for t values between 0.75 and 1.0. When this happens, the structure loses its initial symmetry and becomes tetragonal, rhombohedral, or orthorhombic. This distortion is caused by the reduced size of the A ion, which causes the BX_6 octahedra to be tilted in order to improve the A-X bonding.[13] Perovskite oxides, ABO_3, can be thought of as alternating layers of BO_2 and AO stacked one on top of the other in the direction of [001]. Another method to characterize the A and O ion packing in the ABO_3 structure is to stack close-packed AO_3 layers on top of one another, with the B cations occupying octahedral holes surrounded by oxygen.[14] The most commonly studied perovskite structures include:

Cubic perovskite: This is the basic perovskite structure, where the A, B, and X ions are arranged in a cubic lattice shown in Figure 1.1. Examples of cubic perovskites include $CaTiO_3$ and $SrTiO_3$.[15]

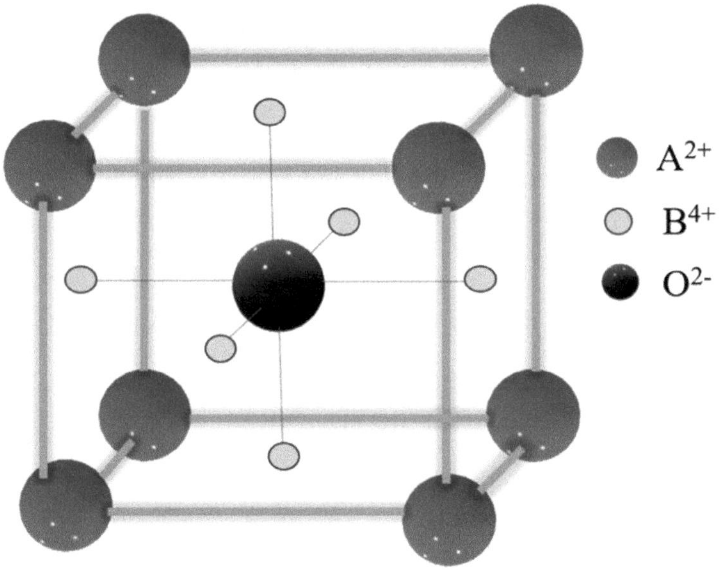

FIGURE 1.1 Illustrates the cubic perovskite.

Orthorhombic perovskite: This is a distorted form of the cubic perovskite structure shown in Figure 1.2, where the A and B ions are shifted along certain directions to form a distorted lattice. Examples of orthorhombic perovskites include $LaAlO_3$ and $LaNiO_3$.[16]

Tetragonal perovskite: This is another distorted form of the cubic perovskite structure shown in Figure 1.3, where the A and B ions are shifted along a different set of directions to form a tetragonal lattice. Examples of tetragonal perovskites include $BaTiO_3$ and $SrTiO_3$.[17]

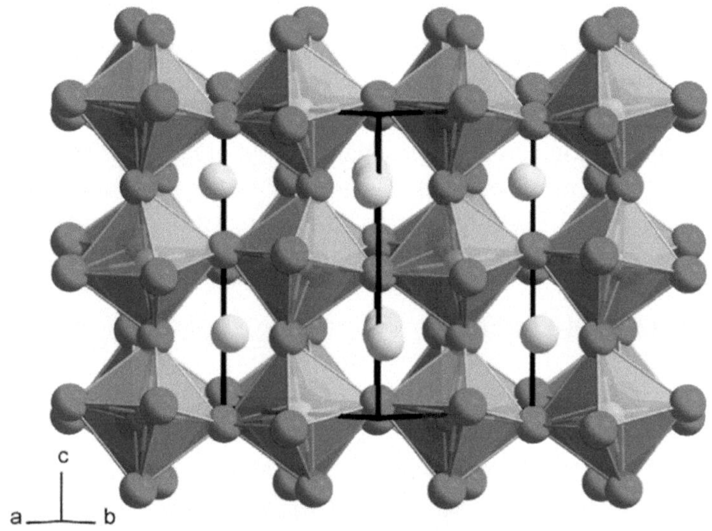

FIGURE 1.2 Illustrates the Orthorhombic perovskite.

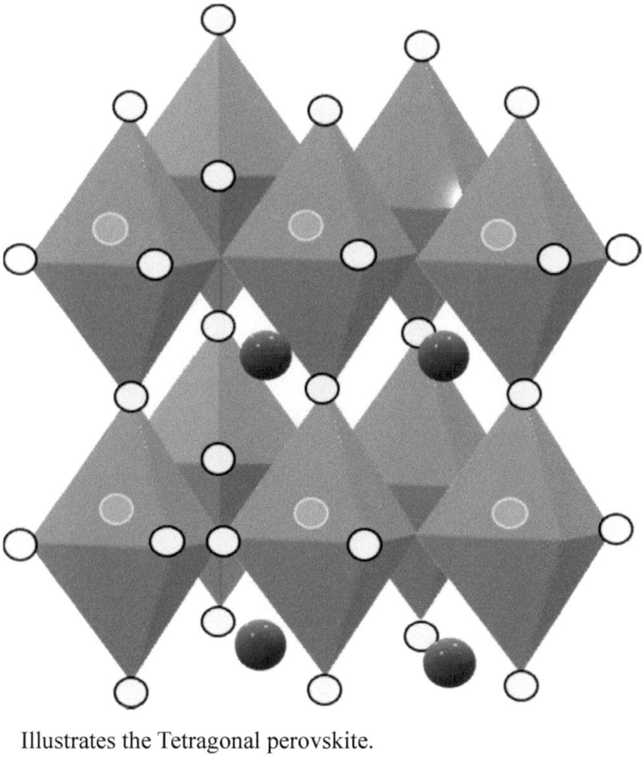

FIGURE 1.3 Illustrates the Tetragonal perovskite.

Ruddlesden-Popper perovskite: This is a layered form of the perovskite structure shown in Figure 4, where the A and B ions are arranged in layers separated by layers of X ions. Unlike conventional three-dimensional perovskites, Ruddlesden-Popper (RP) perovskites have a layered structure where the perovskite layers are separated by organic or inorganic spacer layers.[18] The general formula for Ruddlesden-Popper perovskites is $(A')_2 A_{n-1}B_nX_{3n+1}$, where A is a large cation, B is a metal cation, and n is an integer that determines the number of perovskite layers in the structure.[19] Examples of Ruddlesden-Popper perovskites include $(CH_3NH_3)_2(CH_3NH_2)Pb_2I_7$ and $(CH_3NH_3)_2(CH_3NH_2)\ Pb_2Br_7$.[20]

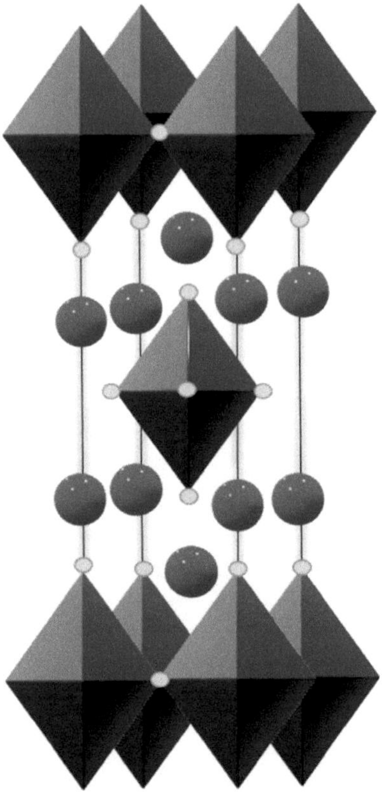

FIGURE 1.4 Illustrates the Ruddlesden-Popper perovskite.

Hybrid organic–inorganic perovskite: This is the most commonly studied form of perovskite shown in Figure 1.5, where an organic cation is incorporated into the perovskite structure. Examples of hybrid organic–inorganic perovskites include methylammonium lead iodide ($CH_3NH_3PbI_3$) and formamidinium lead iodide ($HC(NH_2)_2PbI_3$).[21,22]

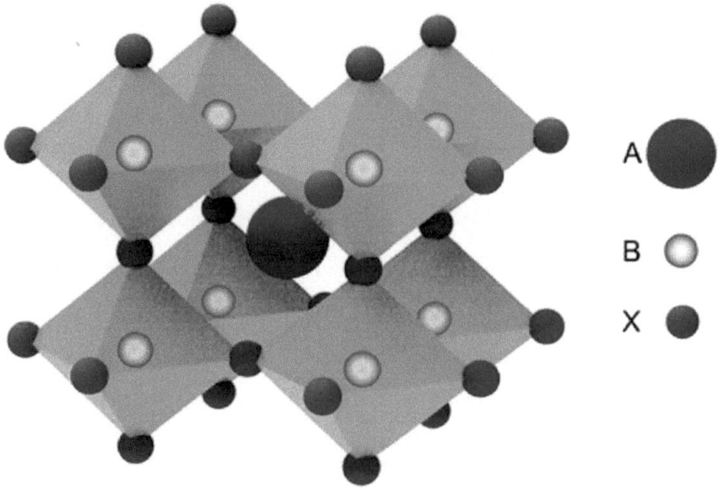

FIGURE 1.5 Illustrates the hybrid organic–inorganic perovskite.

The Goldschmidt tolerance factor is a dimensionless parameter used to predict whether a given compound will form a perovskite structure. It is calculated by dividing the sum of the ionic radii of the A and B cations in the perovskite structure by the square root of 2 times the radius of the oxygen ion and B cation. Mathematically, the formula for calculating the tolerance factor (t) is as follows:

$$t = (R_a + R_o) / (\sqrt{2}*(R_{o+} R_b)$$

where R_a, R_b, and R_0 are the ionic radii of the A cation, B cation, and oxygen ion, respectively.

When the tolerance factor is between 0.8 and 1.0, it indicates that the perovskite structure is likely to be stable and form a cubic crystal structure. Tolerance factors outside this range suggest that the compound may not form a perovskite structure and may adopt a non-perovskite structure instead.[23,24]

1.3 PROPERTIES

Perovskite materials have recently gained a lot of attention because of their outstanding characteristics discussed in Table 1.1, particularly in the field of optoelectronics and photovoltaics. Some of the exceptional properties of perovskite materials include[25]:

TABLE 1.1 Demonstrate the Significant Properties of Perovskite Material.

Property	Symbol	Typical value	References
Bandgap	E_g	1.5–2.3 eV	[32]
Carrier mobility	μ	20–100 cm²/(Vs)	[27]
Carrier lifetime	τ	1–1000 ns	[33]
Exciton binding energy	E_b	10–50 meV	[34]
Dielectric constant	ε	10–40	[35]
Young's modulus	E	10–40 GPa	[36]
Poisson's ratio	ν	0.2–0.4	[37]
Hardness	H	2–4 Mohs	[38]

High absorption coefficient: Perovskite materials have a high absorption coefficient 675 to 1400 nm, which means that they can efficiently absorb light, even with a thin layer of material.[26]

High carrier mobility: Perovskite materials have a high carrier mobility, which means that electrons and holes can move freely within the material, leading to a high charge carrier diffusion length and high device performance.[27]

Multiferroicity: Piezoelectricity is a property exhibited by certain perovskite materials where they can generate an electric charge in response to an applied mechanical stress or strain, or conversely, experience a mechanical deformation when an electric field is applied to them. Perovskite materials that exhibit multiferroicity are typically characterized by their crystal structure, which is composed of a metal oxide framework with an ABO_3 stoichiometry. The A-site cation can be an alkaline earth or rare earth metal, while the B-site cation can be a transition metal ion. The multiferroic behavior is believed to arise from the coupling between the magnetic and electric degrees of freedom within the crystal structure.[28]

Piezoelectricity: Perovskite materials that exhibit piezoelectricity typically have a polar crystal structure that lacks inversion symmetry. The most common piezoelectric perovskite materials are lead-based, such as $Pb(ZrTi)O_3$ (PZT), which is widely used in various electronic devices.[29]

Tuneable bandgap: The ability to easily tune the bandgap of perovskite materials is due to their unique crystal structure, which allows for the incorporation of a variety of elements into the material. By carefully selecting the elements used in the perovskite composition, the bandgap can be precisely tuned over a wide range of energies.[30]

Colossal magneto resistance (CMR): Perovskite materials that exhibit CMR are typically doped with transition metal ions, such as manganese or cobalt, to introduce magnetic ordering into the crystal structure. These materials are also characterized by their complex crystal structures, which can give rise to strong electron correlations that are believed to be responsible for the CMR effect.[31]

1.4 APPLICATIONS

Perovskite materials have shown great potential for various optoelectronic and photonic device applications shown in Figure 1.6 due to their outstanding electrical and optical characteristics. Here are some examples of these applications:

FIGURE 1.6 Illustrates the application of perovskite material in different fields.

Solar cells: Perovskite solar cells have attracted significant attention due to their high-power conversion efficiency (PCE), low-cost fabrication, and versatility in device architecture. Perovskite solar cells can be fabricated using solution-based methods, which makes them easily scalable for commercial production. Perovskite-based solar cells offer a realistic option for achieving sustainable development goals by providing affordable and clean energy. Additionally, perovskite materials can be tuned to absorb different parts of the solar spectrum shown in Figure 1.7, which can lead to higher PCE. The most often employed component in the production of perovskites, particularly in the advancement of perovskite solar cells, is lead (Pb). Perovskite materials made of lead have shown to be highly effective in converting light into electricity, making them a possible replacement for conventional silicon-based solar cells. The first liquid-based dye-sensitized solar cell structure using perovskite materials was discovered in 2009, with a PCE of around 3–4%.[39] However, the stability issues associated with liquid electrolytes limited the practical application of these cells. In 2012, a long-term, stable, and high-efficiency perovskite solar cell was developed by substituting the solid hole conductor with a liquid electrolyte, leading to a rapid increase in efficiency to 18% within just 2 years. PCE values of above 20% are realistically expected with the usage of low-cost organometal halide perovskite materials, further solidifying the potential of perovskite solar cells.[40]

1.4.1 CARBON-BASED PEROVSKITE SOLAR CELL

The fabrication of back electrodes using carbon-based materials such as carbon nanotubes (CNTs), graphite, graphene, and carbon black demonstrates remarkable efficiency and ensures stability better than that of gold-based devices.[42]

- *Graphene-based back electrodes:* Since this initial demonstration, several studies have reported the use of graphene as an electrode in PSCs. These studies have focused on optimizing the synthesis and transfer of graphene, as well as improving the interface between the graphene and the perovskite layer. The resulting device exhibited PCE of 12.81%, which was higher than the PCE of the control device using a conventional indium tin oxide (ITO) electrode.[43] Inverted cell architecture for organic photovoltaic devices using graphene as the transparent cathode was the subject of a 2016 study by Sung et al.[44]

MoO_3 was added to the graphene to enhance the device's performance, by achieving 17.1% power conversion efficiency. MoO_3 increased the band alignment between the graphene and the hole-transporting material (HTM) by raising the work function of the material from 4.23 to 4.71 eV. This increased the open circuit voltage (Voc) of the interface and decreased. The use of MoO_3 was also confirmed to be effective by Yoon et al.,[45] in their study, devices with graphene/MoO_3 as the back electrode showed an average PCE of 15.0%, with a Voc of 0.99 V, a short circuit current density (Jsc) of 21.0 mA cm^{-2}, and a fill factor (FF) of 0.72. Chunyang Zhang developed a revolutionary modular PSC design with a carbon back electrode whose thickness and sheet resistance were significantly reduced by pressing it with another carbon-coated FTO glass. Graphene-based PSCs (G-PSCs) were able to attain PCE of 18.65%, which was among the greatest efficiencies ever recorded for C-PSCs.[46,47]

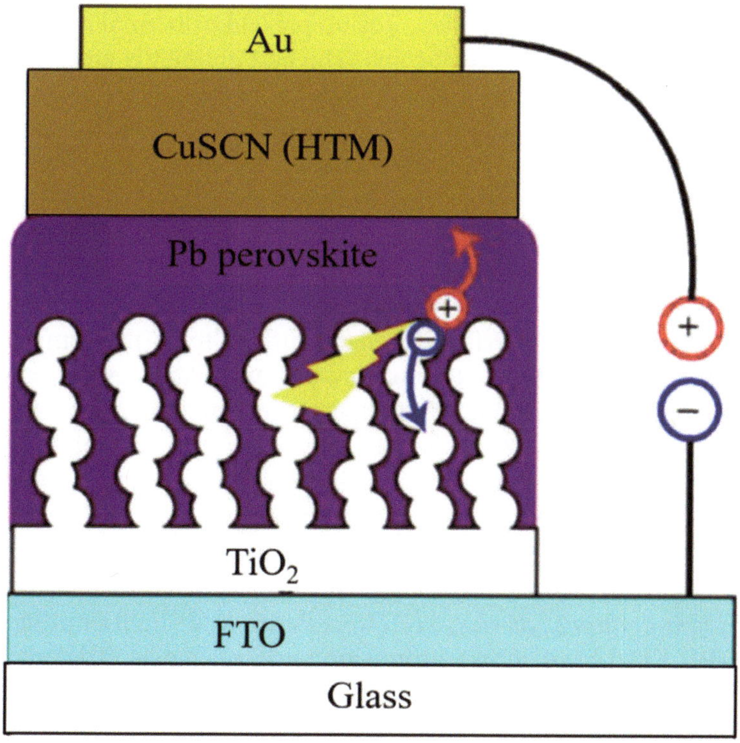

FIGURE 1.7 Illustrates the perovskite solar cell[41].

Source: Reprinted with permission from Ref. [41]. Copyright © 2014, Springer Nature Limited

- *Carbon nanotube (CNT)-based back electrodes:* Due to its distinctive one-dimensional structure, which enables effective electron transfer down the length of the tube, CNTs offer outstanding charge transport properties. They also have a high hydrophobicity, which means they repel water, making them perfect for use in electrical and optoelectronic devices that need to function reliably even when wet.[48] Furthermore, CNTs are extremely stable, withstanding high temperatures as well as chemical and mechanical stresses. After continuous illumination for 1000 h in direct sunlight, the bifacial solar cell displays great stability with only a 5% efficiency loss. A 4-T tandem with a copper indium selenide (CIS) bottom cell and the bifacial device as the top cell produced PCE of 27.1%. Findings imply the possibility of increasing the power output per unit area of perovskite solar cells using bifacial SWCNT-PSCs.[49]
- *Graphite-based back electrodes:* A thin layer of graphite is often deposited onto a substrate, like glass or a polymer sheet, to create graphite-based electrodes. Techniques like thermal evaporation or sputtering can be used for this. The end product is an electrode with a high electrical conductivity that can be employed as a transparent conductive electrode to let light reach the solar cell's active layer. Typically, conductive ink is created by distributing carbon black particles in a polymer matrix to create carbon black-based electrodes.[50] The back electrode can then be created by printing this ink onto a substrate. A new device architecture for HTM-free perovskite solar cells (PSCs) uses printed carbon–graphite back electrodes. The back-electrode interface, though, suffers from performance losses in this architecture. Despite this difficulty, the device nevertheless manages to reach one of the highest reported efficiencies for HTM-free PSCs of 18.5% while also enhancing device stability due to the use of 2D perovskite as an electron-blocking layer (EBL).[51] Graphite is less conductive as compared to other carbon materials. It can only be made conductive by adding different carbon nano-species or other conducting material layering to increase the interface with perovskite.
- *Light-emitting diodes (LEDs):* Perovskite light-emitting diodes (PeLEDs) have shown promising performance as efficient and low-cost light sources. Perovskite LEDs can emit light in a wide range of colors, and their emission can be easily tuned by adjusting the composition of the perovskite material. PeLEDs are based on similar perovskite materials used in perovskite solar cells but are designed

to emit light instead of generating electricity. The development of PeLEDs has been progressing rapidly, and high efficiencies greater than 20% have been achieved.[52] In recent years, PeLEDs based on RP perovskites have demonstrated promising performance. The promise of RP perovskites in the field of optoelectronics is shown by the considerable rise in the external quantum efficiency (EQE) of PeLEDs from 0.1 to 28%.[53] Additionally, perovskite LEDs can be fabricated using solution-based methods, which makes them easily scalable for large-scale production.[54]

- *Lasers:* Perovskite materials have been used to demonstrate efficient lasing action in the visible and near-infrared regions of the spectrum. Perovskite lasers can be fabricated using low-cost solution-based methods, and their emission can be easily tuned by adjusting the composition of the perovskite material.[55] Quantum dots, nanowires, and other reduced-dimensional perovskite structures are essential for creating micro- or nanoscale laser sources for optical devices and high-resolution imaging.[56] Due to their high photoluminescence quantum yield, fast charge carrier mobility, long carrier diffusion length, low Auger recombination rate, and large absorption coefficient, metal halides perovskite materials are proved to be good gain media for laser applications.[57,58]
- *Photodetectors:* Perovskite materials have shown great potential for photodetection applications due to their high sensitivity and low dark current. Perovskite photodetectors can be fabricated using low-cost solution-based methods, and their performance can be easily tuned by adjusting the composition of the perovskite material.[59] Perovskites with a variety of architectures, including classic 3D-2D perovskites, 1D NW(nanowire), and 0D quantum dots, have been created for the photodetector application.[60] Perovskites have a promising potential for using sensitive and quick photodetectors that are intended for image sensing, optical communication, environmental monitoring, or chemical/biological detection. They achieve a combination of strong light absorption across a wide range of wavelengths, efficient generation of photos, and a high level of mobility for charge carriers.[61] Moreover, the combination of $CsPbBr_3$ nanocrystals and $PC_{71}BM$ exhibited outstanding characteristics.[62]
- *5G and wireless communication:* Perovskite materials play a significant role in the development of 5G and wireless communication technologies due to their unique combination of high carrier mobility,

high dielectric constant, and tuneable bandgap.[63] Perovskite filters can offer lower insertion loss, higher selectivity, and better out-of-band rejection. Perovskite modulators can offer low drive voltage, high modulation depth, and low insertion loss. Perovskite photodetectors can offer high sensitivity, high speed, and low noise. The features like lifetime and efficiency can be improved by using Lead halide perovskites' materials. But the restricted use of these material results in poor stability when used under water.[64,65]

1.5 ISSUE OF THE PEROVSKITES

Toxicity: Due to its superior ability in converting sunlight into electricity, lead is a common component in perovskite solar cells. Because of their excellent power conversion efficiency and stability, lead-based perovskite materials are a possible replacement for conventional silicon-based solar cells. Lead is a hazardous substance, and its usage in the production of perovskites raises questions regarding its effects on the environment. If lead is released into the environment, it can have a major negative impact on one's health. Manufacturers must take the necessary safety procedures when producing and handling lead-based perovskite materials in order to reduce the danger of lead exposure.

Charge recombination: Charge recombination occurs when electrons and holes generated by the absorbed light in the perovskite layer recombine before reaching the electrodes, resulting in a loss of efficiency and reduced current density. This issue can be caused by several factors, including defects in the perovskite material, poor interfaces between the perovskite and charge transport layers, and inefficient charge extraction.

Poor interface: The poor interface issue in perovskite solar cells can arise from several factors, including any contamination or impurities on the surface of the perovskite layer, mismatch in energy levels and rough or uneven surface.

1.6 CONCLUSION

Perovskites have become an appealing category of materials for a range of optoelectronic and photonic technological applications because of their exceptional electrical and optical characteristics, clean and sustainable energy, low-cost fabrication, and tunability. Despite their promise, perovskite materials have certain hurdles that must be overcome before they can be extensively used in industrial applications. For example, perovskites can be unstable in the presence of moisture, oxygen, and other environments, which can limit their durability and long-term stability. Moreover, their toxicity due to the presence of lead raises safety concerns in their use. Therefore, research efforts are needed to develop nontoxic lead-free alternatives of perovskites. Researchers are actively working to address these issues to improve the performance and reliability of perovskite-based devices by inventing robust methods of producing high-quality films with high reproducibility, improved charge carrier lifetimes, and reduced trap densities. In summary, perovskites have great potential to revolutionize many aspects of modern technology by inventing new cost-effective, environment friendly methods of producing high-quality, durable, and efficient films.

KEYWORDS

- **perovskite**
- **structure**
- **properties**
- **application**

REFERENCES

1. Cheng, Z.; Lin, J. Layered Organic–Inorganic Hybrid Perovskites: Structure, Optical Properties, Film Preparation, Patterning and Templating Engineering. *CrystEngComm.* **2010,** *12* (10), 2646–2662. DOI: 10.1039/C001929A.
2. Pan, L.; Zhu, G.; *Perovskite Materials: Synthesis, Characterisation, Properties, and Applications*; IntechOpen, 2016. https://books.google.co.in/books?id=SmqQDwAAQBAJ
3. Szuromi, B.; Phillip; Grocholski. Natural and Engineered Perovskites. *Science* **2017,** *358* 732–733.
4. Yang, Y.; Sun, Y.; Jiang, Y. Structure and Photocatalytic Property of Perovskite and Perovskite-Related Compounds. *Mater. Chem. Phys. 2006, 96* (2), 234–239, 2006. DOI: https://doi.org/10.1016/j.matchemphys.2005.07.007.
5. Misra, R. K.; Cohen, B.-E.; Iagher, L; Etgar, L. Low-Dimensional Organic–Inorganic Halide Perovskite: Structure, Properties, and Applications. *ChemSusChem.* **2017,** *10* (19), 3712–3721. DOI: https://doi.org/10.1002/cssc.201701026.
6. Zhu, J. et al., Perovskite Oxides: Preparation, Characterizations, and Applications in Heterogeneous Catalysis. *ACS Catal.* **2014,** *4* (9), 2917–2940. DOI: 10.1021/cs500606g.

7. Vittayakorn, N.; Bongkarn, T; Rujijanagul, G. Phase Transition, Mechanical, Dielectric and Piezoelectric Properties of Perovskite $(Pb_{1-x}Bax)ZrO_3$ Ceramics. *Physica B: Condensed Matter* **2007,** *387* (1), 415–420. DOI: https://doi.org/10.1016/j.physb.2006.04.027.
8. Zhang, S.; Xia, R.; Shrout, T. R.; Zang, G.; Wang, J. Piezoelectric Properties in Perovskite $0.948(K_{0.5}Na_{0.5})NbO_{3-0.052}LiSbO_3$ Lead-Free Ceramics. *J. Appl.Phys.* **2006,** *100* (10), 104108. DOI: 10.1063/1.2382348.
9. Hendriks, K. H.; van Franeker, J. J.; Bruijnaers, B. J.; Anta, J. A.; Wienk, M. M.; Janssen, R. A. J. 2-Methoxyethanol as a New Solvent for Processing Methylammonium Lead Halide Perovskite Solar Cells. *J. Mater. Chem. A* **2017,** *5* (5), 2346–2354. DOI: 10.1039/C6TA09125C.
10. Nedelcu, G.; Protesescu, L.; Yakunin, S.; Bodnarchuk, M. I.; Grotevent, M. J.; Kovalenko, M. V. Fast Anion-Exchange in Highly Luminescent Nanocrystals of Cesium Lead Halide Perovskites ($CsPbX_3$, X = Cl, Br, I). *Nano Lett.* **2015,** *15* (8), 5635–5640. DOI: 10.1021/acs.nanolett.5b02404.
11. Doumbia, Y.; Bouich, A.; Soro, D.; Bernabé, M. S. Improving Stability and Performance of Cesium Mixed Lead Halides for Photovoltaic Applications. *JOM* **2023,** *75* (3), 693–700. DOI: 10.1007/s11837-022-05618-0.
12. Bradley, K. *Crystal Structure Prediction for Complex Modular Materials*; The University of Liverpool, 2017.
13. Chen, Y.; Zhang, L.; Zhang, Y.; Gao, H.; Yan, H. Large-area Perovskite Solar Cells—A Review of Recent Progress and Issues. *RSC Adv.* **2018,** *8* (19), 10489–10508. DOI: 10.1039/C8RA00384J.
14. S. P. N. Nair and Murugavel, P.; 4.03—Oxides: Their Properties and Uses. In: *Comprehensive Inorganic Chemistry II (Second Edition)*, Reedijk, J., Poeppelmeier, K., Eds.; Elsevier: Amsterdam, 2013; pp 47–72. https://doi.org/10.1016/B978-0-08-097774-4.00403-4.
15. Lebedev, A. I. Band Offsets in Heterojunctions Formed by Oxides with Cubic Perovskite Structure. *Phys. Solid State* **2014,** *56* (5), 1039–1047. DOI: 10.1134/S106378341405014X.
16. Zhou, J.-S.; Goodenough, J. B. Intrinsic Structural Distortion in Orthorhombic Perovskite Oxides. *Phys. Rev. B* **2008,** *77* (13), 132104. DOI: 10.1103/PhysRevB.77.132104.
17. Shim, S.-H.; Jeanloz, R.; Duffy, T. S. Tetragonal Structure of $CaSiO_3$ Perovskite Above 20 GPa. *Geophys. Res. Lett.* **2002,** *29* (24), 14–19. DOI: https://doi.org/10.1029/2002GL016148.
18. Chen, Y.; Sun, Y.; Peng, J.; Tang, J.; Zheng, K.; Liang, Z. 2D Ruddlesden–Popper Perovskites for Optoelectronics. *Adv. Mater. 2018*, *30* (2), 1703487. DOI: https://doi.org/10.1002/adma.201703487.
19. Xu, X.; Pan, Y.; Zhong, Y.; Ran, R.; Shao, Z. Ruddlesden–Popper Perovskites in Electrocatalysis. *Mater. Horiz.* **2020,** *7* (10), 2519–2565. DOI: 10.1039/D0MH00477D.
20. Koh, T. M.; Febriansyah, B.; Mathews, N. Ruddlesden-Popper Perovskite Solar Cells. *Chem* **2017,** *2* (3), 326–327. DOI: https://doi.org/10.1016/j.chempr.2017.02.015.
21. Gu, C.; Lee, J.-S. Flexible Hybrid Organic–Inorganic Perovskite Memory. *ACS Nano* **2016,** *10* (5), 5413–5418. DOI: 10.1021/acsnano.6b01643.
22. Berry, J. et al. Hybrid Organic–Inorganic Perovskites (HOIPs): Opportunities and Challenges. *Adv. Mater.* **2015,** *27* (35), 5102–5112. DOI: https://doi.org/10.1002/adma.201502294.
23. Li, Z.; Yang, M.; Park, J.-S.; Wei, S.-H.; Berry, J. J.; Zhu, K. Stabilizing Perovskite Structures by Tuning Tolerance Factor: Formation of Formamidinium and Cesium Lead Iodide Solid-State Alloys. *Chem. Mater.* **2016,** *28* (1), 284–292. DOI: 10.1021/acs.chemmater.5b04107.

24. Rahim, W. et al., Geometric Analysis and Formability of the Cubic A_2BX_6 Vacancy-Ordered Double Perovskite Structure. *Chem Mater.* **2020,** *32* (22), 9573–9583. DOI: 10.1021/acs.chemmater.0c02806.
25. Kreisel, G. j. j.; Glazer, J.; Jones, A. M.; Thomas, G.; Abello, P. A.; Lucazeau, L. An X-ray Diffraction and Raman Spectroscopy Investigation of A-Site Substituted Perovskite Compounds. *J. Phys.: Condensed Matter* **2000,** *12*, 3267. DOI: 10.1088/0953-8984/12/14/305.
26. Barugkin, C. et al., Ultralow Absorption Coe ffi cient and Temperature Dependence of Radiative Recombination of CH_3 NH_3 PbI_3 Perovskite from Photoluminescence, 2015. DOI: 10.1021/acs.jpclett.5b00044.
27. Mukhokosi, E.P.; Maaza, M. Influence of Device Architectures and Mobility on Response/Recovery Time of Metal Halide Perovskites: A Review. *J. Mater. Sci.* **2022,** *57* (3), 1555–1580. DOI: 10.1007/s10853-021-06678-4.
28. Bhatti, H. S.; Hussain, S. T.; Khan, F. A.; Hussain, S. Synthesis and Induced Multiferroicity of Perovskite $PbTiO_3$; A Review. *Appl. Surface Sci.* **2016,** *367*, 291–306. DOI: https://doi.org/10.1016/j.apsusc.2016.01.164.
29. Paik, D.-S.; Park, S.-E.; Shrout, T. R.; Hackenberger, W. Dielectric and Piezoelectric Properties of Perovskite Materials at Cryogenic Temperatures. *J. Mater. Sci.* **1999,** *34* (3), 469–473. DOI: 10.1023/A:1004578225228.
30. Li, H.; Zhang, W. Perovskite Tandem Solar Cells: From Fundamentals to Commercial Deployment. *Chem. Rev.* **2020,** *120*. DOI: 10.1021/acs.chemrev.9b00780.
31. Raveau, B.; Maignan, A.; Martin, C.; Hervieu, M. Colossal Magnetoresistance Manganite Perovskites: Relations between Crystal Chemistry and Properties. *Chem. Mater.* **1998,** *10* (10), 2641–2652. DOI: 10.1021/cm9801791.
32. Farooq, U. et al., Bandgap Engineering of Lead-Free Ternary Halide Perovskites for Photovoltaics and Beyond: Recent Progress and Future Prospects. *Nano Energy* **2022,** *92*, 106710. DOI: https://doi.org/10.1016/j.nanoen.2021.106710.
33. Shi, R.; Fang, Q.; Vasenko, A. S.; Long, R.; Fang, W.-H.; Prezhdo, O. V. Structural Disorder in Higher-Temperature Phases Increases Charge Carrier Lifetimes in Metal Halide Perovskites. *J. Am. Chem. Soc.* **2022,** *144* (41), 19137–19149. DOI: 10.1021/jacs.2c08627.
34. Ruf, F. et al. Temperature-Dependent Studies of Exciton Binding Energy and Phase-Transition Suppression in (Cs, FA, MA)Pb$(I, Br)_3$ Perovskites. *APL Mater.* **2019,** *7* (3), 31113. DOI: 10.1063/1.5083792.
35. Su, R. et al. Dielectric Screening in Perovskite Photovoltaics. *Nat. Commun.* **2021,** *12* (1), 2479. DOI: 10.1038/s41467-021-22783-z.
36. Feng, J. Mechanical Properties of Hybrid Organic-Inorganic $CH_3NH_3BX_3$ (B = Sn, Pb; X = Br, I) Perovskites for Solar Cell Absorbers. *APL Mater.* **2014,** *2* (8), 81801. DOI: 10.1063/1.4885256.
37. Huang, C. W. et al. Abnormal Poisson's Ratio and Linear Compressibility in Perovskite Materials. *Adv. Mater.* **2012,** *24* (3)0, 4170–**4174,** 2012. DOI: https://doi.org/10.1002/adma.201104676.
38. Spina, M. et al. Mechanical Signatures of Degradation of the Photovoltaic Perovskite $CH_3NH_3PbI_3$ Upon Water Vapor Exposure. *Appl. Phys. Lett.* **2017,** *110*, 121903. DOI: 10.1063/1.4978687.
39. Park, N. Perovskite Solar Cells : An Emerging Photovoltaic Technology. *Biochem. Pharmacol.* **2014,** 1–8. DOI: 10.1016/j.mattod.2014.07.007.

40. Zuo, C.; Bolink, H. J.; Han, H.; Huang, J.; Cahen, D. Advances in Perovskite Solar Cells. *Adv. Sci.* **2016,** 1–16, 2016. DOI: 10.1002/advs.201500324.
41. Qin, P. et al., Inorganic Hole Conductor-Based Lead Halide Perovskite Solar Cells with 12.4% Conversion Efficiency. *Nat. Commun.* **2014,** 1–6. DOI: 10.1038/ncomms4834.
42. Fagiolari, L.; Bella, F. Carbon-Based Materials for Stable, Cheaper and Large-Scale Processable Perovskite Solar Cells. *Energy Environ. Sci.* **2019,** 3437–3472. DOI: 10.1039/c9ee02115a.
43. You, P.; Liu, Z.; Tai, Q.; Liu, S.; Yan, F. Efficient Semitransparent Perovskite Solar Cells with Graphene Electrodes. *Adv. Mater.* **2015,** *27* (24), 3632–3638. DOI: https://doi.org/10.1002/adma.201501145.
44. Sung, H. et al. Transparent Conductive Oxide-Free Graphene-Based Perovskite Solar Cells with Over 17% Efficiency. *Adv. Energy Mater.* **2016,** *6* (3), 1501873. DOI: https://doi.org/10.1002/aenm.201501873.
45. Yoon, J. et al. Superflexible{,} High-Efficiency Perovskite Solar Cells Utilizing Graphene Electrodes: Towards Future Foldable Power Sources. *Energy Environ. Sci.* **2017,** *10* (1), 337–345. DOI: 10.1039/C6EE02650H.
46. Zhang, C. et al. Efficient Stable Graphene-Based Perovskite Solar Cells with High Flexibility in Device Assembling via Modular Architecture Design. *Energy Environ. Sci.* **2019,** *12* (12), 3585–3594. DOI: 10.1039/C9EE02391G.
47. Suragtkhuu, S. et al., Graphene-Like Monoelemental 2D Materials for Perovskite Solar Cells. *Adv. Energy Mater.* **2023,** *13* (12), 2204074. DOI: https://doi.org/10.1002/aenm.202204074.
48. Chen, R. et al. Rear Electrode Materials for Perovskite Solar Cells. *Adv. Funct. Mater.* **2022,** *32* (26), 2200651. DOI: https://doi.org/10.1002/adfm.202200651.
49. Zhang, C. et al. CNT-Based Bifacial Perovskite Solar Cells Toward Highly Efficient 4-Terminal Tandem Photovoltaics. *Energy Environ. Sci.* **2022,** *15* (4), 1536–1544. DOI: 10.1039/D1EE04008A.
50. Mo, Y.; Deng, X.; Liu, P.; Guo, J.; Wang, W.; Li, G. Insights into the Application of Carbon Materials in Heterojunction Solar Cells. *Mater. Sci. Eng.: R: Rep.* **2023,** *152*, 100711. DOI: https://doi.org/10.1016/j.mser.2022.100711.
51. Zouhair, S. et al., Employing 2D-Perovskite as an Electron Blocking Layer in Highly Efficient (18.5%) Perovskite Solar Cells with Printable Low Temperature Carbon Electrode. *Adv. Energy Mater.* **2022,** *12* (21), 2200837. DOI: https://doi.org/10.1002/aenm.202200837.
52. Liu, X.-K. et al. Metal Halide Perovskites for Light-Emitting Diodes. *Nat. Mater.* **2021,** *20* (1), 10–21. DOI: 10.1038/s41563-020-0784-7.
53. Chen, W.; Qin, Z.; Neogi, A.; Wang, Z. Composition and Structure Regulation of Ruddlesden–Popper Perovskite for Light-Emitting Diodes Applications. *J. Mater. Chem. C* **2023,** *11* (10), 3448–3458. DOI: 10.1039/D2TC05089G.
54. Lin, K. et al., Perovskite Light-Emitting Diodes with External Quantum Efficiency Exceeding 20 Per Cent. *Nature* **2018,** *562* (7726), 245–248. DOI: 10.1038/s41586-018-0575-3.
55. Stylianakis, M. M.; Maksudov, T.; Panagiotopoulos, A.; Kakavelakis, G.; Petridis, K. Inorganic and Hybrid Perovskite Based Laser Devices: A Review. *Materials* **2019,** *12* (6). DOI: 10.3390/ma12060859.
56. Zhang, Q. et al. Advances in Small Perovskite-Based Lasers. *Small Methods* **2017,** *1* (9), 1700163. DOI: https://doi.org/10.1002/smtd.201700163.

57. Kao, T. S.; Hong, Y.-H.; Hong, K.-B.; Lu, T.-C. Perovskite Random Lasers: A Tunable Coherent Light Source for Emerging Applications. *Nanotechnology* **2021,** *32* (28), 282001. DOI: 10.1088/1361-6528/abe907.
58. Lei, L.; Dong, Q.; Gundogdu, K.; So, F. Metal Halide Perovskites for Laser Applications. *Adv. Funct. Mater.* **2021,** *31* (16), 2010144. DOI: https://doi.org/10.1002/adfm.202010144.
59. Zhang, Y. et al., Lead-Free Perovskite Photodetectors: Progress, Challenges, and Opportunities. *Adv. Mater.* **2021,** *33* (26), 2006691. DOI: https://doi.org/10.1002/adma.202006691.
60. Gu, H.; Chen, S.-C.; Zheng, Q. Emerging Perovskite Materials with Different Nanostructures for Photodetectors. *Adv. Opt. Mater.* **2021,** *9* (5), 2001637. DOI: https://doi.org/10.1002/adom.202001637.
61. Wang, H.; Kim, D. H. Perovskite-Based Photodetectors: Materials and Devices. *Chem. Soc. Rev.* **2017,** *46* (17), 5204–5236. DOI: 10.1039/C6CS00896H.
62. Mei, F. et al. Recent Progress in Perovskite-Based Photodetectors: The Design of Materials and Structures. *Adv. Phys.: X* **2019,** *4* (1), 1592709. DOI: 10.1080/23746149.2019.1592709.
63. Kang, C. H. et al. All-Inorganic Halide-Perovskite Polymer-Fiber-Photodetector for High-Speed Optical Wireless Communication. *Opt. Express* **2022,** *30* (6), 9823–9840. DOI: 10.1364/OE.452370.
64. Xia, M. et al., Ultrastable Perovskite Nanocrystals in All-Inorganic Transparent Matrix for High-Speed Underwater Wireless Optical Communication. *Adv. Opt. Mater.* **2021,** *9* (12), 2002239. DOI: https://doi.org/10.1002/adom.202002239.
65. Shan, Q. et al. Perovskite Light-Emitting/Detecting Bifunctional Fibres for Wearable LiFi Communication. *Light: Sci. App.* **2020,** *9* (1), 163. DOI: 10.1038/s41377-020-00402-8.

CHAPTER 2

Perovskites as a Multifunctional Wonder Material: Categories and Applications

V. J. ANJANA NAIR, S. HEERA, ARYA MOHAN, and K. G. DEEPA

Laboratory for Energy and Environmental Devices, Department of Physics, University of Kerala, Thiruvananthapuram, Kerala, India

ABSTRACT

The chapter discusses the much-attracted perovskites and their variants, which have applications in versatile fields. Perovskites, with their outstanding features such as long-range ambipolar charge transport, high dielectric constant, high absorption coefficient, low exciton-binding energy, and adaptable ferroelectric properties, have revolutionized various aspects of materials science. The perovskites have a generic formula of ABX_3 where A and B are cations. Based on X, the material is divided into oxide perovskite or halide perovskite. Later, more stable, defect variant complex perovskites with formulas $A_2BB'X_6$, $AA'B_2X_{6,}$ and $AA'BB'X_6$ emerged. According to the Goldschmidt tolerance factor, the perovskite structures are stable only if the value is between 0.85 and 1. Ideally, the structure of perovskite is cubic with six-fold BX_6 octahedra which expands, contracts, or tilts according to the size of the A-site cation. Distortion in the octahedra results in the formation of tetragonal, orthorhombic, and hexagonal structures. Perovskites have a surprisingly high flexible structure which offers higher tolerance to structural and compositional changes. The structure can lodge broad varieties of cations having diverse oxidation states and anions vacancies which gives rise to many attractive properties of the material.

While oxide perovskites are mainly used in the areas of photocatalysis, electrocatalysis, LED, LASER, memristive devices, superconductivity,

Advanced Materials for Next-Generation Technologies: Challenges and New Prospects. Shrikaant Kulkarni, Vipul Srivastava, Rabah Khenata, & Yarub Al-Douri (Eds.)

ferroelectricity, and bio-imaging, halide perovskites are used in solar cells, photodetectors, X-ray imaging, gas sensors, etc. Oxide perovskites are wide bandgap semiconductors having a bandgap varied from 2.1 to 4.7 eV. They are the most abundant and naturally occurring type. On the other hand, halide perovskite is utilized in solar cells due to its superior characteristics such as prolonged carrier diffusion length, low-temperature deposition, ultra-high minority carrier lifetime, and extremely high optical absorption coefficient. In photovoltaic applications, organic-inorganic hybrid perovskites are exploited as absorber layers, methylammonium (MA^+), or formamidinium (FA^+) cations occupying the A site. The cations used in this perovskite are highly hygroscopic which demands the inert atmospheric condition for the preparation and characterization. Layered hybrid perovskites have recently been identified as a feasible solution for this issue which acts as an organic spacer and impedes the ion migration and thereby degradation of the organic cations. The interfacial charge transport across the layered perovskite needs to be improved in order to achieve higher efficiency. Another class of oxide perovskites known as hexagonal perovskites exhibit the properties of ionic conduction and they have several advantages in the power generation application apart from being used as sensors and catalysts. These oxide ion conductors have the capability to operate on different fuels with low emission of pollutants resulting in high energy conversion rates. In the meantime, zero-dimensional (0D) perovskite behaves similarly to a molecule having properties such as low mobility and electrical conductivity together with substantial polaron binding energy. Photoluminescence from 0D perovskite with color purity attracts special interest in LEDs. The chapter also enumerates the applications of perovskite materials in solar cells, photodetectors, photocatalysis, gas sensing, memristive devices, bio-imaging, LASER, and LED. The challenges in the stability and commercialization of the devices are mentioned briefly.

2.1 INTRODUCTION

As an obscure mineral obtained from the Akhmatovskaya Kop (Akhmatov's Pit) in the Ural Mountains, perovskite revolutionized itself as a versatile functional material by traveling for more than 180 years. The Prussian mineralogist Gustav Rose discovered perovskite in 1839 which was sent by his friend August Alexander Kämmerer, a mineral collector who was aware of its potential in scientific society. The name "perovskite" was given to the mineral in honor of Napoleonic Wars veteran and an eminent mineralogist Count Lev A. Perovskiy.[1] Initially the name perovskite was coined for calcium titanate mineral ($CaTiO_3$); later it began to be used for all crystalline structure of ABX_3 ionic compounds where A can be an alkaline earth metal/rare earth metal; B can be a transition metal with partially filled d-orbitals; and X can be oxides, halides, sulfides, chlorides, or nitrides.[2]

Structure: The ideal structure of perovskite is cubic (space group Pm-3m) and consists of a 12-fold coordinated A cation in the middle of the cube formed by eight corner-sharing [BX_6] octahedra. The B atom has six-fold coordination with the X atom where the X atom is coordinated to four A-site cations and two B-site cations. The A-site cation is located at the corners of the cube in (0, 0, 0) position although the oxygen atoms are at the position (½, ½, 0), that is, at the face center. The B site occupies the body-centered position (½, ½, ½) which lies within the oxygen octahedral.

The structure as shown in Figure 2.1 can be portrayed as a three-dimensional (3D) matrix of regular corner-connected BO_6 octahedral. There are three types of perovskite materials: the first consists of localized electrons, the second consists of delocalized energy band states, and the third can be a transition between these two types. The perovskite structures exist in a variety of forms including ABO_3-perovskite, $A_2BB'O_6$ double perovskite, $A_2A'B_2B'O_9$ triple perovskite, A_2BO_4 layered perovskite, etc.[3]

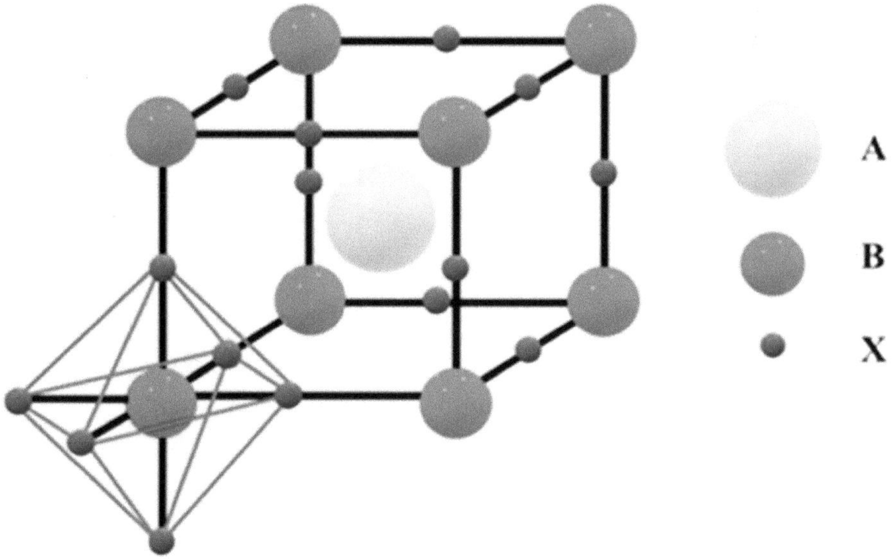

FIGURE 2.1 Structure of ABX_3 perovskite.

Ideally the perovskite structure is inflexible since there is no adjustable atomic position parameter. So, any compositional modification is addressed by the change in lattice parameter. Usually, the mineral perovskite gets slightly distorted, so we could rarely find the perovskite in its ideal structure. The distortion of BX_6 octahedral units, the displacement of B site cation within the octahedra, and BX_6 octahedral tilting are the common types of distortion found in perovskite. The suppleness of the corner-connected BX_6 octahedra network is owing to the structural and compositional flexibility of perovskites.[4–6]

According to Goldschmidt's empirical approach, perovskite is considered as a group of rigid spheres having ionic radii r_A, r_B, and r_X, which determine their sizes. The two dimensionless quantities tolerance factor t and octahedral factor μ are determined by their ionic radii.

$$\text{the tolerance factor } t = \frac{(r_A + r_X)}{\sqrt{2}(r_B + r_X)} \quad (2.1)$$

and

$$\text{the octahedral factor } \mu = \frac{r_B}{r_X} \quad (2.2)$$

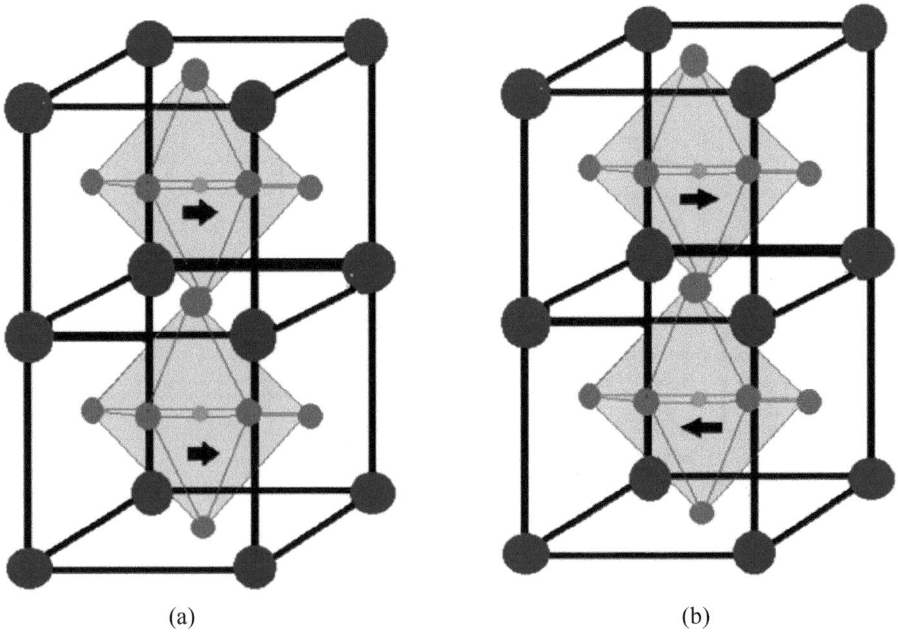

(a) (b)

FIGURE 2.2 Octahedral tilting of perovskite: (a) In-phase tilting and (b) Out-of- phase tilting.

In agreement with "no rattling" hypothesis, the maximum possible number of anions should surround a cation so that all anions are in contact with the cation. For a perovskite, the range of t and μ values are restricted by the "no rattling" principle.[7] The value of the tolerance factor should be close to 1 for the formation of perovskite structure. If the t value lies in the range of 0.9–1.0, a cubic perovskite gets formed. If $t > 1$, the hexagonal packing of the AX_3 layer results, because the A cation is large and the B cation is small. When t lies between 0.71–0.9, the structure distorts to a lower symmetry than cubic which can be a cuboctahedral coordination polyhedron. For chlorides or sulphides, when the t value is near 0.8–0.9, cubic or distorted cubic structure forms, and for t greater than 0.9, hexagonal perovskites form. Now, it is common to use the bond length in place of ionic radii to get an observed tolerance factor which however depends on the anion present in the perovskite structure. The usage of ionic radii is favorable for oxides, reasonable for fluoride, and incompatible with chlorides and sulphides.[8] The octahedral component also contributes to the prediction of the formation of cubic perovskites. The greatest r_B/r_O value is 0.704 for cubic perovskites, and the lowest limit of the octahedral factor for their synthesis is 0.433. Generally, the well-known range of r_B/r_X values for octahedron BX_6 ranges from 0.414 to 0.732. The octahedral factor is an indispensable, but not an adequate, prerequisite for the production of the cubic perovskite structure.[9]

Occasionally, the cation in a perovskite structure simply displaces away from the octahedron's center, because the size of the cation is too small for an octahedral site. It causes the tolerance factor to fall below 1. The trigonal, tetragonal, or orthorhombic structures are developed depending on the magnitude and direction of cation displacement. These displacements give rise to permanent electric dipoles in the unit cell which bring about the ferroelectric, pyroelectric, and antiferroelectric effects. Octahedral tilt or rotation mentioned in Figure 2.2 is typically correlated with A cations which are too tiny for the cuboctahedral cage site. The BX_6 octahedra twist effectively diminishes the cavity dimensions, permitting the structure to once more support values of $t < 1$. Similar to cation displacement, rotation also reduces the crystal's symmetry and has a significant impact on the physical characteristics of these phases.

The BX_6 octahedron is intrinsically capable of distortion, resulting in elongated or flattened octahedra, which in extreme circumstances can result in square planar or square pyramidal coordination. The Jahn–Teller effect, which is a manifestation of the interplay of cation electron orbitals with the surrounding anions, causes these distortions. Cation valence changes including disproportionation can also lead to octahedral deformation.[8]

Properties: Due to the distinctive chemical nature, including non-stoichiometry of the anions and/or cations, the valence mixing electronic structure, the change in cation configuration, and the mixed valence, perovskite materials display a variety of fascinating characteristics. The perovskite exhibits optical properties, colossal magneto-resistance (CMR), dielectric properties, catalytic activity, multiferroicity, etc., as shown in Figure 2.3, since the partial substitution of A-site and B-site cation is possible.

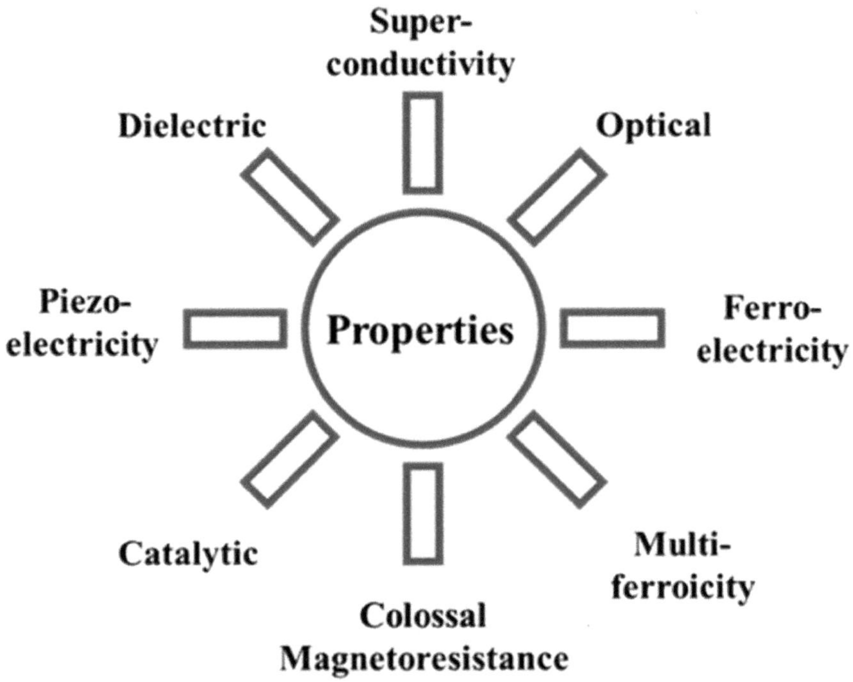

FIGURE 2.3 Properties of perovskite.

Dielectric properties: Dielectric materials are those materials where the electrostatic fields last over a very long time. With the effect of applied direct current voltage, it displayed a strong resistance to electric current and diverged drastically from conductive materials in terms of their basic electrical properties. The materials have applications in capacitors, resonators, switches, etc. High dielectric constant (κ) and low dielectric loss (tan δ) are essential design requirements for modest devices, but it's just as crucial to maintain constant fidelity for the components manufactured under various operational circumstances. Even though there have been many studies done in the past, there is still much to be done to attain higher κ (>1000) and lower temperature coefficient (100 ppm/K) together in a dielectric material. This results from the fast temperature-dependent shift in the dielectric permittivity when it approaches the Curie temperature (T_C).[3]

In $BaTiO_3$, as the temperature rises from the high temperature cubic form, the TiO_6 octahedra are deformed, changing its lattice's crystallographic dimensions. Since the distorted octahedra are interconnected, there is a substantial amount of spontaneous polarization, which causes a substantial temperature and dielectric constant dependency.[10] A group of dielectric materials mentioned as relaxor ferroelectrics, simply relaxors, are acquired with complex perovskites having general formula Pb $(M_xM'_{1-x})$ O_3. They sustain higher κ upon a broad temperature scale due to compositional heterogeneity arising from octahedral cation distribution disorder. Oxynitride perovskite framed from the partial replacement of O^{-2} with N^{-3} shows interesting dielectric properties due to the changed bonding and structure of partially substituted perovskite.[11] Pb $(Zr,Ti)O_3$ (PZT) and $Pb(Mg_{1/3}Nb_{2/3})$ O_3 (PMN) are the commonly used dielectric ceramics which are perovskite ferroelectrics based on lead or relaxor oxides. $SrTaO_2N$ ceramic samples from the oxynitride family were proclaimed to have substantial dielectric permittivity of about *ca.* 2900 with modest temperature dependency across 180–300 K.[12]

Optical Property: Perovskite has a wide range of applications, especially in LEDs, solar cells, and photodetectors. They possess a large absorption coefficient in the order of 10^5 cm^{-1}. Due to the same reason, the thickness of the perovskite layer can be reduced thus making it more economical. According to the calculations by Shockley and Queisser, the maximum solar energy conversion efficiency utilizing single-junction solar cells is almost 33% at an absorber energy bandgap of 1.4 eV. However, its practical achievability is limited by defects, impurities, interfacial barrier, etc. Perovskite bandgaps are easily tunable by compositional variations of A, B, and X ions. The bandgap is also affected by the symmetry of the perovskite structure. Perovskite quantum dots (QDs) display distinctive optical features by making use of defect tolerance, despite the system's numerous fundamental structural defects. As opposed to the other compounds, these defects don't function as traps. Recent years have seen reports of almost every variety of perovskite QD, including all-inorganic and organic-inorganic hybrid materials. Numerous studies reveal the potential of two-dimensional (2D), 3D, and hybrid 2D/3D structures.[13]

Multiferroicity: Single-phase materials possessing more than one primary ferroic order characteristics, such as ferroelectricity, ferromagnetism, and ferroelasticity, are referred to as multiferroics. After discovering the magnetic control polarization in $TbMnO_3$ (TMO) and massive polarization in $BiFeO_3$ (BFO), research on materials that exhibit both magnetism and ferroelectricity simultaneously has gained increased attention for their intriguing physics and possible uses as memory devices. Soft mode theory has long established the root of ferroelectricity in perovskites: polar-phonon mode at the lowest frequency softens on cooling and eventually resorts to zero frequency at T_c, where T_c is the critical temperature at which material changes from a perfect cubic to a lower symmetrical ferroelectric structure. The ferroelectric state arises in perovskite due to both A-site and B-site cation, the long-range coulomb interaction reducing the A–O or B–O distance is the prime supportive force behind ferroelectric characteristics. In contrast to ferroelectricity, quantum-mechanical exchange interactions, such as the double exchange interaction and the superexchange interaction, govern magnetism. In perovskite materials, the ferromagnetic ordering increases the conductivity and impairs the ferroelectricity because long-range coulomb interactions get screened in such perovskites. The perovskites possessing superexchange interaction rather than the double exchange interaction are commonly insulators so they are ferromagnetic and ferroelectric simultaneously. The BO_6 octahedra strongly affects the multiferroic property of the perovskite.[14]

Many multiferroics, such as ferrites and rare-earth manganite, are examples of transition metal oxides with perovskite structures. Multiferroicity can be seen in these materials even at room temperature. Across a large temperature range, significantly higher than room temperature, perovskite compound-bismuth ferrite, which has rhombohedrally deformed perovskite structure (compounds with multiferroics property), exhibits antiferromagnetic behavior in addition to ferroelectric order.[3]

Colossal Magnetoresistance: The property of some materials in which the resistance gets reduced on applying a magnetic field is known as colossal magnetoresistance (MR). With the discovery of significant MR among these materials, manganese oxides having a perovskite structure emerged as the focus of numerous researches. These manganites, having the generic formula $Ln_{1-x}A_xMnO_3$ (A = Ca, Sr, Pb, and Ba), show a resistance drop when a magnetic field is applied close to their T_C, resulting in MR, which is typically characterized as $MR = R/R_0 = (R_H - R_0)/R_0$ to be negative. Subsequent studies have revealed that, for these oxides, MR tends toward 100%; therefore, the effect's amplitude is preferably demonstrated by RR ratio, where it can reach high values of up to 10^{11}. As a result of this, these substances are mentioned as colossal MR (CMR) manganites. The mechanism of CMR effect is the double exchange between Mn^{3+} and Mn^{4+} species.[15] Different magnetic phases are seen depending on the orbital occupancy of the manganese ions and the corresponding orbital order. In these compounds, due to the close coupling between the orbitals and spins, ordering temperatures for both degrees of freedom are identical in magnitude. As correlated electron systems with interactions involving the lattice spin, the Jahn–Taller effect, charge and orbital degrees of freedom, electronic phase separation, charge ordering, etc., manganites are closely connected to the CMR effect.[3]

Catalytic Activity: In addition to the strong catalytic activity, perovskite structures demonstrated high stability there by permitting compound formation from elements having unusual valence states or with a significant lack of oxygen. They are able to function as catalysts in numerous environmental catalytic reactions such as cleaning catalyst, intelligent automobile catalysts, and motor exhaust gas catalyst. Due to oxygen deficiency and the direct removal of the surface as a reaction product, several perovskite materials (containing Cu, Co, Mn, or Fe) demonstrated good catalytic activity for the direct decay of NO at higher temperatures.[3]

A fundamental reason behind the perovskite catalytic activity is metal-oxygen bond which grounds the tuning of the perovskite electronic structure and leads to the controlling of surface-binding energetics. In perovskite, the s and p orbitals are formed as a result of the hybridization of transition metal d and oxygen 2p atomic orbitals. The s* and p* orbitals are otherwise called e_g and t_{2g} orbitals in which the s* orbital has high energy and is unstable. Crystal field stabilization gives the degree of stabilization which reflects the symmetry of atomic orbital overlap. Apical oxygen is removed at the surface of perovskite, so the exposed B site has BO_5 coordination environment which further causes the splitting of e_g and t_{2g} states into distinct energy levels. The chemistry behind surface reactivity and catalytic activity is highly influenced by these orbitals in perovskites. The e_g orbital of B-site ion (transition metal) in BO_5 coordination detains surface absorbate because of their vertical orientation in relation to surface-bound intermediates.[16]

Other Properties—Ferroelectricity, Piezoelectricity, and Superconductivity

The phenomenon of spontaneous electric polarization on applying an external electric field is called ferroelectricity. For a ferroelectric material, dielectric constant is two times larger than the ordinary dielectric material. The discovery of ferroelectricity in $BaTiO_3$ triggered the research in this area.

The material property in which the materials have zero resistance and magnetic field expulsion while cooling under a specific temperature is called superconductivity. In perovskite materials, the superconductivity is first discovered in La-Ba-Cu-O. The perovskites having Cu act as high-temperature superconductors.

The phenomenon of producing electric charge in response to applied mechanical stress is known as piezoelectricity. Piezoelectric ceramics with perovskite crystal structures having general formula $A^{2+}B^{4+}O^{2-}{}_{3}$ is an example of synthetic piezoelectric material. Several useful scientific applications of the piezoelectricity trait include high-voltage power sources, sensors, pick-ups, microphones, force sensors, piezoelectric motors, actuators, loud-speakers, valves, nanopositioning in AFM and STM, energy harvesting, etc.[3]

The perovskite structures are broadly classified into oxide and halide perovskites based on whether the X-anion is an oxide or halide.

2.1.1 OXIDE PEROVSKITES

The chemical formula of the oxide perovskite structure is ABO_3. In oxide perovskite, a 12-fold coordinated A-site cation is larger than six-fold coordinated B-site cation. (Mg, Fe)SiO_3 or $CaSiO_3$ is the oxide perovskite which is commonly found in the earth's crust. The A-site and B-site cation carries different charge combinations such as 1+5,2+4 and 3+3. Some of the oxide perovskite shows small distortion from symmetrical structure and crystallizes in polymorphic structure. Even though the ideal structure of perovskite is cubic, a large number of oxides gets slightly distorted with lower symmetry, and have hexagonal, orthorhombic, and rhombohedral lattices.

Polymorphs of perovskite structures also exist, one of them is Ilmenite structure which has similar compositions but the size of A and B cation being approximately the same. The tolerance factor of an ilmenite structure is lower than that of the ideal value of perovskite. The structure depends on both the size and nature of A and B cations. For example, the tolerance factor, t, of ilmenite $BaGeO_3$ is bordering on 1 but it crystallizes into silicate-related structure rather than in perovskite structure since the preferred coordination number of Ge is 4. The stability of perovskite depends on the tolerance factor which is determined by the ionic size but the chemical nature is determined by the coordination number. Incorporating dopants in B site with large difference in ionic radii leads to the formation of superstructures instead of random ordering of the two types of ions. Some other polymorphs of perovskite are brownmillerite ($A_2B_2O_5$) which is an oxygen deficit and K_2NiF_4 which is a combination of ordered Bsite and oxygen defects.[17]

The ABO_3 perovskites exhibit several intriguing physical characteristics, including ferroelectricity ($BaTiO_3$), weak ferromagnetism ($LaFeO_3$), ferromagnetism ($SrRuO_3$), superconductivity (YBa_2-Cu_3O_7), insulator-to-metallic transitions for thermistor applications ($LaCoO_3$), and large thermal conductivity due to exciton transport ($LaCoO_3$) fluorescence compatible with laser action (La_2CuO_4).

Each oxygen in the ABO_3 structure is shared by two B^{3+} ions in the ideal cubic perovskite, creating an angle B–O–B of 180°. Superexchange interactions between magnetic B^{3+} cations are more advantageous in this configuration. Adjacent magnetic moments are often coupled in an antiparallel manner as a result of this exchange. Further spin configurations are feasible when the B^{3+} ions are accommodated in two sublattices ($A_2BB'O_6$). The B^{3+} ions exhibit antiferromagnetic alignment when B′ is a diamagnetic ion; moreover, the critical exchange mechanism is expected to be a superexchange interaction of longer range through two oxygen arranged in B–O–B′–O–B configuration. In contrast to ideal perovskites, where the B–B separation was 0.4 nm, it is now significantly longer.

The iron moments in $LnFeO_3$ (Ln lanthanide) perovskites are somewhat tilted despite being antiferromagnetically aligned, so they exhibit a weak spontaneous magnetic moment. The iron moments line up in such a manner that magnetization is easily directed along a or c axis of orthorhombic cells. Materials with mild ferromagnetic moments of 0.03-0.07 μB/mol are targeted for memory devices. The $LnMnO_3$, which predominately contain Mn^{3+} or Mn^{4+} ions, exhibits antiferromagnetic properties. At the range of 25–35% Mn^{4+}, ferromagnetic behavior is seen. In addition to a modest magnetic connection between Mn^{3+} ions, there exists a negative interaction between Mn^{4+} ions along with strong positive interactions between Mn^{3+} and Mn^{4+}. When Co^{3+} and Co^{4+} are combined, a similar pattern of behavior was seen, but antiferromagnetic behavior was observed in Cr and Fe compounds.

The outermost electrons, which either can be collective or may be localized at certain atomic locations, determine the electrical activity in perovskites. There is a close relationship between perovskites' electrical and magnetic characteristics because localized electrons have the potential to carry spontaneous moments. Rare-earth perovskites with transition ions display a wide range of electrical characteristics. For instance, a collective d-electron behavior was observed in $LaNiO_3$ and $LaTiO_3$, where the former contains nickel in the low-spin state Ni^{III}, exhibiting Pauli paramagnetism and metallic conductivity. $LaCrO_3$ is a semiconductor in the $LnCrO_3$ (Ln lanthanide) perovskite series. It appears that there are two conductivity areas for the heavier chromites. An increase in activation energy is noticed along the lanthanide series in the low-temperature zone (from 0.27 eV for Dy to 0.37 eV for Yb), whereas it is roughly 0.23 eV in the high-temperature region. High mobility Cr^{4+} ions are thought to be the mechanism of conduction, causing p-type extrinsic conduction. A reduction and an improvement in the covalence of the Cr–O bond and Ln–O bond, respectively, may be responsible for lowering conductivity in heavier chromites. The conductivity and magnetic susceptibility of $LaMnO_3$ perovskite immediately change when the temperature approaches 720 K. The charge carriers' theory of hole hopping between localized levels was used to explain this phenomenon. $LaCoO_3$ behaves in a manner that is considerably more complicated. Up to approximately 400 K, it is a semiconductor, but after that, conductivity increases more quickly till the temperature reaches 823 K. Between 823 and 1200 K, $LaCoO_3$ progress along a broad, flat maximum, and beyond 1200 K, it exhibits metal-like behavior. The low-spin state (Co^{III}) of Co ions is primarily found at lower temperatures; and at higher temperatures, it can also be found in the high-spin paramagnetic Co^{3+} state. With a temperature rise, the Co^{3+} and Co^{III} ion couples change to Co^{2+} and Co^{IV} pairs. As a result, while long-range order alone is present, the proportion of Co^{3+} vanishes at about 1200 K, leading to a first-order transition.

The outer electrons exhibit a change from a localized to a collective behavior in addition to the spin behavior. The lanthanide partially replaced with a divalent ion will increase the conductivity. The Mn ions are hence Mn^{3+} for $x = 0$ in $Sr_xLa_{1-x}MnO_3$. On the other hand, as x rises, Mn^{4+} holes form, resulting in an increase in conductivity. The system shows a metal-semiconductor crossover for substitutions in the range $0.2< x<0.4$ and becomes ferromagnetic. A similar phenomenon was seen in cobaltites that had Sr replaced. Perovskites especially their electrical characteristics got a lot of attention with the discovery of superconductivity at 40 K in cuprates. Those cuprates, which have a mixed copper Cu^{II}-Cu^{III} valence, are hole superconductors. The exception to this rule is Ce-doped Nd_2CuO_4, which has a T_c close to 25 K, is an electron superconductor, and belongs to a separate structural class. The bidimensional nature of the structure, a characteristic shared by all of these compounds, has been proven to be crucial for the presence of superconductivity at high temperatures.

On examining the magnetic ordering of both the transition metal ion in the Bsite and rare earth ion A perovskite oxides (ABO_3) are well suited for infrared, visible, and ultraviolet spectroscopic studies. There is little interaction since the ordering temperatures of the two types of ions are so different. You can examine the absorption spectra of the host ion in a perovskite structure or those produced when a low-concentration "probe" is incorporated into a diamagnetic lattice. Several ABO_3 compounds, where B corresponds to a transition metal ion with either a partially filled or a vacant d-shell have had their optical characteristics investigated. Since an empty d-shell transition ion (Ti^{4+}) is present in $SrTiO_3$ perovskite, the emission spectrum reveals luminescence that is unaffected by sample history or contamination. This luminescence has a band maximum at roughly 500 nm at a lower quenching temperature of 35 K, and a slight Stokes shift (7000 cm^{-1}). In $SrTiO_3$, the titanate octahedra sharing corners are advantageous for the creation of energy bands. Energy for the first absorption transition is significantly reduced as a result of the widening of the titanate octahedron's energy levels and hence observes a mobile titanate excited state. The free excitons coexisting with the delocalized (self-trapped) excitons have been attributed as the cause of this condition. Numerous studies have been conducted on the optical characteristics of metal ions incorporating perovskites that have partially filled d-shells. The optical characteristics of perovskites have been widely studied using ions in the d^3 configuration. In the perovskites $MTiO_3$ (M = Ca, Sr, Ba) and ABO_3 (A = La, Gd, Y; B = Al, Ga), Cr^{3+} spectroscopy has been investigated. The Cr^{3+} ion in these systems has the potential to emit in two entirely distinct ways: narrow-line, spin-forbidden emission (^2E to $^4A_2$) and broad-band, spin-allowed emission ($^4T_2$ to $^4A_2$), where the second occurs for only weak fields. Following the discovery of titanates-based photoelectrochemical water splitting, transition-metal–doped titanates attracted attention. With the doping of transition metal ions, it seems conceivable to make an $SrTiO_3$ electrode sensitive to visible light.[18]

2.1.2 HALIDE PEROVSKITES

Halide perovskites have gained popularity because of their low cost and facile production, multicolor emission, high photoluminescence quantum yields (PLQYs), and superior excitonic and charge carrier characteristics. The basic formula for halide perovskites is ABX_3 same as that of oxide perovskite, where the A-site cations, such as Cs^+, methylammonium ($CH_3NH_3^+$, MA), or formamidinium [$(CH(NH)_2)^{2+}$, FA], fill the voids formed by the BX_6 octahedra's corner-sharing. By changing the halide ion as well as the A- or B-site cation, isomorphs of halide perovskites can be produced. Perovskites with mixed A-site and/or B-site cations are also derived, including $MAFAPbX_3$,

$KCsMAFAPbX_3$, Cs_2AgBiX_6, $Cs_2AgInCl_6$, and $Cs_2AgInBiX_6$. Halide perovskites are useful for engineering products including LEDs, lasers, photodetectors, and solar cells.

The fascinating charge carrier and optical properties of perovskite arises from energy states distribution at the band edge. Halide perovskite possesses direct bandgap with an exception of noncentrosymmetric distribution of valence band maximum (VBM) and conduction band minimum (CBM). The PL emission of the perovskite can be tuned from visible to near-infrared region by changing the halides. Chloride perovskites emit violet-to-blue light, bromine emits green, and iodide emits deep red-to-near-infrared light. The change of band-edge states is created by the hybridized s- and p-orbitals of halides and the halide-dependent tuning of bandgap or emission color is caused by B-site cations. Similar band edge structure is applicable to lead free perovskite. CBM degeneracy is disrupted when spin orbit coupling (SOC) becomes substantial, it separates into spin–orbit split-off doublet and quadruplet states. Halide perovskites display fine optical absorption characteristics by the optically permitted transitions from the VBM to the spin-orbit split-off states and quadruplet states. The optical bandgap is also lowered by a reduction in the conduction band's dispersion, which has an impact on the charge carrier and PL characteristics. In reality, other parameters including composition, octahedral tilt, and temperature are linked to SOC, and controls the perovskite's optical properties. The octahedral tilt emanates as a consequence of (i) the decrease in Pb–X–Pb angle when large A cation is replaced by a smaller one, and (ii) change in crystal structure. Three separate regimes of PL property fluctuation exist in lead iodide perovskites: orthorhombic (low temperature), tetragonal (room temperature), and cubic (high temperature). A perovskite crystal tends to stabilize in the cubic phase as the temperature rises, resulting in a small octahedral tilt and strong SOC. This results in the shrinkage of bandgap and red shifting of PL and absorption spectra. Pressure has a role in changing the SOC by inducing compression of the Pb-X bond, shrinking of the Pb–X–Pb angle, or increasing the octahedral tilt causing amorphization of the crystal lattice.[19]

Halide perovskite shows PL blinking which can be accounted to the photoinduced activation and deactivation of emitting or quenching sites caused by geometric distortions or with the presence of various chemical structures. Charge traps or crystal defects are equivalent to the quenching sites. Alternatively, the creation of quenchers close to the emitting sites or turning the emitting site into a quencher can be related to PL blinking. Quencher switching is thermally activated, and at low temperatures, blinking is suppressed. Because of inadequate quenching site diffusion, there is a suppression of blinking. Quenching by halide or proton migration alone cannot model PL blinking, despite the fact that the thermal barrier for activating quenching sites is in the order of the activation energy for halide migration. The "quenching complex" that combines a shallow localized electron trap and a mobile hole trap present close to the conduction band and valence band, respectively, is therefore used to describe blinking. Such particular shallow defects are however unlikely to result in blinking.[19]

Even though the halide perovskite can be prepared using low-temperature solution chemistry, it has characteristics similar to inorganic photovoltaic material such as long charge carrier lifetime and sharp optical absorption onset. Density, energy, and distribution of the trap states determine how trap-assisted processes in perovskites control monomolecular recombination. Additionally, material purity and crystallinity have an impact on the monomolecular recombination rate. The ratio of metal and halide ions determines how bimolecular recombination occurs in perovskites. For instance, perovskites made of metals (Sn instead of Pb) and lighter halides (Br instead of I) have higher bimolecular recombination rates. Such chemical composition-dependent rate fluctuations are linked to changes in exciton-binding energy and band structure.[19] Halide perovskite has slow recombination which can be attributed to the defect properties. Since the defect states are localized close to the band edges, the material is considered defect tolerant. The presence of defects in halide perovskite, particularly the intrinsic ones, and how they contribute to long lifetimes has recently become a hotspot of research.[21]

Defects or a high density of charge traps, which are abundantly present at the grain boundaries or at the surfaces of $MAPbI_3$ polycrystalline films, typically affect the carrier diffusion length. The diffusion lengths of $MAPbI_3$ polycrystalline thin films are shorter than those of single crystals. In the past, solvent annealing allowed for the creation of $MAPbI_3$ polycrystalline films with carrier diffusion lengths of roughly 1 μm. By adopting a low-temperature solution-growth technique, $MAPbI_3$ single crystals might achieve considerable diffusion lengths of about 175 μm. Longer carrier diffusion lengths are induced by small charge trap densities in single crystals with high carrier mobilities. The mixed halide $CH_3NH_3PbI_{3-x}Cl_x$ has an extended carrier diffusion length of 1 μm which is comparatively longer than the diffusion length of $MAPbI_3$. Electron and hole effective masses influence the carrier mobility. In $MAPbBr_3$, conduction band electrons and valence band holes are found to have effective masses of .15 m_0 and 0.25 m_0, respectively. The inorganic halide perovskite (IHP) has comparatively smaller effective mass than its organic counterparts because of the larger lattice constants of organic cations.[22]

Having a mechanically soft lattice or low phonon (vibrational) energy and elastic constants is an unusual feature of halide perovskites. Soft materials can have elastic moduli in the orders of magnitude lower than halide perovskite, the soft lattice of halide perovskite is neither surprising nor novel in a more general context of materials physics. It can also be explained physically by taking into account the fact that the halide perovskite structure's formal ionic charges are rather low, which leads to a weak electrostatic interaction inside the lattice. This mechanical softness had a crucial role in photostablity of the material.[21]

The characteristics of halide perovskite could be refined by using halide mixing technique. Generally, the halide mixing imparts mainly the following three advantages—(i) enhancing the stability, (ii) improving carrier transport, and (iii) bandgap tuning. The $CH_3NH_3Pb\ (I_{1-x}Cl_x)_3$ shows more stability and effective charge carrier transport than $CH_3NH_3PbI_3$. The bandgap of $CH_3NH_3PbBr_{3-x}Cl_x$ gets increased to 1.5 eV when Cl is incorporated into $CH_3NH_3PbBr_3$ which has a bandgap of 1.3 eV.[20]

2.2 PEROVSKITE VARIANTS

Different modifications in the distribution and arrangement of anions and cations have led to the formation of various perovskite-derived structures as listed later. In addition, perovskites can also be modified in their dimensionality, producing layered (2D) and 0D structures for specific applications.[23]

2.2.1 DOUBLE PEROVSKITES

The substitution of A-site or B-site cation leads to the formation of a doped single perovskite, has a structure $A_{1-x}A'_xBX_3/AA'B_{1-x}B'_xX_3$ where A and B take disordered arrangement When the x value becomes approximately equal to 0.5, the size and charge of A and A′ (B or B′) show substantial difference which causes A-site (B-site) cation ordering. This cation ordering causes the development of double perovskite having the formula $AA'B_2X_6$ (or $A_2BB'X_6$). Three ordering types (layered, columnar, or rock-salt) are possible for A-site, as well as B-site–, ordered double perovskites.[24]

Double perovskite is formed when the metal ions are selected for the B and B′ sites and the alkali, alkaline earth, or rare earth ions are selected for the A site. The B and B′ cations can exist in a wide variety of oxidation states because they contain a divalent A-cation. To build the perovskite phase, these two B cations require a suitable oxidation state. The overall oxidation state of B site in the occurrence of an A^{2+} cation will be four, which may also be adapted to B^{+}/B^{7+}, B^{2+}/B^{6+}, B^{3+}/B^{5+}, and B^{4+}/B^{4+}. The oxidation states in the B site for $A_2^{2+}BB'O_6$ range from 1 to 7. The typical oxidation state of the B site for the A^{3+} cation is three. As a result, the B-site combinations for the A^{3+} cation will be B^{+}/B^{5+}, B^{2+}/B^{4+}, and B^{3+}/B^{3+}. Nonetheless, the ionic radii are enormous for the A^{+} cation. As a result, the oxidation states of the B sites will be B^{3+}/B^{7+}, B^{4+}/B^{6+}, and B^{5+}/B^{5+}.[25]

The Goldschmidt tolerance factor t for double perovskite can be written as

$$t = \frac{(R_{A'} + R_{A''})/2 + R_0}{\sqrt{2}x[(R_{B'} + R_{B''})/2 + R_0]} \tag{3}$$

where $R_{A'}$, $R_{A''}$, $R_{B'}$, RB'', and R0 are the Shannon ionic radii of the constituent ions.

For a double perovskite, B–B' cation radii mismatch has much significance rather than the A–B cation radii variation as in the case of a single perovskite.[25]

Many significant advantages to the physicochemical properties, stability, impacting activity, and effectiveness in pertinent applications can be brought about by the formation of a double perovskite structure. At first, double perovskites enable the elements that are normally unable to exist in single perovskites, such as high valence elements (6^+ or 7^+), expanding the compositional space of the perovskite family. Second, double perovskites have more complicated atomic environments than single perovskites, which might result in electronic structures that are unusual. Owing to the oxygen vacancy localization in the AO plane, A-site–ordered oxide double perovskites ($AA'B_2O_{5+\delta}$), for instance, have alternating $|AO_\delta|BO_2|A'O|BO_2|$ layers and hence have two symmetries for B-site cations (i.e., octahedral and square pyramidal). In double perovskites with B-site ordering, alternate octahedra of BX_6 and $B'X_6$ are anticipated to function in a more complimentary manner, contributing to noticeably enhanced chemistries. Finally, double perovskites show greatly improved chemical stability in catalytic settings, like intensely reducing or oxidizing atmospheres, strongly acidic or basic solutions, and environments with heat, CO_2, humidity, or light.[24]

The lead-free halide double perovskite has unique electronic dimensions and enhanced stability compared to other lead-free perovskites. In optoelectronic applications, where two divalent lead ions are substituted by combining one monovalent and one trivalent ion, or one tetravalent ion, halide double perovskite materials with formulas of $A_2M(I)M(III)X_6$ or $A_2M(IV)X_6$ may be considered as stable and environmentally friendly substitutes. They can be employed as sensors, light-emitting diodes, photocatalytic dye degradation, solar cells, and X-ray detectors.[26,27]

The structure of the A2B′B″O6 double perovskite oxides material can be incredibly varied; metals from every group of the periodic table can combine with rare earth elements to generate this type of double perovskite. The eventual outcome has a great deal of potential for use in a variety of fields, including magnetic, electrical, optical, and catalytic materials. Recently, there has been a lot of interest in the new functionality of rare earth double perovskites, namely their MR, magnetocapacitance, white light and long afterglow luminescence, multiferroics, photovoltaic characteristics, and electrocatalytic or photocatalytic activities.[28]

2.2.2 LAYERED (2D) PEROVSKITES

The layered structure largely confines electric and magnetic moment to layers, which give rise to exceptional physical properties and quantum derived behaviors. 2D perovskites based on all-inorganic perovskite layers and organic-inorganic hybrid perovskite layers are developed at present. 2D hybrid perovskite structure was developed first in which the organic layer is used as a spacer between the layers formed by corner shared $[BX_6]^{4-}$ metal halide octahedra.

Layered perovskites are further classified into three types as low-dimensional perovskites with <100>-, <110> -, and <111> orientations having network structures of the form $BnX3n_{+1}$ ($n \geq 1$), B_mX_{3m+2} ($m > 1$), and B_qX_{3q+3} ($q > 1$), respectively.[32] Layered hybrid perovskites mostly pertain to the <100> series. Two main types of <100> phases are stipulated for layered oxide perovskites.

i) Dion–Jacobson (DJ) phases $A'[A_{n-1}B_nO_{3n+1}]$ have one interlayer cation A′ per formula unit,
ii) Ruddlesden–Popper (RP) series $A2'[A_{n-1}B_nO_{3n+1}]$ have two interlayer cations A′ per formula unit, where A′ cation can correspond to either a monocation or a dication.

Layered hybrid perovskites with <100> orientation are mostly based on organic monocations A'^{+} and are structurally similar to RP phases. Since the structure of DJ phase always contains two cationic sites per formula unit in the interlayer space, hybrid-layered perovskite cannot be interpreted with the DJ phase. It expounds on one category of inorganic-layered perovskites.

The layered structure largely confines electric and magnetic moment to layers, which give rise to exceptional physical properties and quantum-derived behaviors. Studies have revealed that A-site cations have an immense effect on the crystal structure. Bulky cations favors 2D structure whereas too bulky cation occupancy brings about deviation from corner-sharing octahedral structure to edge or face-sharing octahedra within the 2D layers.[29]

The magnetic characteristics of the layered organic and inorganic perovskites initially sparked interest. The layered perovskite structures have been found to contain long-range magnetic ordering when Cu(II), Fe(II), Cr(II), and Mn(II) are used as the metal. Most noteworthy are the 2D ferromagnetic insulators that were originally made possible by the hybrid Cu(II) halides. The magnetism of hybrid perovskites is typically controlled via magnetic superexchange interactions, J, via the B–X–B linkages, as there are no direct metal–metal connections in these materials. The structural parameters, such as bond lengths and angles, are the main variables affecting the magnetic exchange. Moreover, the hybrid compounds' structures and magnetism are affected by the presence of Jahn–Teller active metal cations. The Goodenough–Kanamori (or Goodenough–Kanamori–Anderson) criteria that were created and are frequently used to explain the magnetism of oxide perovskites also hold true for hybrid halide perovskites (HHP). A strong antiferromagnetism is predicted by the Goodenough Kanamori rules when the magnetic orbitals on nearby B atoms are orientated at 180° and B has partially filled d shells, whereas a weak ferromagnetism is predicted when they are at 90°.[30]

On comparing with the 3D hybrid perovskite, in addition to negligible contributions of organic spacers to the frontier orbitals, the limited orbital overlap between inorganic layers leads to the confinement of electron-hole wave functions within the inorganic sublattice, thereby developing a 2D quantum well structure.[31] Practically it can be achieved by substituting the smaller A cationic sites with longer ones while multilayered perovskites are obtained from partial substitution.[32] The larger A-site organic cations function as "spacers" separating the inorganic layers. The general formula for 2D and quasi-2D HHP is expressed as $R_2A_{n-1}B_nX_{3n+1}$, where R, A, B, X, and n denotes larger organic spacers, smaller A-site cations filling the inorganic octahedral voids in quasi 2D HHP, inorganic cations, halogens, and the number of inorganic octahedral sheets held together, respectively. As the value of n increases, a greater number of inorganic layers are separated by the spacers and finally, it approaches the 3D HHP configuration. The layered tin perovskite structure is shown in Figure 2.4. Van der Waals interaction holds the spacers together whereas hydrogen bonding, halogen bonding, and Coulombic interactions act between organic and inorganic layers.[33] This weak interaction between the organic and inorganic layers can be disturbed under extreme parameters thereby tuning the orientation of the organic spacers.

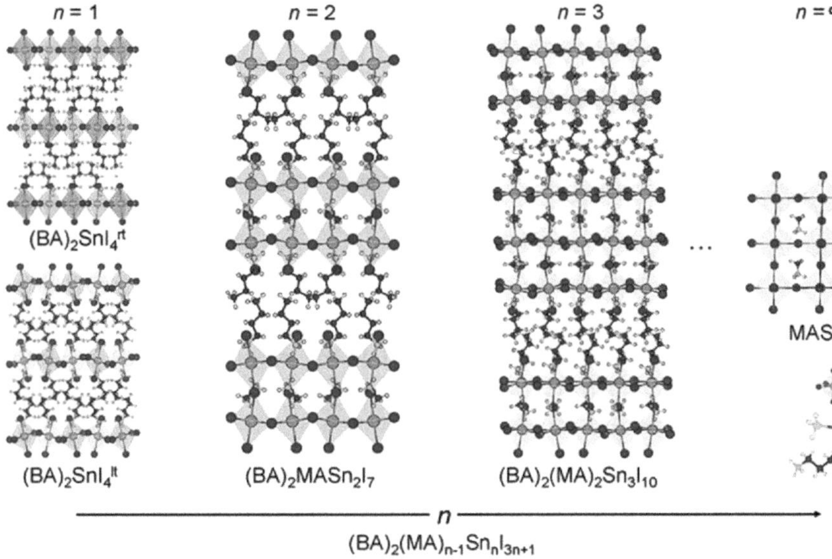

FIGURE 2.4 Layered tin perovskite structure.

Source: Reprinted from Ref. [34]. http://creativecommons.org/licenses/by/4.0.

The organic part in the 2D HHP materials brings forth chemical flexibility, dimensionality tuning, broad photoluminescence, and narrow and color tunable emission.[33] This confinement alters different intrinsic parameters of the material like their bandgap, Coulombic interaction, exciton binding energy, and radiative recombination probability. All the 2D HHP material parameters are dependent on the lattice parameters, permittivity contrast, and octahedral distortions, and hence vary with the choice of organic spacers and inorganic anions.[31] In addition to the elaborated choice of A-site cations, the wider chemical flexibility provided by the long-chain organic spacers makes the material a potential candidate for several optoelectronic applications.

The unique structure of the 2D HHP materials encourages them to be administered in electrochemical applications such as solid-state ion batteries. For a lithiated HHP, the conductivity can reach approximately 10^{-3}Scm^{-1}.[35] This efficient ionic diffusion is assigned to their characteristic structure. The enhanced diffusion property along with higher stability under extreme working conditions promotes them to be used as electrode materials in all solid-state batteries. Moreover, a volume change is associated with the charge–discharge process of the electrode under the presence of organic spacers. Hence, the electrochemical battery performance is enhanced by the chemical structure of 2D HHP and the organic cation volume by promoting interfacial electrode–electrolyte reactions and chemical stability.

With a wider bandgap and higher exciton binding energy, 2D HPP has limited application as a photoactive layer in photovoltaics in comparison with their 3D counterparts. 2D HHP are excellent photodetectors that are capable of detecting dark current in orders of pico amperes (Pa) and with ultrafast response rates. The enhanced ferroelectric behavior of 2D HHP that creates an ultrahigh electric field is responsible for the superior photodetection performance. Ferroelectric memory devices based on resistive switching which in turn is linked to ferroelectric polarization switching can also be developed from 2D HHP materials.[33] Some 2D HHP materials possess excellent two photon absorption coefficients which makes the material a suitable candidate for non-linear and photonic-sensor device applications. Because of its intrinsic environmental stability against external parameters, 2D HHP materials can be designed as suitable power sources for wearable electronics.

In the meantime, research on all-inorganic 2D perovskite material is attaining momentum with the scope of maintaining stability in extreme conditions. While hybrid perovskites are formed by slicing the 3D ABX_3 structure along the <100> directions, inorganic 2D layers are mainly formed by slicing in the <100> and <111> directions. The main orientation of all-inorganic 2D perovskite is along <111> direction, the general formula being $A_{n+1}B_nX_{3n+3}$. Here, $n = 2$ for the bilayer structure, and B sites are occupied by the trivalent cation and vacancy at a ratio of 2:1. B sites are occupied by metal ions like $Bi^{3+}$and Sb^{3+}, resulting in an $A_3B_2X_9$ structure. Surprisingly, depending upon the size of A cation, this structure could be 2D, as well as 0D.[41] Replacing A-site cation with larger size causes the formation of dimers, where BiI_6 octahedra fused together form a face-sharing $Bi_2I_9{}^{3-}$ dimer. These electronically isolated dimers build a 0D crystal structure of $A_3B_2X_9$. <111>-oriented triple-layered perovskites could be formed by mixing bivalent cations so that the vacancy sites of $A_3B_2X_9$ is lowered by 25%. Structures are also formed by mixing the neighboring halides such as $Cs_3Bi_2I_6Cl_3$ which alters the opto-electronic properties. $Cs_3Sb_2I_9$, $Cs_3Bi_2Br_9$, $Rb_3Sb_2Br_9$, $Cs_3Sb_2Br_9$, and $Cs_3Sb_2Cl_9$ are examples of 2D structures along <111> plane.

Apart from this, a few RP phases having the structure A_2BX_4 with orientation along <100> plane are formed in inorganic perovskites. The RP phase can be visualized as two AX rock-salt-type layers sandwiched between *n* numbers of ABX_3 perovskite layers along the *c*-axis. Similar to the phases in the <111> plane, mixed halide structures are also found along the <100> plane. Recently, transition metal-based halides with the formula A_2MCl_4 are also reported where A could be Cs or Rb and M could be Cd and Mn.[36]

With its versatile properties and rapid development, 2D perovskites are undoubtedly the next generation of optoelectronic materials. It is the unique structure of the material that reasons for their various inherent properties. Thus, an depth understanding of the cationic composition and individual structural nature of perovskite constituents is essential for advanced technological developments of the perovskite material for next-generation device fabrication.

2.2.3 HEXAGONAL PEROVSKITES

In a common perovskite structure, the BX_6 octahedra are rigid entities that tilt around the shared corners and facilitate the accommodation of A-site ions with different sizes.[37] In the former case, tilting of the corner-sharing BX_6 octahedra is responsible for filling up the vacant space created in the structure bringing about orthorhombic, monoclinic, tetragonal, and rhombohedral symmetries. But for the latter, the lattice strains produced by higher cationic size mismatch forces to form face-sharing octahedra.[38] This leads to the formation of hexagonal perovskites. Since they are deviated from the ideal BX_6 corner-sharing perovskite structure, hexagonal perovskites are generally referred to as perovskite-related hexagonal polytypes. That is, consecutive AX_3 layer's stacking determines the material's structure as perovskite for corner-sharing BX_6 octahedra network, hexagonal for face-sharing octahedral chains, and hexagonal perovskite polytypes for those having a combination of face- and corner-sharing octahedral. In general, if the structure possess edge- or face-sharing network apart from corner-sharing mode, the structure is referred to as halometalate network with a perovskite sub-network or "perovskite like" or "perovskitoid." In effect, hexagonal perovskites are not really of the perovskite type.[38]

The possibility of octahedral tilting, a combination of atoms, structural distortions, and physical properties advocates the development of enormous sets of variants. In the case of an oxide perovskite framework possessing cubic symmetry, the effect of tilting is visible near the associations between corner-sharing MO_6 octahedra, where it causes a change in the degree of overlap between the metal *d* and oxygen *p* orbitals.[37] As a result, the magnetic and electronic properties get significantly changed.

The marked difference between the cationic sizes in traditional 3D perovskites compels some of the corner-sharing BO_6 octahedral units to share faces leading to the formation of hexagonal polytypes. They are characterized by intermediate bandgaps (between 1.9 and 2.3 eV), which are tuned by the structural proportion of corner-to-face–sharing octahedra.[39] Numerous hexagonal perovskite derivatives can be formulated from different combinations of corner- and face-sharing octahedra, which are capable of accommodating cationic and anionic vacancies.[40] Variation of synthesis conditions enables some cations to stabilize in more than one crystal structure.

Apart from having corner-sharing geometry, they facilitate face sharing of MO_6 octahedra or A- or M-centered trigonal prisms in the structural framework. This results in the presence of dimers, trimers, tetramers, or longer fragments of chains of face-sharing octahedra in the structure. As a result of face sharing, octahedral structures not only reduce the metal–metal distances and metal–oxygen–metal bond angles but also dramatically alter the quantum properties.[37] Depending on the electronic properties, semiconducting, metallic, and dielectric hexagonal perovskites are developed. Different structural principles in hexagonal perovskites deviates the Goldschmidt tolerance factor from 1, which is close to one for cubic perovskites. The unique properties of the materials make them potential candidates for quantum-spin-liquid-state at low temperature.

The versatility, excellent electrical and chemical stability of the materials, and their ability to accommodate oxygen vacancies and intrinsic cations promote their utilization in various fields of application, especially as ion transport materials. The superior performance of the material as an electrolyte can be grounded to the low ohmic resistance without external gas humidification along with good proton and oxide ion conduction.[37] Perovskite oxide structure with reduced symmetry usually exhibits lower proton mobility, but surprisingly hexagonal perovskite derivatives are an exception. It is revealed that lattice disorderness promotes protonic defect delocalization, leading to higher proton mobility. The transport mechanism is favored along two dimensions. The hexagonal perovskite materials with cation-deficient layers or palmierite-like motifs are highly suitable for oxide and proton conduction.

Besides various structural aspects of hexagonal perovskites, one of the fascinating facets that enable their multifunctional application is their ability to sculpt chain fragments of varying lengths from face-centered transition metal octahedra. The chains then interact with triangular prisms by face sharing and are capable to shrink or stretch along their prismatic lengths to accommodate ions with suitable M−O bond lengths.[37] Ba_2CoTeO_6 is an example of hexagonal perovskites where the crystal structure comprises dimers occupied with ordered face sharing CoO_6 and TeO_6 octahedra, which are disembodied by single CoO_6 and TeO_6 octahedral layers.

Double hexagonal perovskites are a different category of hexagonal perovskites possessing multiple chemically distinct M-site ion in the MO_3 framework. The M–M' structure can be either ordered or disordered. The M ions in the double hexagonal perovskites can occupy two distinct octahedral sites;

i) Favoring only corner-sharing with the octahedral, forming a separate triangular plane
   or
ii) Favoring corner-sharing and face-sharing with different sets of (M, M') O_6 octahedra.

The basic formula for double hexagonal perovskites, with two octahedra in the face-sharing substructure, is $A_3MM'_2O_9$ and is generally known as dimer-based hexagonal perovskites. Here, M and M′ can be the same or different with M ions forming triangular planar lattices of MO_6 octahedra whereas M′ ions form a dimer array on a triangular lattice. Further, subgroup classification of the material can be made by the difference in the totality of electrons on the M'_2O_9 dimer. M can have oxidation states varying from 1^+ to 4^+ and as a basic requirement of maintaining neutrality, as the M-site ion charge increases, the oxidation state of the M′ ion decreases. Hexagonal perovskites become unstable when the A-site ionic size becomes very small. Dimer-based hexagonal perovskites may or may not exhibit structural disorderness and can be exploited in quantum materials research.

Trimer-based hexagonal perovskites are variations of hexagonal double perovskites which are formed when the ratio of M to M′ ions is 1:3 instead of 1:2 and have a general formula $A_4MM'_3O_{12}$. These materials have three $M'O_6$ octahedra that share corners with a triangular lattice of MO_6 octahedra and face each other. When the central octahedron occupies different ions in it other than the two outer octahedrals, such materials are described as hexagonal triple perovskites, with the general formula $A_4MM'_2M''O_{12}$. The excellent magnetic behaviur of the material is one of the superior material qualities.[37]

10H-$Ba_5Fe_4NiO_{13.5}$ is an example of a material having both dimers and trimmers in the hexagonal oxide perovskite structure. They crystallize in the P63/mmc hexagonal space group and are synthesized by nitrate salt decomposition. The crystal structure is composed of stacked trimer cluster corners shared with dimer clusters. Studies revealed the existence of various other substructures in hexagonal perovskites such as the one having both tetrahedral, as well as octahedral, sites in the same structure and so on.

Complex structured hexagonal perovskites can be developed from creating oxygen deficiency, where the vacancies can be either ordered or random. Like other materials, in hexagonal perovskites also different polytypes can be created, especially with pressure. Based on the conditions of crystal growth, different structures of oxide perovskites, such as oxyhalide perovskites, can be developed. Besides having oxide layers, this material comprises halide layers between the face sharing building blocks. Of these, oxychlorides are the most common. The ability of the material to be synthesized as good single crystals and their potential applications as cathode materials and photocatalysts promote their exploration.

Corner-sharing produces extremely dispersive electronic conduction and valence bands by promoting destabilization of the ns^2 lone pair on the M^{2+} sites and also enhances charge carrier mobility. In the case of edge- or face-sharing octahedral configurations, lone pair energy is lowered, hence creating flat electronic bands and wide bandgaps.[39] Thus, the materials need to be suitably selected for each application.

In a nutshell, because of the wide varieties of crystal structures, hexagonal oxide perovskites and their distorted derivatives are appreciably substantial than their perovskite analogs. Though various materials are identified and studied, they are not much utilized for quantum material applications. The complexity of the structure, difficulty in synthesis, and the lack of knowledge hinder the exploration of the material category. Future investigations of the materials can provide us with plausible candidates for low-temperature quantum spin-liquid states.

2.2.4 PEROVSKITE NANOSTRUCTURES (0D)

When the materials change their phase or reduce their dimensionality, their overall properties vary. The increased surface area and the excitonic nature of the dimensionally reduced materials (like nanorods, nanowire, etc.) make them a hot topic of discussion.[41] As for example, nanowires and nanorods of $MAPbX_3$ exhibits high PL Quantum Yields and long PL time compared to their 3D analogue. Cs-based inorganic perovskites with colloidal nanocube structure exhibit halide composition-based visible range bandgap tunability. Several other interesting properties can be observed when the dimensionality is reduced.

Based on the corner-sharing octahedral connectivity, there exist four distinguished crystal structures.

i) 3D lattice: Octahedras form a 3D network.
ii) 2D lattice: Octahedra has a layered arrangement with organic or inorganic cations separating the layers.
iii) 1D lattice: Octahedral chains are formed with cationic isolation.
iv) 0 D lattice: Complete individual octahedral isolation from each other in the crystal lattice in all dimensions. Not a single octahedra is connected to any other directly in any dimension.

Cs_4PbBr_6 is an example of 0D perovskite demonstrating superior PLQY in both powder and single crystal forms.[42]

The requirement for reducing the dimensionality of a semiconductor lies in confining the excitons in a volume of radius less than or comparable with the Bohr radius.[43] This could be accomplished by following either of the two mechanisms described below.

i) When the exciton diameter is more than its particle size, 0D perovskite materials can be formed.
   or
ii) By the complete isolation of BX_6 octahedral units, within the bulk crystal lattice, by a structural barrier.

Since the former ones are similar to metal chalcogenide QDs, they are entitled as perovskite QDs. The latter has the benefit of occupying several millimeters or larger physical dimensions. However, the intrinsic confinement behavior of the material makes them act like a bulk quantum material. As they do not have any corner-sharing connectivity, they are termed 0D perovskite-related materials (0D PRMs).[43] The unique feature of such materials is that though they can have diverse sizes and morphology, their properties resemble that of a single-constituent octahedron.

The crystal phases, like cubic, tetragonal, orthorhombic, etc., vary with the octahedral network stress, which in turn depends on the cationic size. $[BX6]^{4-}$ octahedral units control the frontier energy levels, where B-metal (s) and halogen (p) valence orbitals make up the highest occupied valence electronic state and the lowest unoccupied conduction band state is composed of B-metal outer p-orbital, suggesting that A-site cations do not directly intervene in determining the frontier electronic states.[43] Since excitons are generated in individual octahedrons with respect to photoexcitation, the octahedral connectivity greatly affects the overall bandgap and optical properties of these materials.

Hot injection method or ligand-assisted antisolvent precipitation methods are the main approaches for the synthesis of hybrid and all-inorganic metal trihalide perovskite QDs.

The most noted material in this category is Cs_4PbBr_6. Various other materials with Bi, Sn, and Sb instead of lead can also be formed, with versatile optical properties. The composition of 0D PRM and their isolated structural unit nature are determined by the central metal ion valency. This makes it difficult to have a general structural formula.

Most of the lead free 0D PRM feature low PLQY and long absorption tail, and exhibits fluorescence with a broad PL spectrum. The PL emission of such materials results from electron-phonon coupling.[43] Strong electron-phonon coupling in the lattice results in the generation of polarons, which bind charge carriers resulting in excitonic self-trapping. When such self-trapped excitons recombine, they exhibit PL. Another reason for PL emission is the radiative recombination from intrinsic trap states or can be the combination of both.

In conclusion, some theories proposed for PL emission in 0D perovskites are:

i) From some kind of emissive impurities present in the material. However, studies revealed that the emissions do not occur from selected localized domains but from the entire 0D material.
ii) Presence of intrinsic defect states, which act as recombination centers. This is mostly favored because in a large bandgap material such an emission is possible only in the presence of a mid-gap state.
iii) Increased binding energy of self-trapped excitons due to structural distortion.[44]

In some materials, like CsPbBr3, the materials can be fabricated in different forms, for instance, thin films, single crystals, and nanocrystals. The different perovskite phases can be attained by simply altering the crystal growth conditions, like controlling the precursor ratios. As the PL emission in 0D perovskites is insensitive to the growth conditions, especially temperature and synthesis methods, the materials of the same form can be easily reproduced and are highly stable for over a year. The materials experience strong quantum confinement from the complete octahedral isolation which favors strong exciton–phonon interactions and photoexcitaton can result in small-polaron generation.[45] The photoexcitation process perturbs individual octahedral structure and the structural deformation potentials induce the formation of polaronic states having short life time and limited transport. All these fascinating optical features distinguish 0D from their low-dimensional perovskites.

One of the main applications of perovskites is as LEDs. Apart from other materials in this field, they have the benefits of color purity, efficiency, and solution-processed color tunability. However, it faces the problem of instability, which is its main drawback. A solution for these limitations can be made by 0D perovskites, which are highly stable with excellent PLQY.[45] When hybrid organic–inorganic lead halide perovskites are prominent as light harvesters, all-inorganic perovskites in nanocrystalline form are preferred for photon-emitting applications. Since the exciton-binding energies of traditional 3D perovskites are lower, they dissociate easily even at room temperature thus decreasing the efficiency of devices based on 3D perovskites that work based on photon emission.[43] This limitation of 3D perovskites is sorted out by their 0D analogs because the latter possess high excitonic binding energies.

Confinement makes the optical properties of 0D perovskites similar to that of individual octahedrals and increases the effective mass.[46] So, the carriers are now heavy charge carriers. The isolated structure determines small-polaron generation, exciton localization and self-trapping, and optical and photo-physical properties of the 0D structure.[44]

In 0D perovskites, the optical properties can be modulated in a number of ways including high-pressure application, chemical substitution, and thermal treatment.[47] Cs_4PbBr_6 with the same structural formula, the material sometimes behaves as emissive and sometimes as non-emissive. The reason for PL emission is defect states which usually arise from the decrease in unit cell volume along c-axis, which is caused by the metal–halogen bond length contraction. It is also found that green light emitters exhibit paramagnetic behavior while nonemitters exhibit diamagnetic behavior. It is believed that paramagnetism is a result of increased defect concentration. The optical, structural, and magnetic behaviors of the material have close relation with the defect states.[47]

Electrochemical capacitors are a good choice of energy storage devices. Although perovskites have various properties, like high ionic conductivity, beneficial for working as a good capacitor electrode, they lack the feature of high-power density.[48] $(CH_3NH_3)_3Bi_2I_9$ is an example of an electrode material, having 0D perovskite structure. All the benefits of the material make them an excellent material for EDLC application.

Despite that the octahedral units are electronically decoupled in 0D perovskites, the materials exhibit efficient carrier transport and hence can be utilized as stable and nontoxic photovoltaic materials.[43] The high bandgap favors its application in tandem solar cells. Crystal defects have a great influence on the properties of such materials. This is observed when the presence of defects turns the 0D PRM with an insulator bandgap to a highly luminescent material. Having high PLQY in the solid state with enhanced environmental stability, 0D Cs_4PbBr_6 has promising applications as solid-state luminophores. Since Cs_4PbBr_6 materials are efficient green emitters, they have been applied in the fabrication of LED devices, where they act as a phosphor material.[45] These materials have the advent of easy tailoring of color-rendering index (CRI) and correlated color temperature (CCT). It is believed that further studies of the material will enable us to tune the Pb^{2+} ion emissions by changing the 0D perovskite host lattices and promote their application as fast scintillators and as near–UV light sources.

Since the materials are in their stage of development, much information regarding the material, such as exciton formation, defect chemistry, and dissociation mechanism, are yet to be identified and confirmed.[43] This impedes the exploration of these materials as functional devices. Even though it is a cumbersome task to prepare high-purity 0D perovskites, their novel optical properties nurture their further research.

2.3 APPLICATIONS OF PEROVSKITES

The outstanding properties and the tolerance to structural deviation allow multifunctional application of perovskite materials. A few of them are detailed in the following sections.

2.3.1 SOLAR CELLS

The rapid hike in the energy necessities of the world demands alternate sources since fossil fuel gets depleted day-by-day. The earth gets $2.9*10^{15}$ KW of energy every day from the sun which is 100 times more than the world's energy requirement; thus, solar energy is a promising preference over other nonrenewable sources.[49]

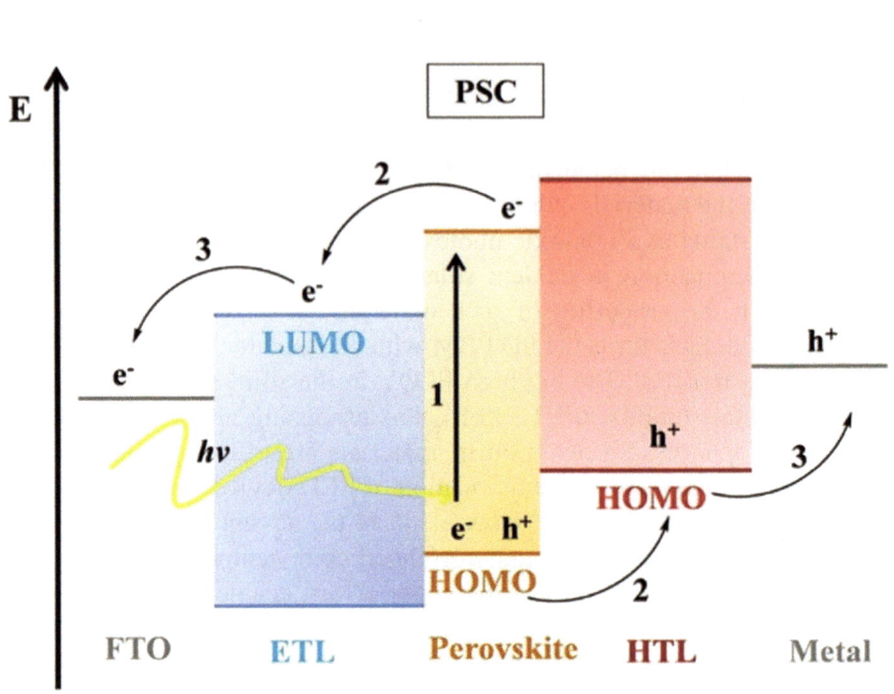

FIGURE 2.5 Working of perovskite solar cell.

Source: Reprinted from Ref [50]. http://creativecommons.org/licenses/by/4.0.

A solar cell is a device that employs photovoltaic effects or photochemical reactions to transform light energy directly into electrical energy. Perovskite has emerged as a game changer in the field of photovoltaics. Being a heterojunction device, a perovskite solar cell (PSC) uses electron transport layer or hole transport layer (or both) for selective harvesting of charge carriers at respective electrodes in order to produce a photovoltage. The mechanism of the formation of electron-hole pairs and their separation is mentioned in Figure 2.5. Despite some early debate, it has been determined that these interfaces are not necessary to break up an exciton, which is an electron-hole pair produced by a photon because the exciton's binding energy is close to the thermal energy being used (a few tens of millielectron volts). This feature of perovskite grounds for excellent internal quantum efficiencies even for the first-reported PSCs, along with a relatively long diffusion length and suitably high electron and hole mobilities.[51]

The power conversion efficiency (PCE) of PSCs rises from 3.8% to 25.7% over a decade. In 2009, Miyasaka and co-workers incorporated perovskite in dye-sensitized solar cell (DSSC) as a thin layer on titanium oxide (TiO2).

Even though the stability was low because of the liquid electrolyte, it shows a PCE of 3.8% and it was considered as a milestone in the journey of PSC.[52] Now, the efficiency gets triggered to 25.7%[53] followed by curious research over structure design, materials chemistry, process engineering, and device physics which is attributed to the renaissance in the field of next-generation solar energy harvesters.

A number of fascinating characteristics of perovskite materials greatly aid its application in solar cells. The materials have great photoelectric performances as well as high optical absorption coefficients (up to 10^4 cm^{-1}) and decreased exciton-binding energies. Furthermore, the light-absorbing layer made of perovskite effectively absorbs solar energy. Moreover, the electrons and holes may be transferred and collected efficiently because of the materials' high dielectric constant. The simultaneous transmission of electrons and holes is possible at distances of up to 100 nm and even more than 1 μm. If perovskite is used in solar cell devices, these properties result in a high open-circuit voltage (V_{oc}) and a short-circuit current density (J_{sc}).[54]

The perovskite system is classified into an inorganic oxide perovskite system and halide perovskite system. The inorganic oxide system is again classified into intrinsic and doped perovskite systems. Recently, halide perovskite witnessed potential growth in the field of photovoltaics which again brought back perovskite to the limelight. They are also classified into alkali halide and organometal perovskite systems.[49]

In most cases, perovskites adopt a 3D crystal structure made up of three main ions with the stoichiometry ABX_3. Cesium is the A-cation, lead or tin is the B-cations, and halides of such as chlorine, bromine, or iodine serve as the X anion in all inorganic compounds. Methylammonium (MA) or formamidinium (FA) serve as the A cations in the "hybrid" perovskite compounds, which have so far been identified as a potential candidate. Different ion combinations can be seen in the A and X sites of efficient and stable compounds. Perovskites and related compounds are receiving increasing attention as a result of their wide range of compositions and several research avenues available to them.[55] 2D and 2D/3D perovskites have received a lot of interest recently because of their superior moisture resistance, higher degree of chemical freedom, and higher formation energy compared to 3D perovskites, which have improved the stability of devices.[56]

The BX_6 octahedra would not be able to preserve their corner-sharing geometry of the ABX_3 3D perovskite, leading to the creation of a layered 2D A'_2BX_4 structure. This might happen if the size of the A-site cation exceeds the volume of the cuboctahedral site generated by the corner-sharing BX_6 network. The family of 2D perovskite crystals includes these compounds. In contrast to 3D perovskite materials, 2D perovskite materials often contain large organic cations with long alkyl chains or aromatic rings, making them less hygroscopic. The charge-carrier transport inorganic network, however, may be hampered by the large organic cations, and as a result, their charge-carrier transport capabilities fall short of those of 3D perovskite materials. Quasi-2D perovskite materials have been used as light-absorbing layers since they have exceptional stability. The $A'_2A_{n-1}B_nX_{3n+1}$ quasi-2D perovskite is created by combining 2D and 3D perovskite materials, with a bilayer of large A′ cations separating the *n* layers of the 3D perovskite. Although they possess exceptional environmental and operational stability, the quasi-2D perovskite materials were detrimental to the PCE of the devices because their bulk crystals had insulating organic layers.[57]

Moving beyond single-junction PV cells to more sophisticated concepts will result in significant efficiency improvement. Using multi-junction principles, where perovskite films are coupled with silicon or other materials to broaden the absorption spectrum range and convert the solar photons into electrical potential energy at a higher voltage, is a short-term way to achieve higher efficiencies. The ability to tune the bandgap of the perovskite layers, ideally from 2 to below 1.2 eV, depending on the absorption characteristics of the materials they are coupled with, is crucial to the success of this strategy. Although bandgap adjustment has been proven by altering the chemical composition or via quantum confinement, only cells with a gap in the 1.5–1.7 eV range give adequately high efficiency and stability.[55]

Even though the perovskites possess remarkable optoelectronic characteristics according to their structural composition, it also makes them unstable in the presence of air. Its stability may be impacted by a number of variables, including heat, light, ion migration, oxygen (O_2), and moisture (H_2O).[58] Another concern about PSC is its toxicity regarding the presence of lead. Although a novel class of air stable perovskite emerged recently, which resolved the instability and toxicity issues, research needed to be conducted to improve its efficiency to make it comparable with lead halide perovskite.

2.3.2 PHOTODETECTOR

The photodetector is a device that detects and measures the optical signal with a wide range of incident photon flux density and exhibits it as a photocurrent. It has extensive applications in science and industry including medical imaging, optical communication, chemical/biological sensing, environmental monitoring, automated production, and military detection.

The simple, low-cost, flexible organic materials, nanomaterials, and nanocomposites replace the traditional photodetector having complex and expensive manufacture, mechanical inflexibility, and high driving voltage. However, the poor charge carrier mobility impedes the overall performance of the photodetector. The emergence of perovskite in the field of optoelectronics resolved the existing complications because it has high charge carrier mobility, effective light absorption, and facile solution processability. The perovskite materials are compatible to use as photodetectors though they have improved responsivity, detectivity, noise equivalent power, linear dynamic range, and response speed.[59,60]

The two primary types of perovskites are single crystals and polycrystalline films. A precise control of the crystallization process is needed for film formation in order to regulate the film morphology, including the homogeneity and surface coverage, which ultimately dictate the performance of the device. The methods for producing defect-free high-quality perovskite can follow the relatively advanced techniques for solar cells. Perovskite single crystals have advantages over perovskite polycrystalline films that suffer from morphological disorders, including low trap density, low intrinsic carrier concentration, high mobility, and long diffusion length. This makes the material perfect for achieving both quick and sensitive photodetection. A variety of perovskite single crystals with exceptional structural and photoelectric capabilities, including 3D large-sized crystals, 2D nanosheets, one-dimensional (1D) nanowires, and 0D QD, show great potential for high-performance photodetectors.[60]

The major perovskite-based photodetectors are classified into the following four categories: photodiodes, photoconductors, phototransistors, and photomultiplication (PM)-type photodetectors according to their operational mechanisms and photoelectric performances. PM (or termed gain)-type photodetectors and photodiodes are vertically structured photodetectors while photoconductors and phototransistors are of lateral design. Photodiodes and photoconductors are two-terminal, dual-electrode devices. Photoconductors can be built with either a horizontal or vertical structure. Due to their built-in amplification function, phototransistors with source, gate, and drain terminals have a more complicated structure and are therefore able to multiply their internal photocurrent gain.[60,61]

The photodiodes have a configuration similar to solar cells but they work with a reverse applied field with low dark currents which promotes fast charge carrier collection. This category includes the most current perovskite-based photodetectors, which include traits like high linearity and minimal noise. The presence of Pb, Br, and I in perovskite polycrystalline films or single crystals helps in detecting X-rays, gamma rays, in addition to regular ultraviolet-visible and near-infrared light. Photodetectors are classified as broadband and narrowband depending on the range of spectral response. The halide-based perovskite is more investigated as broadband photodetector since it enables superficial bandgap tuning from ultraviolet to near infrared. In narrowband photodetectors, only a selective spectral region is needed to be detected. So, it is necessary to suppress usual background or environmental radiation. In recent past, the charge collection narrowing concept has been employed to develop filterless narrowband photodetector.[60]

The photoconductors exhibit an intrinsic photocurrent amplification (photoconductive gain) facilitated by an incident photon which generates an electrical current flow through a number of electrons. Using a vertical or lateral device configuration, photoconductors-based perovskites showed remarkable responsivity. In vertical structure photoconductors, significantly smaller terminal spacing (<500 nm) causes considerable gain with low working voltage and this can be ascribed to perovskite's high charge carrier mobility. The perovskite film devices utilize the vertical design for producing high responsiveness >200 AW^{-1} at low bias (<1 V). This high responsiveness is a result of the accumulation of photogenerated charges at the electrode/perovskite interface that modifies the electrode work function and enables the charge injection into perovskite while illuminated. Relatively high-working voltage is needed for lateral-structure devices for perovskite films with large electrode spacing (>5 μm), which is related to the perovskite film quality. This is owing to the fact that film defects can more easily trap charge carriers during long-distance transit, leading to photocurrent losses. Perovskite nanosheets, wires, and particles make it simple to create lateral-structure devices. Certain photoconductors built on perovskite single crystals with low surface trap density displayed ultrahigh responsivities above 10^3 AW^{-1}.[60]

A high internal photocurrent gain is frequently seen in perovskite-type phototransistors because of their built-in amplification function. In comparison to perovskite photodiodes and photoconductors, it may have an optical response several orders of magnitude larger due to its unique geometry. The incident light can be employed as a supplementary gate to change optical properties. By integrating plasmonic nanoparticles or hybridizing with 2D-layered materials, the phototransistor performance can be further enhanced.[61]

In inorganic PM-type photodetector, photomultiplication occurs by impact ionization triggered by hot carriers whereas the working mechanism of organic PM-type photodetector materials is interfacial trap-assisted charge tunneling injection. The working mechanism similar to organic PM-type photodetector is successfully introduced in the perovskite photodetectors. In perovskite photodetector with a vertical structure, photocurrent multiplication can be achieved by one of two mechanisms: ferroelectric polarization process and interface-barrier-modification mechanism via field-driven ion migration.[62]

Perovskite photodetectors will experience many of the same difficulties as PSC in terms of commercialization, including stability and toxicity concerns as well as competition from more established competing technologies. In comparison to relatively stable inorganic semiconductors like silicon, the long-term stability of perovskites is a significant problem. While some current research aims at using interface engineering with buffer layers, controlling perovskite crystallization morphologies, passivating perovskite surface defects, and endorsing sandwiched device architectures to upgrade the operated stability in the photodetection, it is much more crucial to fabricate chemically stable perovskite materials themselves, which aids to solve the complications of perovskite-based photodetectors in terms of reliability.[60]

2.3.3 PHOTOCATALYSIS

The process of stimulating a photochemical reaction with the help of a catalyst is called photocatalysis. The realm of photocatalysis is governed by semiconductor oxides (e.g., TiO_2, ZnO, etc.) and metal chalcogenides (e.g., CdS and CdSe) till the emergence of perovskite. The facile preparation method and compositional tuning of bandgap enable the perovskite an apt candidate for photocatalysis.[63]

The principle behind photocatalysis is the charge carrier's photogeneration, the separation of charge carriers followed by their diffusion to the photocatalyst surface, and the oxidation-reduction reaction on the surface of the catalyst which is shown in Figure 2.6. When the catalyst absorbs a photon with energy greater than its bandgap, charges are generated and separated since the electrons get transferred through the valence band to the conduction band along with the creation of a hole in the valence band. The photoinduced h+ being an oxidizing agent, directly oxidizes the pollutant directly or surface hydroxyl (OH^-)/water to •OH radicals whereas e- being a reducing agent, reduces the adsorbed oxygen into superoxide radical $O_2\bullet^-$.[65]

In order to increase the photocatalytic activity, two main strategies are decreasing the bandgap of the photocatalysts so that they can absorb more incident light and modify the energy levels within VB and CB to maximize reduction and oxidation ability. Moreover, it is important to reduce surface and bulk charge recombination. The low-defect concentrations and high crystallinity eliminate the recombination centers. By reducing the particle size and shortening the charge diffusion length, charge recombination can be lowered. Another useful method for preventing surface recombination is to separate anode and cathode reactions, particularly in water splitting. However, because superoxide radicals have a low diffusion coefficient, this method is ineffective for photocatalytic degradation. Substantial surface area and the predominance of active facets are necessary to reveal more active sites. It is also possible to use scavengers to fuel the reactions.[66]

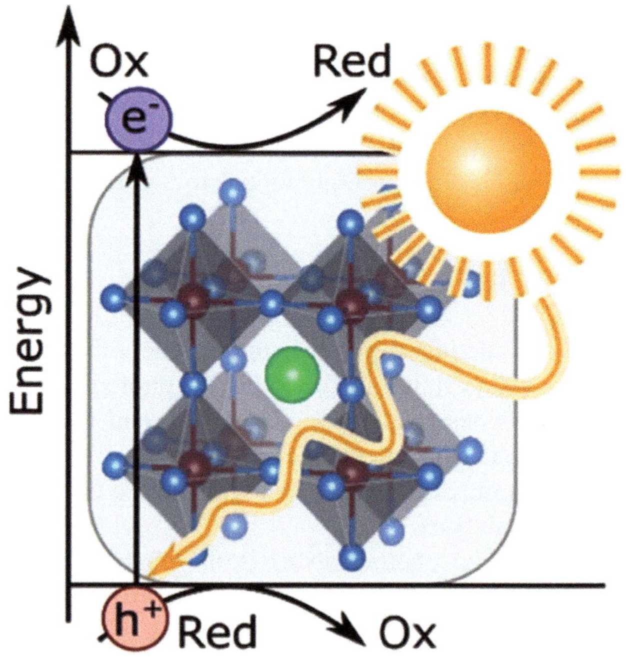

FIGURE 2.6 Mechanism of photocatalysis.

Source: Reprinted with permission from Ref. [64]. © 2020 ACS. https://doi.org/10.1021/acsenergylett.0c00058.

Since the perovskite has a stable framework and considerable space for elemental combination, they construct electronic bands for photocatalytic activities. For photocatalysis, the valency of elements plays a vital role though they determine the light responsivity and photogenerated carrier's lifetime. In photocatalysts, the metal cations' valence state control can be ensured by using perovskites. perovskites. The B-site-substituted metal ions of perovskite, such as Fe^{3+}, Co^{3+}, Ni^{2+}, Cr^{3+}, Ca^{2+}, and Mg^{2+}, can act as extrinsic acceptors, while A-site substituted ions, such as La^{3+}, Sm^{3+}, Nd^{3+}, Pr^{3+}, and Mn^{2+}, can act as extrinsic donors. Furthermore, several metal ions, including Sb^{3+}, Dy^{3+}, Er^{3+}, Ho^{3+}, $Pb^{2+/4+}$, and $Sn^{2+/4+}$, can be positioned at both A and B sites.[65] The photocatalyst's bandgap should be lower than 3 eV. For effective photocatalytic activity, the CB energy level must be more negative compared to the reduction half reaction's potential, and compared to the potential of the oxidation half reaction, the VB energy level must be more positive. In perovskite, doping of foreign elements causes the generation of new intermediate bands between VB and CB, and also the positions of the CB and VB are adjusted by the orbital hybridization of the host and dopant. B-site substitution would be the best way to regulate the bandgap energy based on the contributions of the elements to the CB and VB of perovskites. Doping in A site can also change the bandgap of perovskites and increase carrier mobility because the A-site ions' different sizes can cause the BX6 octahedra to deform, which can change the bandgap and other photoelectronic characteristics. The energy level of the VB rises when the dopant X's electronegativity exceeds that of the X anion already present in the perovskite, enhances the oxidizing power, and widens the bandgap. In the majority of pollution degradation events and in photocatalytic water splitting, due to the intense oxidizing power of oxide perovskite, the anions with lower electronegativity such as N_3 and S_2 are used as dopants for lowering the bandgap. The efficient photoactivity of the mixed-anion perovskites is not guaranteed by a lowered bandgap, as the small bandgap resulting from a negative shift of the VB may lead to insufficient overpotential for the redox processes to occur at the surface.[66]

The A-site doping in perovskite influences the size and morphology. When the particle size gets reduced, bulk recombination also gets reduced due to the shortened diffusion distance and the surface area increases thereby increasing the active site and reactant contact. However, the doping above a particular limit may produce more recombination centers which adversely affect the photocatalysis.[66]

Oxide perovskites serve as a potential photocatalyst for the treatment of wastewater since it has high chemical stability and suitable optical as well as electronic properties. For perovskite photocatalytic reactivity, optical characteristics including light absorption coefficient, scattering coefficient, and refractive index are crucial because they control how light is used and transferred in the material. These factors determine the thermodynamic viability of photooxidation and photoreduction reactions for water purification under light irradiation by getting influenced by the lifetime and destiny of photogenerated electrons and holes.[67]

Halide perovskites have numerous advantages in photocatalysis. Many photocatalytic reactions can be driven by the CBM and VBM locations. Their chemical compositions, ferroelectric, and piezoelectric phenomena enhance photocatalysis. The stability of halide perovskite can be increased by replacing organic MA or FA by Cs^+ ions, by developing heterojunction for extracting photogenerated electrons and holes, and by encapsulating perovskite to avoid its direct contact with the degradable environment.[68]

Some of the layered perovskites are Dion–Jacobson phases (generic formula $A_{n-1}B_nO_{3n+1}$), Ruddlesden–Popper phases (generic formula $an_{+1}B_nO_{3n+1}$), and the Aurivillius phases (generic formula $(A_{n-1}B_nO_{3n+1})_2$) in which the *n* gives the number of perovskite-like layers. Aurivillius phases have been deemed the most effective in producing notable photocatalytic outcomes among these perovskite-related classes.[65]

Light is used in photocatalysis to start chemical reactions such as water splitting, CO_2 reduction, pollutant degradation, etc.[66] Inorganic perovskites, such as $CsPbBr_3$, are employed in the photocatalytic conversion of CO_2 into CO, HCOOH, HCHO, CH_3OH, and CH_4. When transition metal complexes, such as $[Ni(terpy)_2] (PF_6)_2$ (terpy = 2,2′ :6′,2″terpyridine), were immobilized on the perovskite, the photocatalytic performance of $CsPbBr_3$ nanocrystals towards CO_2 reduction to CO/CH_4 was improved (26 times greater compared to pure $CsPbBr_3$). The nanoplatelets of $Cs_2AgBiBr_6$, a lead-free double perovskite structure, were effectively included in the photocatalytic reduction of CO_2 reduction and it has a longer lifespan than its lead counterparts. A highly stable combination of reduced graphene oxide and MA lead iodide perovskite, $MAPbI_3$/rGO, has demonstrated exceptional visible-light activity for H_2 evolution in aqueous HI solutions. In the photocatalytic activity of $MAPbI_3$/rGO, the photogenerated electrons in $MAPbI_3$ get transported to the rGO sites, where the protons get reduced to H_2. Lead-free, non-toxic substitutes have also been suggested. Non-toxic lead-free perovskite 0D-Cs_2SnI_6/FTO films are utilized in PEC water splitting. Despite 70 h of repeated hydrogen evolution devoid of any degradation or oxidation, 0D- $(CH_3NH_3)_3Bi_2I_9$ was utilized because of good cyclic stability for photocatalytic hydrogen creation by HI breakdown. $CsPbBr_3$ QDs that can be used to degrade organic contaminants like 2-mercaptobenzothiazole through photocatalysis. The materials $CsPbCl_3$ and $CsPbBr_3$ QDs exhibit high photocatalytic activity and can completely discolor methyl orange solution in just 100 min. The photocatalytic degradation of organic contaminants was enhanced by mixing $CsPbBr_3$ QDs with 2D MoS_2 and GO materials. The dyes completely mineralize in the case of $Cs_2AgBiBr_6$, and superoxide radicals are the predominant active species.[69]

2.3.4 GAS SENSING

Numerous pollutants are present in the atmosphere which are hazardous not only to humans but also to the environment itself. The emissions occurred most often from automobiles, soil manure, industrial activity, and much more. So, it is very important to constantly monitor the concentration of these gases around us for a healthy environment.[70] This suggests the importance of fabricating highly responsive, low-cost, non-toxic, environmentally friendly gas sensors. Various materials have proved to be good sensors but the growing demands force us to further elaborate the search for future materials that can act as good gas sensors. With the low cost, flexible, simple, and compact fabrication, perovskite materials are a good choice for these materials. The best sensors are those with low working temperature, high selectivity, good sensitivity, fast response- recovery dynamics, long-term stability, lightweight, compact, cost-effective, and can work under large shape deformation. They can have applications in medical fields, the agricultural industry, indoor air safety supervision, the automobile industry, and environmental research. Though it is very difficult to achieve novel morphologies for satisfying all the desired properties in a single material device, through proper material analysis best results can be achieved.

The principle behind sensing oxidizing as well as reducing gases lies in the conductivity change of the material associated with the presence and absence of target gases.[71] The target gas fills in the vacancies of the sensing material and exhibits an increase in its conductivity. The reverse happens when inert gases knock the target gas molecules out of the lattices to create vacancies and hence conductivity gets reduced.

Concerning the gas detection principles, gas sensors can be classified into electrochemical types, catalytic types, metal oxide types, and infrared types. Present-day technology demands gas sensors based on small device size and low power consumption. But due to the extreme working conditions, in most cases, it becomes very difficult to combine both requirements in a single device.[71] Depending on the device prerequisites, different types of materials are being used in gas sensing applications. Though there are many metal oxide semiconductors, such as TiO_2, ZnO, and SnO_2, used as sensing materials in gas sensors, they have the disadvantage of high-temperature operation. The high working temperature of most of the materials is the main hindrance. It increases power consumption at the same time lowers sensitivity. The thicker structure and lower carrier mobility of the metal oxide-based sensors limit them from being used in flexible, wearable, and transparent low-temperature applications. So, further studies are focussed on developing materials favoring room temperature operation.

Perovskite nanomaterials can function as gas sensors in both solid and liquid form.[70] Various all-inorganic perovskites, modified halide perovskites having thin film structures, nanocrystal (NC) structures can all function as gas detectors for recognizing various gases like NO_2, acetone, ethanol, oxygen, etc. In the case of perovskite-based gas sensors, the sensitivity of the material is greatly influenced by the composition and structure of the perovskite. Since it is the grain boundary that permits the penetration of target gas to perovskite thin films, smaller grain sizes are preferred. This provides a high surface-to-volume ratio, fast response, and higher sensitivity.[70]

Metal oxide semiconductors and perovskites differ in their gas detection principle. In a metal oxide semiconductor, the conductivity of the material decreases with gas sensing.[71] When the material absorbs oxygen molecules, they attract electrons from the metal oxide and form O^{2-} ions. This forms a depletion layer, which lowers the conductivity.

For a perovskite of the form $MAPbI_3$, iodine vacancies are the most common. The vacancies are mostly found at the interfacial sites, at the surface, and at the grain boundaries. They act as trap states in the material.[71] In the presence of these trap states, electrons can make transitions through the conduction band to the trap states, making low conductivity. When gas molecules are passed over the film samples, iodine vacancies get passivated and trapped charges become free, so that they can move to the conduction band. This increases the material conductivity. The reverse happens when an inert gas is supplied. They expel the target molecules and restore the crystal defects. The ethanol gas sensor setup based on $MAPbI_3$ perovskite film and the response time graph of the sensor are shown in Figures 2.7 and 2.8, respectively.

Consider the oxide perovskite general formula ABO_3. For oxide perovskites with oxygen vacancies, the general formula changes to $ABX_{3-\delta}$, where the stoichiometry deviates from 3, where delta is the oxygen non stoichiometry. These intrinsic oxygen vacancy sites act as positive charge centers and influence conductivity and hence sensitivity.[72] It can be further enhanced with substitution by a heterovalent cation to create vacancies, heterojunction formation, and catalysts-assisted functionalization.

Oxide perovskite is a good choice of resistive type gas sensors because of (i) stability of constituent metallic elements, (ii) structural and chemical composition tunability with cationic substitution, and (iii) good charge mobility. Oxide perovskites have the ability to host anionic, as well as cationic, vacancies, and have excellent thermal and mechanical stability which influences their gas-sensing property and applications.

FIGURE 2.7 Gas sensing setup based on $MAPbI_3$ perovskite film for ethanol gas sensing.
Source: Reprinted from Ref [72]. http://creativecommons.org/licenses/by/4.0.

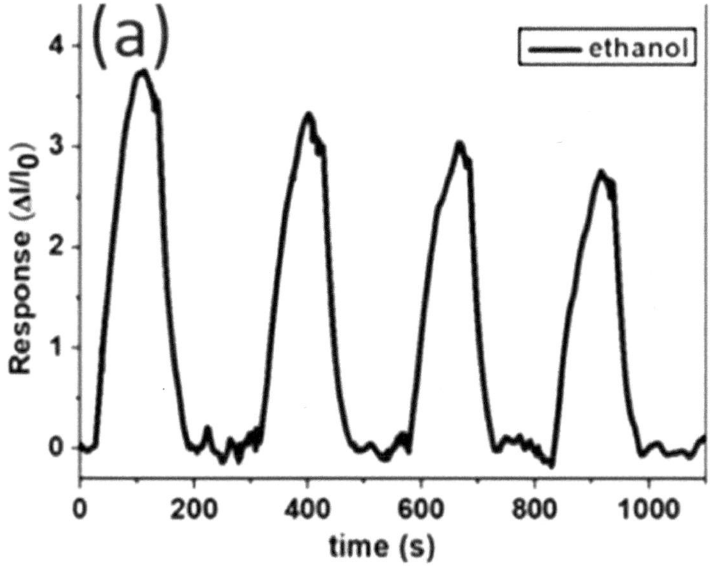

FIGURE 2.8 Response-time graph of the sensor with respect to ON and OFF states of ethanol gas.
Source: Reprinted from Ref [72]. http://creativecommons.org/licenses/by/4.0.

The humidity and interfering gas effects of oxide perovskites can be reduced by the incorporation of suitable chemical or physical fillers on sensing materials. On comparing with metal oxides, the response of oxide perovskite-binary metal oxide composite, with the heterojunction structure displays enhanced sensitivity. In addition to doping and composite formation in oxide perovskites, catalyst inclusion on the perovskite greatly improves oxygen adsorption sites and hence the sensing ability. The most common catalysts are Pd, Au, Ag, etc. The nature of catalysts, synthesis conditions, and catalyst-loading method on host and host–catalyst interactions determine the material responsivity.[72] Oxide perovskites can also act as catalysts and can improve the sensing capacity of the perovskite sensors. The formation of heterojunction and vacancies with the addition of catalysts are the reason for improved sensing activity. Heterojunction formation can create band bending and interfacial charge carrier depletion, which results in increased base resistance. When a material becomes porous the surface area will be higher and hence the concentration of active sites. This favors gas sensing than their dense counterparts.

The sensitivity of the perovskite film is affected by its thickness. With an escalation in thickness, sensitivity is lowered.[71] The following are the two mechanisms that were found to explain the phenomena.

i) With an increase in thickness, only the upper part of the film gets accommodated by the gas molecules
ii) Thicker films have bigger grains and hence lower grain boundary density.

The strong response of semiconductor perovskite sensors is primarily dependent on tunable bandgap, broad absorption, and excellent charge-transport characteristics. Regardless of the benefits of high sensitivity, rapid response speed, and wide-range gas sensing at low working temperature, the limitations of low selectivity, poor stability, irreversibility, and high toxicity cripple their implementation.[73] The aforementioned limitations of hybrid materials suggest 2D perovskite semiconductors with nanosheet structure as the suitable alternative. The 2D layer structure, diverse material composition, and catalytic sensitization of the perovskites provide them with outstanding features like abundant metal oxide compositions, single-crystalline texture, very lower thickness, eminent surface to volume ratio, remarkable anisotropic feature for high-performance, and energy-efficient gas-sensing applications.

In comparison with the hybrid perovskites, inorganic perovskites exhibit high stability against heat, illumination, and moisture.[74] Inorganic perovskite-based NC can be developed into smooth and high-quality thin film structure. As they have applications in LED and solar cells, the large specific surface area of NC enables them to be used for sensing applications.

Among many other volatiles, $CsPbBr_3$ NCs show very high selectivity toward electronegative NO_2. Pristine $CsPbBr_3$ NCs were initially synthesized by hot-injection method and resulted in the cubic crystal structure of $CsPbBr_3$. The low conductivity issue of $CsPbBr_3$ NCs hinders its use in electronics. The low conductivity is a result of the insulation of perovskite NCs by organic capping ligands and defect traps. The large specific surface area of NCs enables selective adsorption of gas molecules on their surface, enabling gas sensing. The prime reason for gas selection and sensing lies in the p-type behavior of the perovskite material. After all, NO_2 is a strong electron-withdrawing gas; when it gets adsorbed on the material's surface, it drags electrons from the material creating holes and hence improves the conductivity of the material which can be sensed and measured. The sensing response of the material is in linear proportion to the target gas concentration. The NCs can detect concentrations as low as 0.4 ppm, and thus are able to function as active materials for NO2 sensing. This explains the fact that why the material provides limited sensitivity for neutral or electron-donating gases. It shows the response to 0_2 gas also.

It is observed that perovskites show higher sensitivity toward polar gases than non-polar ones. $MAPbI_3$ perovskite gas sensor greatly senses VOCs in most of the commercial products, it can function as a low-power, low-cost daily life VOC detector. But because of the lower sensitivity values, the films are not up to the mark for various other applications.[71]

Nanostructured materials with lower dimensions, in comparison with bulk, are more effective for sensing because of the large surface-to-volume ratio, thereby providing numerous active sites for gaseous adsorption. Time-resolved PL studies reveal that Ligand engineering approaches like soft soaking treatment and Ostwald ripening can be carried out to improve conductivity.[71]

2D $NbWO_6$ perovskite, an *n*-type semiconductor, has high sensitivity for H_2S gas. The response-recovery speed is decided by the surface reactions, gas adsorption/desorption behaviors and carrier mobility. The ultra-thin thickness and improved crystallinity of 2D $NbWO_6$ perovskite nanosheets greatly enhance the response-recovery speed than the metal oxide-based sensors. Lower thickness provides higher surface area facilitating diffusion, adsorption-desorption, and transmission for high selectivity gas while the crystalline nature promotes charge transport. The materials exhibit lower sensitivity with humidity. As the response reduces with humidity, a reasonable response can be obtained.[73]

Many mesoporous perovskite nanoparticles, like $BaSnO_3$ nanoparticles, can also efficiently function as gas sensors. The mesoporous structure provides superior specific surface area with 3D interparticle connectivity which enhances the charge collection efficiency and gas sensing.[75]

Approaches to Improve Sensing

- Inclusion of catalysts
  Catalysts enhance carrier movements, promote adsorption sites, catalyze redox reactions, and hence provide better device efficiency. Noble metals were used as catalysts for a long time but their less significant catalytic activity, cost, and agglomerating tendency hinder their wide application and promote the quest for better materials.[77]
- Improving the surface area
  As the specific surface area is larger, it engenders more adsorption sites and hence more sensing.[72]

Effect of Temperature on Gas Sensing

Working of gas sensors is greatly influenced by temperature. The working temperature range needs to be correctly optimized for efficient device performance. The effect of temperature on gas sensing can be explained as follows. When the working temperature is low, the target gas molecules do not receive enough energy to quickly interact with the surface-adsorbed oxygen. This makes low response values. On the other hand, when is temperature is optimum the target molecules get enough interaction energy for complete electron transfer making better responsiveness. But with very high temperatures, molecules gain high energy and vibrate continuously. This makes the surface adsorption of the gas molecules very difficult and hence lowers the sensing.[77]

The chemical instability of some oxide perovskites limits their application as catalysts. Since most of them are synthesized under very high temperatures ultra-small-sized particles are very difficult to be obtained. Synthesizing materials with high porosity is the main challenge in achieving better perovskite materials for gas sensing.[77]

2.3.5 BIOIMAGING

Bioimaging refers to understanding the biochemical activity by introducing chemically inert nanoparticles into the metabolic fluids and thereafter observing the major fluxes in them. Luminescence, magnetic behavior, and high electron density of the materials are the main characteristics required for this application.[78]

Having the advantage of enjoying a stable ABO_3 crystal structure that favors doping with different cations, perovskites are a good choice as nanoparticulate contrasting agents. Doping provides the host with essential features for acting as contrasting agents. Barium titanate-based perovskites are widely used for this purpose. Doping the A site with rare earth atoms gives luminescence while doping the B site with high-valent heavy metal cations contributes them with high electron concentration.

Perovskite QDs with most of the interesting features as that of perovskites have negligible chance of generating deep defects. This happens because the vacancy defects of the material fall well inside the conduction band and the valence band, thereby promoting the PLQY.[79]

As a notion to improve the implementation of PQDs in biotechnology, lead free perovskite based QD are developed. One such example of lead-free perovskites is the bismuth-based perovskite QDs (BQDs). Higher defect tolerance, better PLQY, low toxicity, and water-resistant nature make them a suitable substitute for lead-based perovskites. PLQY of perovskites can be altered by doping with suitable elements. Doping changes, the energy band structure of perovskites thereby introducing luminescent centres and can also influence electronic transitions for reducing non-radiative transitions. The material possesses multiphoton emission and when are doped with Ce^{3+}. The fluorescence peak intensity upgrades and obtains a narrower FWHM. A doping concentration of 30 mol% is considered optimum for the case, as greater and lower doping results in reduced PLQY.[75] Enhanced PLQY in the material also resulted from the reduction in the probability of nonradiative recombinations. The reduction in probability can be attributed to the poor thermal vibration of the lattice at lower temperatures. The excellent stability and biocompatibility of the material promotes them to be used as UC materials for diagnosing human liver hepatocellular carcinoma (HepG2) cells. $Cs_3Bi_2Cl_9$ QDs-based bioimaging is shown in Figure 2.9.

In addition to the enhancement in PLQY and reduced toxicity, doping with Ce^{3+} can improve cellular uptake and biocompatibility and rarely nucleophilicity for targeted applications. This makes the material suitable as up-conversion materials for bioimaging.

IHP lack intermediate energy levels and have low multiphoton efficiency. This makes nonlinear up-conversion emissions (UC) very difficult to achieve in IHPs. Up-conversion process refers to the conversion of two or more photons in to one with a higher energy. On comparing with IHPs, certain inorganic nanoparticles doped with two or more lanthanide ions possess higher stability and enhanced multiphoton absorption efficiency.[76] When Ln NPs are excited with an NIR source it can yield UC emissions in the NIR to

UV range. The higher penetration depth of NIR light in to biosystems leads to their application in phototherapy and bio-imaging. However, employing such superior qualities needed for physiological applications, the presence of limited energy levels of lanthanide ions results in fixed wavelength-assisted fluorescence emissions thereby nontunable emission colors. This limits them to be explored in multiplexed bioimaging.

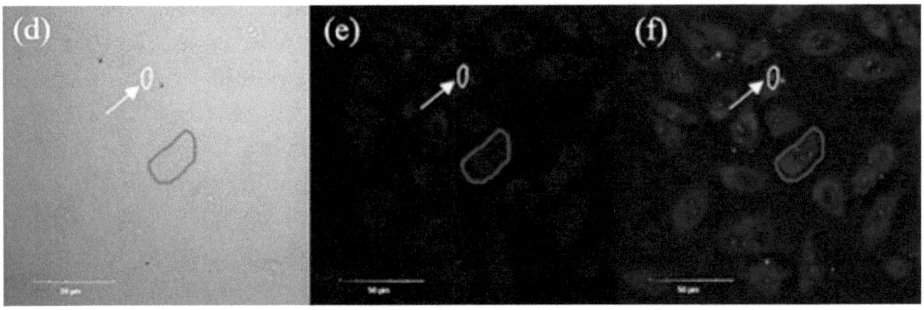

FIGURE 2.9 $Cs_3Bi_2Cl_9$ QDs-based HepG2 cells imaging; brightfield imaging and fluorescence imaging under blue and green channels, respectively.

Source: Reprinted from Ref [80]. http://creativecommons.org/licenses/by/4.0.

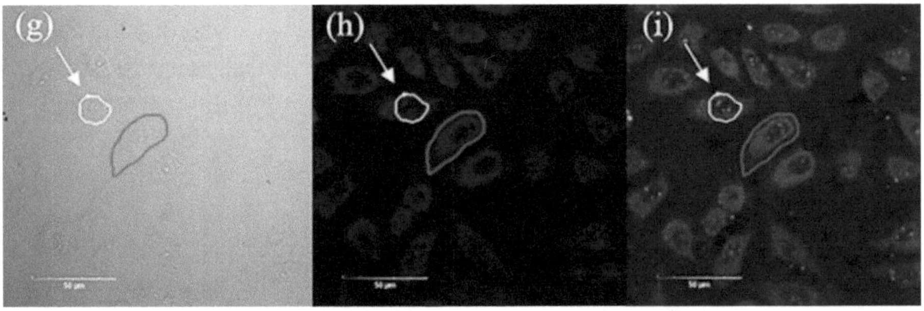

FIGURE 2.10 Ce^{3+} doped $Cs_3Bi_2Cl_9$ QDs based HepG2 cells imaging; Brightfield imaging and fluorescence imaging under blue and green channels, respectively.

Source: Reprinted from Ref. [81]. http://creativecommons.org/licenses/by/4.0.

Halide perovskites suffer from instability and vulnerability due to decomposition and degradation when they come into contact with light, water, and heat. The poor aqueous stability of the material limits their exploration in biological applications.[82] Several strategies are developed to enhance the aqueous stability of perovskite nanocrystals. One of the approaches includes encapsulation of perovskite NCs into polymer matrices where the polymers act as a protective layer against the harsh environment which is capable of decaying the NCs. But for effective encapsulation of NCs for biological applications choice of polymers sets a crucial role. Polymers with low molecular weight need to be selected otherwise the resulting composites achieve sizes in the microscale range. Embedding $CsPbX_3$ NCs into polystyrene microspheres and in polyvinylidene fluoride (PVDF) results in highly aqueous stable NCs but with massive sizes. The former possesses size in the 3–5micrometer range and the latter has an average size of 614 nm. For cellular uptake of nanoparticles by endocytosis, the optimum NP size needs to be below 100 nm in dimension.[82]

The instability problem can also be addressed by coating the NCs with TiO_2 and Al_2O_3, but the complex coating procedures, larger particle sizes, poor aqueous dispersion, and high-cost demand further advancements in procedures.

Encapsulation of $CsPbBr_3$ NCs in silica shells of varying thicknesses and into micelles based on copolymers enhances the procedures by forming composites with in the desirable size range and hence can be effectively utilized for drug delivery and cellular-imaging applications. This mechanism can be resolved by the encapsulation of $CsPbBr_3$ NCs (QDs) by PEG–phospholipid micelles. The so-formed composites take advantage of both PEG and phospholipids. PEG minimizes undesired biomolecule interactions, enables enhanced cellular uptake by target molecule functionalization, and is less cytotoxic whereas the amphiphilic nature of phospholipids imparts the essential water solubility of NCs. This enables the encapsulation of hydrophobic QDs initially in a phospholipid core and thereafter in a PEG-grafted hydrophilic shell, where the shell is functionalized with targeting biomolecules of our requirement for targeted intracellular delivery. This material permits bioimaging by making use of multiphoton excitation. Muliphoton excitation using a femtosecond (fs)-pulsed laser as the light source demonstrates high advantages in bioimaging. The enhanced peak power, small repetition period, and lower average power result in reduced phototoxicity along with insignificant thermal effect. Since the focus area is highly specified, photobleaching of the sample signal can be eliminated and also imparts a deeper penetration depth.[82]

$CsPbBr_3$ NCs with a SiO_2 layer are encapsulated within PEGylated phospholipid (mPEG-DSPE) micelles which surpass the innate instability and vulnerability of bare NCs. The aqueous stable NCs are functionalized with a tumor recognition unit, DSPE–PEG–folate (FA), and are further used for one- and two-photon targeted cellular imaging, especially of cancer cells.

Having the benefit of a facile approach, here the NCs are fabricated under supersaturated recrystallization technique. Besides improving the aqueous stability of NCs such an encapsulation boosts half-life, brings down aggregation, reduces immunogenicity, and improves the pharmacokinetics. Such an encapsulation was found to have an average size of ~30 nm, which is very much suitable for cellular uptake. Both the encapsulated NCs and the bare ones have comparable absorbance and PL emission suggesting the minimal effect of encapsulation on the optical properties of NCs.[82]

The material has also proved to be stable in different pH solutions and biological buffers. Folic acid attachments function as targeting moieties, improving the cellular uptake of the NCs. Multiphoton absorption is associated with deeper penetration depth. For a single-photon imaging, UV or visible light source is used for the excitation of fluorophores. But in two-photon and three-photon imaging, light sources having a wavelength of 700–900 or 1000–1350 nm and wavelength of 1000–1350 nm are used, respectively. Though the cells can be imaged successfully, short wavelength light has the disadvantage of high phototoxicity which may damage the targeted cells and also have short penetration depth. On the contrary, long wavelength light provides lower photodamage, lesser out-of-focus background, higher signal-to-noise ratio, and deeper beam penetration depth.[82] This favors the use of 800 nm fs laser for two-photon excitation processes based on encapsulated $CsPbBr_3$ NCs.

All inorganic $CsPbBr_3$ NCs exhibit better stability by encapsulation in a suitable block polymer like polystyrene-block-poly (acrylic acid), thereby preserving their emission properties with an efficiency of 60% for more than two years in water. The synthesis protocol is insusceptible to surface modification of the NCs and hence can be experimented with for various surface agents. The material in encapsulated form demonstrates stable emission in an acidic environment and can be an alternative for fluorescent dyes, as PL probes for bioimaging. Also, at a dose rate of 0.3 μg_{Pb}/mL, the material exhibits no cytotoxicity.[83] This enables the utilization of these capsules as imaging tools for cellular inspection and biomedical applications, especially in mapping tumoral tissues. Such materials with superior structural and optical stability are of great interest for being used in biological media without any extra treatments. Thus, $CsPbBr_3$ NCs with a high PLQY of approximately 60%, improved water stability, scalable fabrication, robustness, and insignificant toxicity can indisputably be explored for bioimaging.

The ionic nature of perovskites leads to their degradation on aqueous exposure and hence become toxic for biological samples. Though stability issues and toxicity of perovskites hinder such applications, their superior optical properties encourage them to be explored in this field. The difficulty in integrating the materials in biomolecules during circulation, their toxic nature on degradation, stability issues, and their epigenetic interactions need to be mastered before their application.[81]

In the case of all inorganic metal halide perovskites nanocrystals based on Cs^+ at A sites and divalent metal cations at B sites, apart from having high PLQY, tunable bandgap, tunable color emission, and defect tolerance with respect to varying halide composition along under simple low temperature synthesis methods permit their fabrication in variable shapes, sizes, and ligand coating. So, several strategies like shelling, encapsulation, reduction in dimensionality, doping, and using bulky ligands are all employed for achieving longer stability.[83] But the main hurdle is to simultaneously conserve flexible synthesis ways along with high PLQY and stability. The flexible synthesis makes the perovskite NCs superior to metal chalcogenide NCs and fluorescent proteins, having complex synthesis.

2.3.6 MEMRISTIVE DEVICES

The drastic technological developments of our digital world steered the evolution of extremely productive system memory and memory storage. Memory storage technology basically includes optical discs, random access memory (RAM), flash drives, read-only memory (ROM), and solid-state devices. Volatile memory and non-volatile memory, which preserve information even in OFF state, have matured along with these technological advancements.[84]

Current electronic gadgets, such as NAND flash memory, utilize electrical states for data storage. Contemporary memory technologies demand lower costs, better density, and faster writing speeds and research have been done accordingly. Resistive RAM (ReRAM) is an example of cost effective, highly efficient, and high-performance memory storage device.

One possible rival is resistive switching devices, where data is stored between low- and high-resistance states. In RS memory's architecture, in between the bottom and top electrodes, there is a capacitor-like structure having a metal-insulator-metal (MIM) switching layer. Both active electrodes (AEs) like Ag, Cu, and Ni, as well as inert electrodes like Au, Pt, and W, were used. These electrodes function as counter electrodes for reduction and transport channels for the carriers.[85] However, active metals significantly affect RS behavior through electrode electrochemical processes associated with the development and rupturing of the conducting filament (CF).

The top electrode is electrochemically active Ag in the vast majority of the studies. Consequently, applying an electric bias causes Ag cations to migrate and a conducting bridge to develop in the middle of the Ag top and bottom electrodes. When a strong electric field or current comes into play, a dielectric's resistance suddenly changes resulting in RS.

Currently, the hybrid nanocomposites-based nonvolatile memristive devices gained much significance due to their fascinating benefits including facile fabrication, low power consumption, low cost, and their application in logic computing, artificial learning, and memory devices.[86]

Among the different kinds of materials investigated for RS devices, nano-scaled perovskite currently gained relevance. The higher performances of halide perovskites in PV technology are investigated in RS devices for any enhancements in their performance. Applications of perovskite materials in memristor devices emerged initially with oxide perovskites which later on extended to halide perovskites (HPs). Higher PCE, intrinsic hysteresis, ferroelectric properties, and generous charge migration pathways make them to be administered in memristor applications. Though various perovskite structures like all-inorganic perovskites, HPs, lead-free perovskites, and 2D organic–inorganic perovskites are employed for various applications the numerous intrinsic features of HPs like richer dynamics in switching behavior, coexistence, and coupling of ionic and electronic degrees of freedom, tunability of majority carrier concentrations with light and electricity attracted the community.

Compared to oxide perovskite-based RS devices, HP RS devices have benefits in terms of flexibility, simplicity, and low processing temperatures. Further developments are toward engineering lead-free (Pb-free) perovskite RS devices in various fields.

The memristive characteristics of the devices were shown to be influenced by the chemical makeup of the perovskite active layer, where restricted endurance and retention characteristics can result from mobile MA+, Br, and I. Enhanced filament stability with higher endurance and retention properties can be achieved by incorporating multiple cations, such as Rb, in a suppressed MA/Rb ratio.

HPs have generated a lot of attention for use in RS devices due to their remarkable mechanical features, low operation voltage, low manufacturing cost, and high ON/OFF ratio. According to reports, the electroforming origin of resistive switching in perovskites is ascribed to the development and breakage of a conductive filament in the middle of the bottom and top electrode, or the switching mechanism is interface-dominated.

As an interesting application of perovskite in memristive devices, perovskite materials with proper device design can simultaneously harvest solar energy and also exhibit memristive functionalities. Stable resistance switching features along with high PCE can be achieved with proper memristive channel formation for multiple switching cycles. Further improvements in device performance can be achieved by analyzing the impact of each layer on device performance, thereby allowing structural modifications. The various device interfaces have great role on upgrading the overall device performance of MemPVCell device.[87]

Flexible and ultra-high density-based next-generation memory applications of nonvolatile memory devices favors lower-dimensional hybrid perovskite nanocomposites in a polymeric matrix, where the nanomaterials functions as charge storage sites. $CsPbCl_3$ QDs embedded in PMMA polymer matrix is an example of such material. The highly noticeable insulator and dielectric properties of PMMA are reasonable for their selection as the matrix. Since PQD acts as charge storage sites, optimum PQD mass fraction is essential for the best electrical behaviors of memristive devices.[88]

Though with all the advancements, HP-based RS memory devices still encounter a number of difficulties as a consequence of the short lifespan and unstable nature of the HP films.

2.3.7 LIGHT EMITTING DIODE

Being a semiconductor device, light emitting diode (LED) emits light when current flows through it. While passing current through this p-n junction diode, electrons and holes recombine releasing photons and this effect is termed electroluminescence. The energy bandgap of the selected semiconductor determines the color of the light. LEDs are extensively used in everyday lighting and displays. Perovskite materials are anticipated to be crucial in enhancing LED performance and extending their capabilities. Since they are economical, simple-to-manufacture, and can yield light of different colors, direct bandgap semiconductor perovskites could be viable replacements for other types of direct bandgap materials for applications like color displays. It is easy to tune the light emitted by perovskites which makes them potential candidates for optoelectronic applications.[89] Perovskite LED efficiencies are now on pace with some of the best quantum-dot LEDs and

OLEDs, with external quantum efficiencies reaching 20%.[90] That is the electroluminescence (EL) efficiency of perovskite LEDs (PeLEDs) has been exceptionally upgraded in a relatively short span.

PeLEDs have a typical hole-injecting layer (HIL), an electron-injecting layer (EIL), and a perovskite emission layer in between. Electrons and holes are primarily injected into the HIL and EIL, respectively, on applying forward bias. Then, under the applied field, carriers drift into the perovskite layer. As a result of the radiative recombination of the carriers in the perovskite layer, electroluminescence occurs. Luminance, EQE, and current efficiency are the key measures for determining the PeLED performance.[91]

LEDs based on metal halide perovskite materials find use in solid-state lightning and flat panel displays, as they can produce tunable colors, have narrow emission line widths, high PLQY, and are due to inexpensive solution processing.[92] Other desirable optoelectronic properties of metal halide perovskites include high absorption coefficient, high carrier mobility, easily adjustable emission wavelengths, broad color gamut, etc.

Advancement in the EL efficiency of perovskite emitters is required to facilitate their use in practical devices. To do so, one has to suppress the non-radiative recombination rate and at the same time, the radiative recombination should be increased. The former can be done by passivating defects while the latter by confining the charge carriers spatially.[93]

Nonradiative recombination and light trapping often limits the efficiency of LEDs. About 70–80% of light created from emitters/emission layer in planar LEDs, like organic LEDs, is trapped in the gadget. To extract the light trapped in the device, several techniques have been used. However, these techniques usually demand complex fabrication procedures and can alter the directionality and light output spectrum. Since the perovskites have a much greater refractive index than the organic materials, light trapping can be more challenging in perovskite LEDs when compared to organic LEDs. Cao et al. demonstrated an efficient way of extracting trapped light from perovskite LEDs with perovskites that impetuously form sub-micrometer scale structures. To the precursor solutions, amino acid additives are simply added to yield the perovskite. The formation of distinct sub-micrometer-structured perovskite can be made evident from the STEM image of the emitting layer. In a similar fashion, the homogeneous distribution of sub-micrometer structure on the substrate can be confirmed by the SEM and AFM studies. Low trap-assisted nonradiative recombination is reported which can be assigned to the presence of amino acid additive. Moreover, it does the crucial function of passivating the surface defects and improving perovskite emission characteristics. The peak EQE reported is 20.7%, which sets a record efficiency for perovskite LEDs. The primary reason for such an excellent EQE can be attributed to the improved light outcoupling efficiency of the spontaneously formed sub-micrometer structures. Outcoupling efficiency and PLQE can be improved even more with the application of more effective additives.[94]

In most cases, the EQE of red and green light PeLEDS lags far behind the organic LEDs as well as inorganic QD LEDs. By adopting a novel approach to managing the compositional distribution within the device, balanced charge injection as well as high luminesce can be obtained. In this method, perovskite light-emitting polycrystals constitute the bottom part of the quasi-core/shell structure. A defect-passivation layer covering the bottom part passivates the grain boundaries. On the other hand, the top portion consists of passivating layer which balances the charge injection into the device as well. Fabricating compositionally graded structure of CsPbBr3 perovskite and MABr passivant on the bottom and [(onto this defect-passivated perovskite layer (CsPbBr3/MABr)] MABr electrical passivating layer on top yielded an EQE of 17%. Here, the charge injection is balanced by MABr capping layer and the nonradiative defect sites in CsPbBr3 crystals are passivated by MABr shell. Efficiency rose to 20.3% with the further addition of a poly(methyl methacrylate) (PMMA) insulating layer between the electron transfer layer and the perovskite layer. A perovskite layer with small roughness and large surface coverage provides high performance LEDs. However, improving the stability is a major concern.[92]

Recently, a solution process for creating extremely efficient perovskite LEDs employing nanometre-sized crystallites has been developed. The use of small grain sizes or ultra-thin layers of perovskites promotes bimolecular-radiative recombination by spatially confining the electrons and holes. A drastic decrease in the film roughness and impediment of the growth of 3D perovskite grains are observed in the inclusion of long-chain ammonium halides like n-butyl ammonium halides [BAX, X=I, Br]. There was a tremendous increase in EQE from 1% to 10.4% for iodide perovskites, with the addition of BAI. Likewise, with the incorporation of BABr, EQE changed from 0.03% to 9.3%, for bromide perovskites. Similar improvement can be witnessed in terms of current and power efficiency. Since stability is essential for perovskite LEDs, shelf stability is investigated. The unencapsulated LEDs are stored in N2 and are observed over time. Subsequently, the perovskite LEDs with BAX-incorporated layers showcased stable EQE for a longer period with no signs of degradation. There were no traits of degradation for more than eight months for I-perovskite LEDs and for more than four months for the bromine ones. On the contrary, those without BAX displayed a rapid fall in EQE within two days. Also, employing several strategies like using more bulky organic ligands and reducing turn-on voltage helps to enhance the operational stability.[95]

The electroluminescence performance of the very first hybrid MAPbBr3 perovskite NCs-based LEDs was very poor/low. Later, all inorganic CsPbX3 NCs have also been investigated for LED application. CsPbX3 QDs-based devices were fabricated and by altering the anion composition of QDs orange, green, and blue devices were created. Nonetheless, the external quantum efficiency and electroluminescence were relatively low. Besides this, perovskite NCs can be utilized for white light emission. CsPbX3 QDs/polymethyl methacrylate composites were constructed for devices emitting white light. However, the device performance of LEDs based on perovskite QDs must be enhanced before commercialization.[96]

Perovskite nanocrystals make excellent light-emitting diodes as thin films can be created from earth-abundant precursors utilizing solution techniques. These perovskite nanocrystal-based LEDs have attained an EQE of 20%.[92] Notwithstanding these benefits, stability is of major concern. Perovskite nanocrystals as a host in metal-organic frameworks (MoFs) is recently studied. This material is promising due to the integration of MoFs' porous structure and perovskite nanocrystals' optoelectronic features. The use of perovskite- MoF (PeMoF) thin films as an emission layer created stable and bright LEDs. By the excitation of moderately high-power lasers, these thin films of perovskite nanocrystals can yield a high PLQY of over 70%, under optimized processing circumstances. Even for conditions like, electrical stress, heat, and constant-UV exposure, PeMoF films exhibit stable emission features. The device is devoid of nanocrystal merging, ion migration, and interface corrosion, for which the MoF matrix's protection is responsible. The operational stability of LED is strongly determined by the PeMoF processing conditions. Similarly, the conversion yield is directly linked to the nanocrystal distribution and size. All the enhanced stability can be attributed to the protection provided by MoF matrix.[97]

Engineering of interfacial layers, perovskite film morphologies, electrodes, dimensional control, and stoichiometry of precursors and additives have significantly increased the EL efficiency of PeLEDs. Perovskite LEDs' emission strongly corresponds with self-trapping of an exciton, the interaction between exciton and phonon, and the crystalline system's dimensionality. So, by modifying the trap states and improving the crystalline morphology, the device emission and quantum efficiency can be boosted.[99]

2.3.8 LASERS

Being a coherent light source based on stimulated emission, laser offers high color purity, coherence, output light intensity, and good directivity. As a result of the narrow luminescence peak at the bandgap energy and extremely efficient photoluminescence, perovskites have been identified as appealing and potentially useful materials for lasting applications.[99] Perovskite lasers have limitless potential for use in optical interconnects, optical storage, on-chip photonic communications, high-quality displays, and other areas.[100]

In laser, population inversion within the gain media is used to generate stimulated emission. After the excited carriers are radiatively relaxed back to the ground state, a photon triggers the emission of another photon, which increases the intensity of the pumping light.[101] So, the work basically involves the use of a pump source, with adequate intensity, to excite the gain medium, the creation of population inversion and ASE, and the resonant cavity deciding the frequency and propagation direction of the photon.[100] On the contrary, there exist lost mechanisms like nonradiative processes, reabsorption of photons, and scattering. Once the gain exceeds the total losses, lasing happens.

For creating lasers, a variety of effective gain media are offered by metal halide perovskites. When served as a gain medium, perovskites often show an elevated optical gain coefficient. Furthermore, perovskites' lower defect density helps in lowering the optical losses thereby paving the way for low-threshold-lasing behavior. They have also exhibited broad wavelength tunability and high-quality factors. In general, all the remarkable LASER performance can be attributed to factors like minimal nonradiative recombination rates, long carrier lifetimes, high optical gain, ease of bandgap engineering, good PLQY, and above all, the facile material synthesis.[99] By utilizing the thin film or low-dimensional structure (nanoplatelet, nanowire nanocrystal) of perovskites as active materials, numerous beneficial high-performance applications have been developed, such as edge-emitting amplified spontaneous emission (ASE), vertical cavity surface emitting lasers (VCSELs), distributed feedback (DFB) lasers, microlasers, etc.

As a result of their lifetime, which is fairly long, low electric injection power, compact volume, and with an established semiconductor technology resulting in good integration, semiconductor lasers belong to the most significant and widely used lasers. Even so, the gain materials for commercially applicable semiconductor lasers are created using energy-intensive and expensive techniques such as molecular beam epitaxy and metal-organic chemical vapour deposition. Subsequently, significant crystal disorder and subpar electrical characteristics restricted the development of affordable, low-temperature solution-processable organic lasers. In this context, metal halide perovskites are extremely promising because of their exceptional optoelectronic features, such as high absorption/emission efficiency, high defect tolerance, broadband emitted color, bipolar long diffusion distance, and most importantly, their low-temperature solution-processable synthesis.[101]

By substituting the monovalent cation or halides, perovskites' emission spectra can be modified, indicating a wide range of SE spectra for the perovskite lasers. The compatibility of perovskites with various optical microcavities is evident with the development of several laser modes including Fabry–Perot mode, distributed feedback mode, whispering gallery mode, random lasing, etc. Similarly, progress can also be observed in terms of development from pulsed optically pumped lasers to continuous-wave (CW) optically pumped lasers and now leading up to perovskite electrically pumped lasers. Electrically pumped perovskite lasers are still in their initial phases, although pulsed and CW optically pumped perovskite lasers have greatly advanced. Now, based on the type of optical microcavity, perovskite lasers fall into two distinct classes; active cavity lasers and passive cavity lasers. Autologous cavity or active cavity lasers' flat smooth crystal surface serves as reflector mirrors and offer optical feedback. This includes nanowire (NW) lasers and WGM lasers. The passive cavity or exogenous cavity lasers that use polycrystalline films or nanocrystals as their gain medium require the coupling of additional optical resonators to accomplish lasing oscillation. Examples are DFB lasers and distributed Bragg reflector (DBR) lasers.[100]

Compared to organics and colloidal QDs, perovskites have high carrier mobility allowing rapid carrier transport and minimal heat loss, which in turn facilitates the development of high current density and raises the possibility of electrically pumped lasers. Toward realizing the high stability, high power, and low threshold electrically pumped perovskite lasers, gain medium has to meet certain conditions including balanced charge carrier mobility, slow nonradiative decay, sharp band tails, and high gain cross-section at the lasing wavelength. All these features, apart from high carrier density and stability, have been exhibited by perovskites. Electric field-induced ion migration is the dominant issue affecting stability. Various strategies like adding ionic additives, using bulky organic cations, small organic molecules or ligands for covering grain boundaries, etc., have been employed for suppressing ion migration and thereby enhancing stability.[100]

Several perovskite lasers are found to function better in terms of high coherence, low threshold, multicolor tunability, and cost-effectiveness. However, a number of challenges, such as their stability in ambient conditions, also limit the practical applications of perovskite lasers.

2.4 CHALLENGES

Even though perovskites exhibit versatile properties such as a high absorption coefficient, excellent compositional tuning, optimum bandgap, high charge carrier mobility, high diffusion length, defect tolerance, etc., some factors challenge the practical application of perovskites. The main intrinsic factors affecting the stability of perovskites are chemical degradation, phase transformation, segregation of perovskite, and ion migration.

In perovskite film having A-site organic ammonium ions, its chemical reactivity gets boosted by polar surroundings of C–N or N–H bonding that adversely affects halide perovskites' decomposition rate at raised temperatures. For the most part, water intrusion expands the unit cells rather than changing the crystal structure. It is observed that chemisorbed water molecules on the surface of perovskite do not alter the material's bandgap and optical properties, because they interact with the perovskite surface electrostatically rather than by creating chemical bonds. A strong hydrogen bond is created between organic cations and iodides by the penetration of water, causing a weak bond between PbX_6 octahedron and the cation. Additionally, the protonation of iodide by water molecules results in the production of volatile hydroiodide acid (HI), which accelerates the breakdown of perovskite into its corresponding PbX_2 precipitate, halide anion, and cations.[102]

The distortion of cubic structure developed by the incompatibility in the size of A-cation leads to the phase transformation from a photo-active α-phase to photo-inactive δ-phase. The distortion also occurs when the high electronegative ion is incorporated within the cubic framework because it creates hydrogen bonding with PbX_2 octahedron, also leads to the breakage of the protonated cation-hydrogen original bond. Beyond adhering to the tolerance rule, adjusting the electronic coupling between the molecular A-cation and the BX_6 octahedral of the perovskite structure is another potential method for stability. For example, $MAPbI_3$ is a widely used absorber that has electronegative N-atom which is stronger than I^- anion. The N-atom induces the H-atoms to MA^+ cations, leading to weak interaction with PbI_6 octahedron. An unstable $MAPbI_3$ structure is a result of an imbalance in the electron coupling between the protonated cations and the halides in the octahedron. By the density functional theory approach, the stability can be improved by balancing the electron coupling between the protonated cations and halides in the octahedral.

Another issue that has recently come out and received a lot of attention in this perovskite community is phase separation. Halide segregation is often triggered by carrier gradient created during the illumination of the surface since I^- migrates much more slowly than Br^-, I^- enriched zones arise as Br is exhausted from the surface. The four factors leading to the selective halide ions migration are: (i) cation or anion hybrid perovskite's lattice mismatch upon photo-irradiation; (ii) the presence of an electric field; (iii) vacancy-impacted halide ion migration; and (iv) the dissimilar affinities of Pb^{2+} toward Br^- and I^-.

Ion migration is a critical and well-recognized problem with perovskite materials, impeding their practical use in photovoltaics and posing a threat to their long-term stability. During operation, the mobile ions deteriorate perovskites and produce halogen segregation, which alters the local bandgap of perovskites. Ion migrations might get speeded up by voltage bias and light irradiation. The kinds of defects and densities of perovskite films are highly correlated with ion movement. Ion migrations are more likely to be reduced in high quality perovskites with fewer defects. High efficiency and stable operation are produced using perovskite films with improved crystallinity and quality.[103] Being responsible for the J-V curve hysteresis and causing irretrievable structural destruction in ambient air, ion immigration causes significant harm to the PSC and photodetectors.[104]

The major environmental conditions deteriorating the perovskites are moisture/water, oxygen, light, and temperature. The creation of powerful hydrogen bonds between the organic amines within the perovskites and the surrounding water molecules is the root cause of this deterioration. In addition to compositions, microstructures like grain size are reported to have an impact on perovskite devices' moisture stability. These studies provide credence to the idea that larger grains lead to smaller grain borders, which correlate with increased stability.

The photoelectrons form an O_2^- by reacting with oxygen, and this O_2^- is then energetically favored to bind to the iodine vacancy, reviving the entire octahedral coordination of Pb_2^+. The perovskite is then decomposed into PbI_2, water, and other by-products as a result of the photoelectron's reaction with the organic component in $MAPbI_3$. By switching the MA component in $MAPbI_3$ to a species without acid protons, the tolerance to oxygen may be enhanced. An extensive analysis of compositional influences on the stability of perovskites in exposure to oxygen and light further corroborated this. The deterioration brought on by oxygen started at grain boundaries and later spread inside the grains.[103]

The lattice expansion, defect formation, and annihilation are found under short-time illumination whereas the degradation of perovskites is noticed upon long-time illumination. The photostability can be improved by the passivation of the surface and grain boundary which served as a defect reservoir. The thermal instability of perovskites is yet another feature to progress. Since annealing is required for perovskite film fabrication, they will likely be exposed to high temperatures. Even in the inert atmosphere, $MAPbI_3$ degrades at a temperature of 85 °C due to its limited thermal conductivity. The problem caused by the thermal effect is even more regrettable for the PSC since the heat created by the irradiation cannot be easily eliminated. Because it is volatile by nature, the organic cation component tends to decrease at elevated temperatures. At 85 °C, heating in a nitrogenous atmosphere causes the perovskite layer to break down into PbI_2, completely removing the organic component (MA). The experiment was carried out for 24 h in pure dry oxygen and pure dry nitrogen at ambient pressure and 50% relative humidity. This fact was further supported by testing the thermal stability of $MAPbI_3$ films at three different temperatures: room temperature, 100 °C, and 200 °C. The $MAPbI_3$ to PbI_2 ratio was calculated using the hard X-ray photoelectron spectroscopy (HAXPES) study, and it was found to be 85:15 to 73:30, 0:100, respectively. This provided more evidence that $MAPbI_3$ begins to convert to PbI_2 as the temperature rises. Thus, it is required to maintain thermal stability up to 85 °C for a stable PV cell. A constant exposure to 85 °C degrades $MAPbI_3$ since the formation energy range of $MAPbI_3$ lies between 0.11 to 0.14 eV.[104]

2.5 SUMMARY

The chapter discusses the basic structure of the perovskite ABX_3 and its derivatives. The two major parameters, that is, tolerance and octahedral factors which depend upon the ionic radii, determine the stability and the structure of the perovskite. Mobility of the BX_6 octahedra network enables the architectural and compositional adaptability of the perovskites. The material exhibits outstanding qualities such as dielectric properties, multiferroicity, colossal magneto-resistance, high absorption, fluorescence, piezoelectricity, superconductivity, catalytic, and photovoltaic activity which triggers the multifunctionality of the material. Lower-dimensional structures also have shown amazing characteristics. The distinguishing properties of halide and oxide perovskites are analyzed in detail. The importance of the HHP is also emphasized in the report. Application of the perovskite material in the fields of solar cells, photodetector, photocatalysis, gas sensing, memristive devices, bio-imaging, LED, and LASER are described with examples. Apart from the advantages and significance of perovskite materials, the challenges in scaling up the technology are also discussed briefly, the major one being the instability. Future research must take two strategies into account for increasing the stability: (1) appropriate encapsulation and (2) enhanced inherent moisture resistance. Encapsulation will enable the technology to get closer to the optimal lifetimes that were calculated in controlled circumstances, and increased perovskite stability will hasten the technological progress toward commercialization. It is envisaged that materials suitable for commercial mass manufacturing must possess excellent intrinsic stability and tolerance to operational stress factors.

KEYWORDS

- **perovskites**
- **tolerance factor**
- **stability**
- **absorption coefficient**

REFERENCES

1. Chakhmouradian, A. R.; Woodward, P. M.; Celebrating 175 Years of Perovskite Research: A Tribute to Roger H. Mitchell. *Phys Chem Miner*. **2014,** *41*, 387–391.
2. Katz, E. A.; Perovskite: Name Puzzle and German-Russian Odyssey of Discovery. *Helv Chim Acta*. **2020,** *103*.
3. Assirey, E. A. Perovskite Synthesis, Properties and Their Related Biochemical and Industrial Application. *Saudi Pharm. J.* **2019,** *27*, 817–829.
4. Johnsson, M.; Lemmens, P.; Crystallography and Chemistry of Perovskites. In: *Handbook of Magnetism and Advanced Magnetic Materials*, 2007.
5. Islam, M. A.; Rondinelli, J. M.; Spanier, J. E. Normal Mode Determination of Perovskite Crystal Structures with Octahedral Rotations: Theory and Applications. *J. Phys. Condens. Matter.* **2013,** *25*, 175902.

6. Howard, C. J.; Stokes, H. T. Group-Theoretical Analysis of Octahedral Tilting in Perovskites. *Acta Crystallogr. B.* **1998,** *54*, 782–789.
7. Filip, M. R.; Giustino, F. The Geometric Blueprint of Perovskites. *Proc. Natl. Acad. Sci. India Sect. B Biol. Sci.* **2018,** *115*, 5397–5402.
8. D, T. R. J. *Perovskites Structure-Property Relationships*; Wiley: Chichester, 2016.
9. Kumar, A.; Verma, A. S.; Bhardwaj, S. R.; Prediction of Formability in Perovskite-Type Oxides. *Open Appl. Phys. J.* **2008,** *1*, 11–19.
10. Wang, H.-L. *Structure and Dielectric Properties of Perovskite -Barium Titanate ($BaTiO_3$)*; Dissertation, 2002.
11. Kim, Y.-I.; Woodward, P. M.; Baba-Kishi, K. Z.; Tai, C. W. Characterization of the Structural, Optical, and Dielectric Properties of Oxynitride Perovskites AMO_2N (A = Ba, Sr, CA; M = Ta, NB*). Chem. Mater.* **2004,** *16*, 1267–1276.
12. Zhang, Y.-R.; Motohashi, T.; Masubuchi, Y.; Kikkawa, S. Sintering and Dielectric Properties of Perovskite $SrTaO_2N$ Ceramics. *J. Eur. Ceram. Soc.* **2012,** *32*, 1269–1274.
13. Arul, N. S.; Nithya, V. D. *Revolution of Perovskite: Synthesis, Properties and Applications*; Springer: Singapore, 2020.
14. Liu, H.; Yang, X. A Brief Review on Perovskite Multiferroics. *Ferroelectricals* **2017,** *507*, 69–85.
15. Raveau, B.; Maignan, A.; Martin, C.; Hervieu, M. Colossal Magnetoresistance Manganite Perovskites: Relations between Crystal Chemistry and Properties. *Chem. Mater.* **1998,** *10*, 2641–2652.
16. Hwang, J.; Rao, R. R.; Giordano, L.; Katayama, Y.; Yu, Y.; Shao-Horn, Y. Perovskites in Catalysis and Electrocatalysis. *NY Sci. J.* **2017,** *358*, 751–756.
17. Ishihara, T. *Perovskite Oxide for Solid State Fuel Cells*; Springer, 2009.
18. Peña, M. A.; Fierro, J. L. Chemical Structures and Performance of Perovskite Oxides. *Chem. Rev.* **2001,** *101*, 1981–2018.
19. Lata, C.; Sushant G.; Challapalli S.; Tsutomu M. Vasudevanpillai, B. Synthesis, Optoelectronic Properties and Applications of Halide Perovskites. *Chem. Soc. Rev.* **2020,** *49*, 2869–2885.
20. Yin, W.-J.; Yang, J.-H.; Kang, J.; Yan, Y.; Wei, S.-H. Halide Perovskite Materials for Solar Cells: A Theoretical Review. *J. Mater. Chem. A.* **2015,** *3*, 8926–8942.
21. Egger, D. A.; Bera, A.; Cahen, D.; Hodes, G.; Kirchartz, T.; Kronik, L.; Lovrincic, R.; Rappe, A. M.; Reichman, D. R.; Yaffe, O. What Remains Unexplained About the Properties of Halide Perovskites. *Adv. Mater.* **2018,** *30*, 1800691.
22. Kim, H.; Han, J. S.; Choi, J.; Kim, S. Y.; Jang, H. W. Halide Perovskites for Applications Beyond Photovoltaics. *Small Methods.* **2018,** *2*, 1700310.
23. Nguyen, L. T.; Cava, R. J. Hexagonal Perovskites as Quantum Materials. *Chem. Rev.* **2020,** *121*, 2935–2965.
24. Xu, X.; Zhong, Y.; Shao, Z. Double Perovskites in Catalysis, Electrocatalysis, and Photo (Electro) Catalysis. *Trends Analyt Chem.* **2019,** *1*, 410–424.
25. Afroze, S.; Karim, A.; Cheok, Q.; Eriksson, S.; Azad, A. K. Latest Development of Double Perovskite Electrode Materials for Solid Oxide Fuel Cells: A Review. *Frontiers* **2019,** *13*, 770–797.
26. Dave, K.; Fang, M. H.; Bao, Z.; Fu, H. T.; Liu, R. S. Recent Developments in Lead-Free Double Perovskites: Structure, Doping, and Applications. *Chem. Asian J.* **2020,** *15*, 242–252.

27. Chu, L.; Ahmad, W.; Liu, W.; Yang, J.; Zhang, R.; Sun, Y.; Li, X. A. Lead-Free Halide Double Perovskite Materials: A New Superstar Toward Green and Stable Optoelectronic Applications. *Nanomicro. Lett.* **2019,** *11*, 1–18.
28. Chen, X.; Xu, J.; Xu, Y.; Luo, F.; Du, Y. Rare Earth Double Perovskites: A Fertile Soil in the Field of Perovskite Oxides. *Inorg. Chem. Front.* **2019,** *6*, 2226–2238.
29. Stoumpos, C. C.; Mao, L.; Malliakas, C. D.; Kanatzidis, M. G. Structure–Band Gap Relationships in Hexagonal Polytypes and Low-Dimensional Structures of Hybrid Tin Iodide Perovskites. *Inorg. Chem.* **2017,** *56*, 56–73.
30. Saparov, B.; Mitzi, D. B. Organic–Inorganic Perovskites: Structural Versatility for Functional Materials Design. *Chem. Rev.* **2016,** *116*, 4558–4596.
31. Kandada, A. R. S.; Silva, C.; Exciton Polarons in Two-Dimensional Hybrid Metal-Halide Perovskites. *J. Phys. Chem. Lett.* **2020,** *11*, 3173–3184.
32. Lekina, Y.; Shen, Z. X. Excitonic States and Structural Stability in Two-Dimensional Hybrid Organic-Inorganic Perovskites. *J. Sci.: Adv. Mater. Dev.* **2019,** *4*, 189–200.
33. Hou,Y.; Wu,C.;Yang, D.; Ye, T.; Honavar, V. G.; Duin, A. C. T. V.; Wang, K.; Priyanka, S.; Two-Dimensional Hybrid Organic–Inorganic Perovskites as Emergent Ferroelectric Materials. *J. Appl. Phys.* **2020,** *128*, 060906.
34. Wang, Z.; Ganose, A. M.; Niu, C.; Scanlon, D. O. First-Principles Insights into Tin-Based Two-Dimensional Hybrid Halide Perovskites for Photovoltaics. *J. Mater. Chem. A.* **2018,** *6*, 5652–5660.
35. Fujii, Y.; Ramirez, D.; Rosero-Navarro, N. C.; Jullian, D.; Miura, A.; Jaramillo, F.; Tadanaga, K. *ACS Appl. Energy Mater.* **2019,** *2*, 6569–6576.
36. Zhang, Y.; Saidaminov, M. I.; Dursun, I.; Yang. H.; Murali, B.; Yengel, E.;Alarousu, E.; Alshankiti, B. A.; Bakr, O. M.; Mohammed, O. F. *J. Phys. Chem. Lett.* **2017,** *8*, 961–965.
37. Nguyen, L. T.; Cava, R. J. Hexagonal Perovskites as Quantum Materials. *Chem. Rev.* **2020,** *121*, 2935–2965.
38. Mercier. N. Hybrid Halide Perovskites: Discussions on Terminology and Materials. *Angew. Chem. Int. Ed.* **2019,** *58*, 2–8.
39. Stoumpos, C. C.; Mao, L.; Malliakas, C. D.; Kanatzidis, M. G.; Structure–Band Gap Relationships in Hexagonal Polytypes and Low-Dimensional Structures of Hybrid Tin Iodide Perovskites. *Inorg. Chem.* **2017,** *56*, 56–73.
40. Tarasova, N.; Animitsa, I.; Galisheva, A.; Medvedev, D. Layered and Hexagonal Perovskites as Novel Classes of Proton-Conducting Solid Electrolytes. A Focus Review. *Electrochem. Mater. Technol.* **2022,** *1*, 20221004.
41. Giorgi, G.; Yamashita, K. Zero-Dimensional Hybrid Organic–Inorganic Halide Perovskite Modeling: Insights from First Principles. *J. Phys. Chem. Lett.* **2016,** *7*, 5, 888–899.
42. Zhang, Y.; Saidaminov, M. I.; Dursun, I.; Yang. H.; Murali, B.; Yengel, E.; Alarousu, E.; Alshankiti, B. A.; Bakr, O. M.; Mohammed, O. F. *J. Phys. Chem. Lett.* **2017,** *8*, 961–965.
43. Seth, S.; Samanta, A. Photoluminescence of Zero-Dimensional Perovskites and Perovskite-Related Materials. *J. Phys. Chem. Lett.* **2018,** *9*, 176–183.
44. Yin, J.; Yang, H.; Song, K.; El-Zohry, A. M.; Han, Y.; Bakr. O.; Bredas. J. L.; Mohammed, O. F. Point Defects and Green Emission in Zero-Dimensional Perovskites. *J. Phys. Chem. Lett.* **2018,** *9*, 5490–5495.
45. Almutlaq, J. M.; Yin, Y.; Mohammed, O. F.; Bakr, O. M. The Benefit and Challenges of Zero-Dimensional Perovskites. *J. Phys. Chem. Lett.* **2018,** *9*, 4131–4138.
46. Yin,J.; Maity, P.; De Bastiani, M.; Dursun,I.; Bakr, O. M.; Brédas, J. L.; Mohammed, O. F.; Molecular Behavior of Zero-Dimensional Perovskites. *Sci. Adv.* **2017,** *3*, 1701793.

47. Cha, J. Y.; Lee, H. J.; Kim, S. H.; Ko, K. C.; Suh, B. J.; Han, O. H.; Jung, D. Y.; Superparamagnetism of Green Emissive Cs4 PbBr6 Zero-Dimensional Perovskite Crystals. *ACS Energy Lett.* **2020,** *5*, 2208–2216.
48. Pious, J. K.; Lekshmi, M. L.; Muthu, C.; Rakhi, R. B.; Vijayakumar, C.; Zero-Dimensional Methylammonium Bismuth Iodide-Based Lead-Free Perovskite Capacitor. *ACS Omega* **2017,** *2*, 5798−5802.
49. Thomas, S.; Thankappan, A. *Perovskite Photovoltaics Basic to Advanced Concepts and Implementation*; Academic Press: London.
50. Kheralla, A.; Chetty, N. A Review of Experimental and Computational Attempts to Remedy Stability Issues of Perovskite Solar Cells. *Heliyon* **2021,** *7*.
51. Correa-Baena, J.-P.; Saliba, M.; Buonassisi, T.; Grätzel, M.; Abate, A.; Tress, W.; Hagfeldt, A. Promises and Challenges of Perovskite Solar Cells. *Science* **2017,** *358*, 739–744.
52. Kojima, A.; Teshima, K.; Shirai, Y.; Miyasaka, T. Organometal Halide Perovskites as Visible-Light Sensitizers for Photovoltaic Cells. *J. Am. Chem. Soc.* **2009,** *131*, 6050–6051.
53. Best Research-Cell Efficiency Chart. https://www.nrel.gov/pv/cell-efficiency.html (accessed on 30 Mar 2023).
54. Zhou, D.; Zhou, T.; Tian, Y.; Zhu, X.; Tu, Y. Perovskite-Based Solar Cells: Materials, Methods, and Future Perspectives. *J. Nanomater*. **2018,** 1–15.
55. Snaith, H. J. Present Status and Future Prospects of Perovskite Photovoltaics. Nat. Mater. **2018,** *17*, 372–376.
56. Dai, X.; Xu, K.; Wei, F. Recent Progress in Perovskite Solar Cells: The Perovskite Layer. *Beilstein J. Nanotechnol*. **2020,** *11*, 51–60.
57. Kim, J. Y.; Lee, J.-W.; Jung, H. S.; Shin, H.; Park, N.-G. High-Efficiency Perovskite Solar Cells. *Chem. Rev.* **2020,** *120*, 7867–7918.
58. Spanopoulos, I.; Ke, W.; Kanatzidis, M. G.; In Quest of Environmentally Stable Perovskite Solar Cells: A Perspective. *Helv. Chim. Acta*. **2020,** 104.
59. Mei, F.; Sun, D.; Mei, S.; Feng, J.; Zhou, Y.; Xu, J.; Xiao, X. Recent Progress in Perovskite-Based Photodetectors: The Design of Materials and Structures. ADV PHYS-X. **2019,** *4*, 1592709.
60. Wang, H.; Kim, D. H. Perovskite-Based Photodetectors: Materials and Devices. *Chem Soc Rev.* **2017,** *46*, 5204–5236.
61. Li, C.; Ma, Y.; Xiao, Y.; Shen, L.; Ding, L. Advances in Perovskite Photodetectors. *InfoMat* **2020,** *2*, 1247–1256.
62. Miao, J.; Zhang, F. Recent Progress on Highly Sensitive Perovskite Photodetectors. *J. Mater. Chem. C.* **2019,** *7*, 1741–1791.
63. Schanze, K. S.; Kamat, P. V.; Yang, P.; Bisquert, J. Progress in Perovskite Photocatalysis. *ACS Energy Lett.* **2020,** *5*, 2602–2604.
64. Huang, H.; Pradhan, B.; Hofkens, J.; Roeffaers, M. B.; Steele, J. A.; Solar-Driven Metal Halide Perovskite Photocatalysis: Design, Stability, and Performance. ACS Energy Lett. **2020,** 5, 1107–1123.
65. Kong, J.; Yang, T.; Rui, Z.; Ji, H.; Perovskite-Based Photocatalysts for Organic Contaminants Removal: Current Status and Future Perspectives. *Catal. Today.* **2019,** *327*, 47–63.
66. Mai, H.; Chen, D.; Tachibana, Y.; Suzuki, H.; Abe, R.; Caruso, R. A.; Developing Sustainable, High-Performance Perovskites in Photocatalysis: Design Strategies and Applications. *Chem Soc Rev*. **2021,** *50*, 13692–13729.

67. Wei, K.; Faraj, Y.; Yao, G.; Xie, R.; Lai, B. Strategies for Improving Perovskite Photocatalysts Reactivity for Organic Pollutants Degradation: A Review on Recent Progress. *J. Chem. Eng.* **2021,** *414*, 128783.
68. Luo, J.; Zhang, W.; Yang, H.; Fan, Q.; Xiong, F.; Liu, S.; Li, D. S.; Liu, B.; Halide Perovskite Composites for Photocatalysis: A Mini Review. *EcoMat.* **2021,** 3.
69. Kaltzoglou, A.; Falaras, P. Recent Developments on Hybrid Perovskite Materials for Solar Energy Conversion and Environmental Protection. *Curr. Opin. Chem. Eng.* **2021,** *33*, 100708.
70. Chen, T. W.; Ramachandran, R.; Chen, S.-M.; Anushya, G.; Ramachandran, K. Graphene and Perovskite-Based Nanocomposite for Both Electrochemical and Gas Sensor Applications: An Overview. *Sensors.* **2020,** *20*, 6755.
71. Nur'aini, A.; Oh, I.; Volatile Organic Compound Gas Sensors Based on Methylammonium Lead Iodide Perovskite Operating at Room Temperature. *RSC Adv.*, **2020,** *10*, 12982–12987.
72. Nur'aini, A.; Oh, I. Volatile Organic Compound Gas Sensors Based on Methylammonium Lead Iodide Perovskite Operating at Room Temperature. *RSC Adv.* **2020,** *10*, 12982–12987.
73. Bulemo, P. M.; Kim, I. D. Recent Advances in ABO_3 Perovskites: Their Gas-Sensing Performance as Resistive-Type Gas Sensors. *J. Korean Ceram. Soc.* **2020,** *57*, 24–39.
74. Wang, J.; Ren, Y.; Liu, H.; Li, Z.; Deng, Y.; Fang, X. Ultrathin 2D $NbWO_6$ Perovskite Semiconductor Based Gas Sensors with Ultrahigh Selectivity Under Low Working Temperature. *Adv. Mater.* **2022,** *34*, 2104958.
75. Kim, j.; Hu, l.; Chen, H.; Guan, X.; Anandan, P. R.; Li, F.; Tang, J.; Lin, C. H.; Zadeh, K. K.; Tricoli, A.; Wu, T. P-type Charge Transport and Selective Gas Sensing of All-Inorganic Perovskite Nanocrystals. *ACS Mater. Lett.* **2020,** *2*, 1368–1374.
76. Wang. G.; Bai. J.; Shan, C.; Zhang, D.; Lu, N. C.; Liu, Q.; Zhou, Z. X.; Wang, S. G.; Liu, C. Synthesis and Ethanol Gas Sensing Properties of Mesoporous Perovskite-Type $BaSnO_3$ Nanoparticles Interconnected Network. *Mater. Lett.* **2017,** *205*, 169–172.
77. Qin, W.; Yuan, Z.; Gao, H.; Zhang, R.; Meng, F.; Perovskite-Structured $LaCoO_3$ Modified ZnO Gas Sensor and Investigation on Its Gas Sensing Mechanism by First Principle. *Sens. Actuat. B: Chem.* **2021,** *341*, 130015.
78. Seisenbaeva, G. A.; Kessler, V. G.; Pazik, R.; Strek, W. Heteroleptic Metal Alkoxide "Oxoclusters" as Molecular Models for the Sol–Gel Synthesis of Perovskite Nanoparticles for Bio-Imaging Applications. *Dalton Trans.* **2008,** 3412–3421
79. Xiao, L.; Huang, L.; Su, W.; Wang, T.; Liu, H.; Wei, Z.; Fan, H. Efficiency Enhancement Strategies for Stable Bismuth-Based Perovskite and Its Bioimaging Applications. *Int. J. Mol. Sci.* **2023,** *24*, 4711.
80. Xiao, L.; Huang, L.; Su, W.; Wang, T.; Liu, H.; Wei, Z.; Fan, H. Efficiency Enhancement Strategies for Stable Bismuth-Based Perovskite and Its Bioimaging Applications. *Int. J. Mol. Sci.* **2023,** *24*, 4711.
81. Arumugam, G. M.; Karunakaran, S. K.; Galian, R. E.; Pérez-Prieto, J. Recent Progress in Lanthanide-Doped Inorganic Perovskite Nanocrystals and Nanoheterostructures: A Future Vision of Bioimaging. *Nanomaterials* **2022,** *12*, 2130.
82. Chan, K. K.; Chan, D.; He. H.; Sum. T. C.; Yong, K. T.; Water-Stable All-Inorganic Perovskite Nanocrystals with Nonlinear Optical Properties for Targeted Multiphoton Bioimaging. *ACS Appl. Nano Mater.* **2021,** *4*, 9022–9033.

83. Avugadda, S. K.; Castelli, A.; Dhanabalan, B.; Fernandez, T.; Collantes,C.; Silvestri, N.; Baranov, D.; Imran, M.; Manna, L.; Pellegrino, T.; Arciniegas, M. P. Highly Emitting Perovskite Nanocrystals with 2-Year Stability in Water Through an Automated Polymer Encapsulation for Bioimaging. *ACS Nano* **2022,** *16*, 13657–13666.
84. Lee, Y. H.; Kim, D. H.; Wu, C.; Kim, T. W. Memristive Devices with a Large Memory Margin Based on Nanocrystalline Organic-Inorganic Hybrid $CH_3NH_3PbBr_3$ Perovskite Active Layer. *Organ. Electron.* **2018,** *62*, 412–418.
85. Thien, G. S. H.; Rahman, M. A.; Yap, B. K.; Tan, N. M. L.; He, Z.; Low, P. L.; Devaraj, N. K.; Osman, A. F. A.; Sin, Y. K.; Chan, K. Y. Recent Advances in Halide Perovskite Resistive Switching Memory Devices: A Transformation from Lead-Based to Lead-Free Perovskites. *ACS Omega*. **2022,** *7*, 39472–39481.
86. Bagdzevicius, S.; Maas, K.; Boudard, M.; Burriel, M. Interface-Type Resistive Switching in Perovskite Materials. *J Electroceram*. **2017,** *39*, 157–184.
87. Rogdakis, K,; Loizos, M.; Viskadourosa, G.; Kymakis, E.; Memristive Perovskite Solar Cells Towards Parallel Solar Energy Harvesting and Processing-Inmemory Computing. *Mater. Adv*. **2022,** *3*, 7002.
88. An. H.; Kim, W. K.; Wu, C.; Kim, T. W. Highly-Stable Memristive Devices Based on Poly(Methylmethacrylate): $CsPbCl_3$ Perovskite Quantum Dot Hybrid Nanocomposites. *Organ. Electron.* **2018,** *56*, 41–45.
89. The Perovskite Experts. https://www.perovskite-info.com/
90. Jeong, S.-H.; Park, J.; Han, T.-H.; Zhang, F.; Zhu, K.; Kim, J. S.; Park, M.-H.; Reese, M. O.; Yoo, S.; Lee, T.-W. Characterizing the Efficiency of Perovskite Solar Cells and Light-Emitting Diodes. *Joule* **2020,** *4*, 1206–1235.
91. Ren, Z.; Wang, K.; Sun, X. W.; Choy, W. C.; Strategies Toward Efficient Blue Perovskite Light-Emitting Diodes. *Adv. Funct. Mater.* **2021,** *31*, 2100516.
92. Lin, K.; Xing, J.; Quan, L. N.; de Arquer, F. P.; Gong, X.; Lu, J.; Xie, L.; Zhao, W.;Zhang, D.; Yan, C.; Li, W.; Liu, X.; Lu, Y.; Kirman, J.; Sargent, E. H.; Xiong, Q.; Wei, Z.; Perovskite Light-Emitting Diodes with External Quantum Efficiency Exceeding 20 Per Cent. Nature. **2018,** *562*, 245–248.
93. Pacchioni, G. Highly Efficient Perovskite Leds. *Nat. Rev. Mater.* **2021,** *6*, 108–108.
94. Cao, Y.; Wang, N.; Tian, H.; Guo, J.; Wei, Y.; Chen, H.; Miao, Y.; Zou, W.; Pan, K.;He, Y.; Cao, H.; Ke, Y.; Xu, M.; Wang, Y.; Yang, M.; Du, K.; Fu, Z.; Kong, D.; Dai, D.; Jin,Y.; Li, G.; Li, H.; Peng, Q.; Wang, J.; Huang, W. Perovskite Light-Emitting Diodes Based on Spontaneously Formed Submicrometre-Scale Structures. *Nature* **2018,** *562*, 249–253.
95. Xiao, Z.; Kerner, R. A.; Zhao, L.; Tran, N. L.; Lee, K. M.; Koh, T.-W.; Scholes, G. D.;Ran, B. P. Efficient Perovskite Light-Emitting Diodes Featuring Nanometre-Sized Crystallites. *Nat. Photon.* **2017,** *11*, 108–115.
96. Fu, P.; Shan, Q.; Shang, Y.; Song, J.; Zeng, H.; Ning, Z.; Gong, J. Perovskite Nanocrystals: Synthesis, Properties and Applications. *Sci. Bull.* **2017,** *62*, 369–380.
97. Tsai, H.; Shrestha, S.; Vilá, R. A.; Huang, W.; Liu, C.; Hou, C.-H.; Huang, H.-H.; Wen, X.; Li, M.; Wiederrecht, G.; Cui, Y.; Cotlet, M.; Zhang, X.; Ma, X.; Nie, W. Bright and Stable Light-Emitting Diodes Made with Perovskite Nanocrystals Stabilized in Metal–Organic Frameworks. *Nat. Photon.* **2021,** *15*, 843–849.
98. Yang, D.; Zhao, B.; Yang, T.; Lai, R.; Lan, D.; Friend, R. H.; Di, D.; Toward Stable and Efficient Perovskite Light-Emitting Diodes. *Adv. Funct. Mater*. **2021,** *32*, 2109495.
99. Fan, J.; Du, W.; Mi, Y.; Liu, X. Perovskite-Based Lasers. *Nanoscale Semiconductor Lasers* **2019,** 41–74.

100. Liu, A.; Guan, G.; Chai, X.; Feng, N.; Lu, M.; Bai, X.; Zhang, Y. Metal Halide Perovskites toward Electrically Pumped Lasers. *Laser Photonics Rev.* **2022,** *16*, 2200189.
101. Zhang, Q.; Shang, Q.; Su, R.; Do, T. T.; Xiong, Q. Halide Perovskite Semiconductor Lasers: Materials, Cavity Design, and Low Threshold. *Nano Lett.* **2021,** *21*, 1903–1914.
102. Teo, S. W.; Ng, C. H.; Ng, Y. H.; Islam, A.; Hayase, S.; Taufiq-Yap, Y. H. Resolve Deep-Rooted Challenges of Halide Perovskite for Sustainable Energy Development and Environmental Remediation. *Nano Energy* **2022,** *99*, 2211–2855.
103. Zhang, S.; Han, G. Intrinsic and Environmental Stability Issues of Perovskite Photovoltaics. Progress in Energy **2020,** *2*, 022002.
104. Roy, P.; Sinha, N. K.; Tiwari, S.; Khare, A. A Review on Perovskite Solar Cells: Evolution of Architecture, Fabrication Techniques, Commercialization Issues and Status. *Solar Energy*. **2020,** *198*, 665–688.

CHAPTER 3

Perovskites are the Future for Upcoming Technologies

PARVESH K. DEENDYAL[1,2], SHWETA DHAKLA[1], ANKUR TAYA[3], SARVESH KUMAR[4], and MANISH K. KASHYAP[1]

[1]*School of Physical Sciences, Jawaharlal Nehru University, New Delhi, India*

[2]*Government Polytechnic for Women, Faridabad, Haryana, India*

[3]*Department of Physics, Mukand Lal National College, Yamunanagar, Haryana, India*

[4]*Inter-University Accelerator Centre, Aruna Asaf Ali Marg, New Delhi, India*

ABSTRACT

Perovskite is a naturally occurring mineral of calcium titanate ($CaTiO_3$) which was first discovered by German mineralogist Gustav Rose in 1839 and was named in honor of the Russian mineralogist Lev Perovski (1792–1856). In general, all the materials having the same crystal structure of $CaTiO_3$ are known as perovskite materials. Interestingly, the research in perovskite materials has gained unprecedented attention owing to its favorable properties. Thus, the research in perovskite materials is the hot topic among physicists, chemists, and material scientists for finding their effective uses in novel technologies. Due to its inherent properties such as large carrier diffusion length, tunable band gap, high absorption coefficient, and high charge carriers' mobility, the power conversion efficiency (PCE) of these emerging solar cells showed a steep growth progress from 3.8% in 2009 to 25.7% in 2022. At present, the hybrid halide perovskites are considered the most promising candidate for the next generation of photovoltaics, capable of replacing commercially demanding polycrystalline silicon materials. In addition, metal halide

Advanced Materials for Next-Generation Technologies: Challenges and New Prospects. Shrikaant Kulkarni, Vipul Srivastava, Rabah Khenata, & Yarub Al-Douri (Eds.)

perovskites have proved to be potential materials for optoelectronic devices like light-emitting diodes (LEDs) due to their tunable bandgap, high external quantum efficiency of over 20%, high carrier mobility, and narrow emission width. In this chapter, all the existing applications of perovskite materials have been reviewed and an emphasis on their upcoming technological applications has been given.

3.1 INTRODUCTION

Human society relied intricately on the energy for its evolution throughout the centuries. But most recently, given its importance in nearly every aspect of economy, environment, and policy-making, it has become a matter of critical choice. Till this date, most of the energy demands are being fulfilled by fossil fuels but due to numerous issues linked with these, focus is now being shifted toward renewable energy resources and solar energy is emerging as the most reliable source for catering future demands.[1] Sun globally provides 2.9×10^{15} kW of energy per day, which is approximately 100 times of the net energy consumption of the world in a year. Due to this huge potential, the demand and development of solar energy are escalating at a steep rate. According to Global Status Report of Renewables 2022, power generation from solar photovoltaic (PV) stands at a historic value of 1003 TWh, which is massive increase from mere 63 TWh in 2011. As far as electricity generation is concerned, solar PV shows a contribution of 3.6% globally.

Based on the superior technologies, technical attributes, and net price reduction, there are two main PV technologies, that is, wafer based and thin film based which further are classified into three main generations.[2] Solar cells having high efficiency alongside high production cost are put under first generation. Fabricated from pure Si, these solar cells make use of a single junction to drive the energy obtained from photons. Since the theoretical maximum efficiency in these cells approaches 33%, these are the most efficient solar cells so far. Although solar cells from second generation are much thinner than crystalline solar cells, these have not been able to take over the commercial market from first-generation solar cells yet. These are manufactured just from few micrometers thick semiconducting material layer, less manufacturing and material cost enables the solar panels manufactured from this technology to be procured and sold at cheaper prices. These undergo intense development during the 1990s and the early 2000s and are based on amorphous silicon, copper indium gallium diselenide (CIGS), and cadmium telluride (CdTe).[3] The maximum power conversion efficiency

(PCE) achieved by CdTe/CIGS is 22.1%/23.4%.[4] Third-generation solar cells are constituted from emerging thin-film technologies. These are known to produce low-cost and high-efficiency cells by utilizing the materials' properties at nano-structural and molecular levels to generate photovoltaic effect by emphasizing on state-of-the-art technologies. Although results are encouraging and promising yet most of these technologies are still under development phase and thus confined to laboratories and have not been commercialized till date.

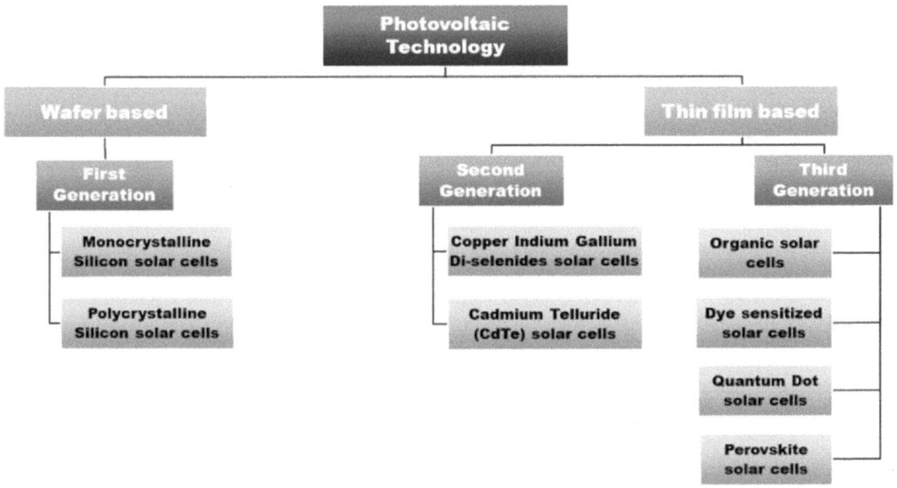

FIGURE 3.1 Types of photovoltaic technologies discovered so far.

With an aim to attain the soaring efficiencies in 30–60% range using novel approaches, the expected attributes of these thin-film solar cells from third generation are colorfulness, lightweight, flexibility, adaptability, and environmental friendly methodologies. Thin-film solar cells such as dye sensitized solar cell (DSSC), organic bulk heterojunction solar cell, quantum dot solar cells, and perovskite solar cells fall under this category.[5] DSSCs use a thin layer of photosensitive dye to absorb sunlight and convert it into the electricity. They are cheaper to manufacture than traditional silicon-based solar cells and can be used in a wide range of applications, whereas organic solar cells are made of organic materials that can be dissolved in solvents and printed onto the flexible substrates. Being lightweight and flexible, they can be produced in large quantities using roll-to-roll manufacturing processes. To absorb and convert sunlight into electricity, tiny semiconductor particles are used in quantum dot solar cells. They have shown high efficiencies and can be manufactured using inexpensive materials. Perovskite solar cells use a hybrid organic–inorganic material as the light-absorbing layer. They have shown high efficiencies at low manufacturing costs, making them a promising candidate for commercialization. The details of perovskite materials are discussed in the next section.

3.2 PEROVSKITES

Perovskite is any material that has a crystal structure similar to structure of a mineral called calcium titanium oxide ($CaTiO_3$) which was first identified by Gustav Rose in 1839. Over the period of time, material scientists showed enormous interests in perovskites due to their abundance availability and richness in behavior. They can be used in a variety of applications including solar cells, LEDs, and lasers. There are several types of perovskite materials available for technological applications which are as follows:

1. Hybrid perovskites: These materials are made up of a combination of organic and inorganic molecules. They are the most commonly used perovskite materials in solar cells and are known for their high efficiency and low cost. Some examples for this purpose are Methy-ammonium lead iodide ($CH_3NH_3PbI_3$), Formamidinium lead iodide ($CH(NH_2)_2PbI_3$) etc.
2. Inorganic perovskites: These materials are made up entirely of inorganic elements such as tin, lead, or cesium ($CsPbBr_3$, $CsPbI_3$ etc.). They are known for their stability and high melting points and being studied for their use in optoelectronic applications like LEDs.
3. Ruddlesden-Popper perovskites: These materials have a layered crystal structure and are being studied for use in solar cells and other optoelectronic devices. They have the potential to be more stable and efficient than the traditional perovskites. The examples for this type are $(CH_3NH_3)_2(CH_3NH_2)_2Pb_3I_{10}$ and $(C_6H_5CH_2NH_3)_2(CH_3NH_3)Pb_2I_7$ etc.
4. Double perovskites: These materials contain two different cations in the perovskite structure, which can result in unique electronic and magnetic properties. They are being studied for use in spintronics and other advanced electronic applications. $Cs_2NmPnCl_6$, and $Cs_2BiAgBr_6$ etc. are the examples of double perovskites.

These are just few examples of the many types of perovskite materials that exist. Researchers are incessantly exploring novel materials and upgrading the existing ones to develop more efficient and cost-effective optoelectronic devices.

Hybrid halide perovskite (HHP)-based solar cells are a promising new technology in the field of photovoltaics. These solar cells use a perovskite material that is a combination of organic and inorganic components which results in a unique combination of properties that make it highly efficient for converting sunlight into electricity. Their unique crystal structure plays a crucial role in their remarkable photovoltaic properties. The basic structure of HHP consists of a 3D framework of metal halide octahedra, surrounded by organic cations that occupy the space between the octahedra. The generic chemical formula for HHP is ABX_3, where A and B are the organic and metal cations, respectively, such that size of cation A is greater than that of cation B while X is an halide ion as shown in Figure 3.2. The metal halide octahedra form a network of corner-sharing octahedra, creating a continuous lattice of tetrahedral spaces.

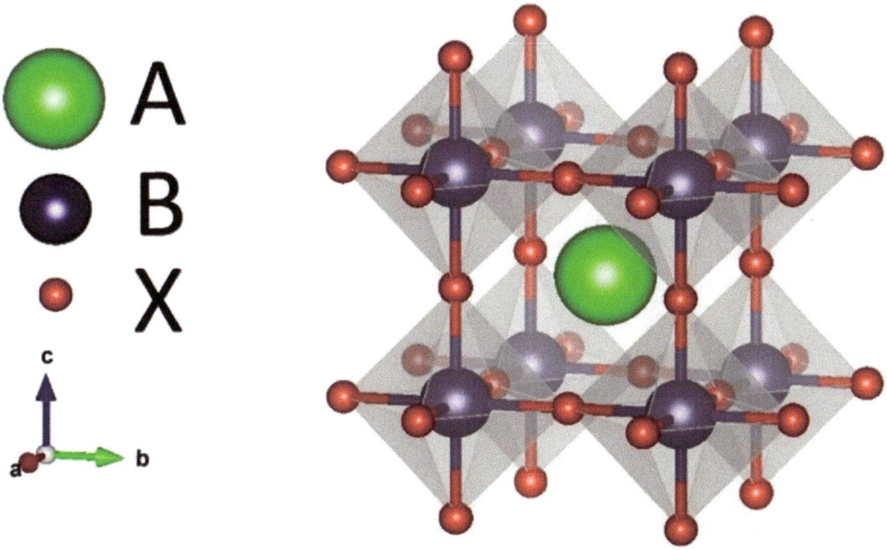

FIGURE 3.2 Structure of archetypical HHP. The octahedral coordination of metallic ions is highlighted here.

The organic cations, such as methylammonium ($CH_3NH_3^+$), ethylammonium ($C_2H_5NH_3^+$), or formamidinium ($HC(NH_2)_2^+$), occupy the tetrahedral spaces between the metal halide octahedra. This results in a crystal structure that has a high degree of structural flexibility, allowing it to adapt to different chemical environments. The organic cations can undergo dynamic changes in orientation and position within the crystal structure, which can help to enhance the performance of the solar cell. For example, when the solar cell is exposed to sunlight, the organic cations can reorient themselves to create pathways for efficient charge transfer, leading to improved photovoltaic performance. Owing to these properties of HHPs, in just a few short years, the researchers have been able to achieve efficiencies of over 25%, which is comparable to traditional silicon solar cells. Also, unlike traditional silicon solar cells, which require expensive and energy-intensive manufacturing processes, HHP solar cells can be made using facile and low-cost solution-based fabrication processes. This makes them highly attractive for large-scale commercial production.[6]

However, there are also some challenges associated with HHP solar cells. One major challenge is their stability, as they are known to degrade quickly under ambient conditions and another is the toxicity of Pb in mostly used $MAPbI_3$.[7] But, a large number of research efforts are actively going on to improve their stability and eliminate toxicity so that HHP solar cells can be made more durable, reliable, and commercially viable.

Double perovskite materials have a unique crystal structure consisting of two interpenetrating perovskite lattices. The general chemical formula for a double perovskite material is $A_2BB'X_6$, where A site is typically occupied by a smaller cation such as Ba^{2+}, Sr^{2+}, or Ca^{2+}, while the B and B' sites are occupied by a larger cation such as Fe^{3+}, Mn^{3+}, or Co^{2+}. One of interpenetrating perovskite lattices consists of the B cations whereas other is made up of the B' cations, while A cations occupying the interstitial sites between the two lattices as shown in Figure 3.3.

The high degree of structural flexibility in double perovskite materials gives rise to some interesting physical properties. For example, these materials have been found to exhibit photovoltaic, magnetoresistance, magnetocaloric effects, and multiferroic behavior.[8] Since the A/B/B'-site cations can be substituted with other suitable elements to tune the physical properties of the material. This makes double perovskites highly versatile and attractive for various other technological applications.

3.3 TECHNOLOGICAL APPLICATIONS

Future technological advancements may be fueled by the continual development of new materials with improved properties. In this regard, cutting-edge research by material scientists has unlocked the magical features of perovskite materials, revealing their potential applications in emerging tsechnologies.

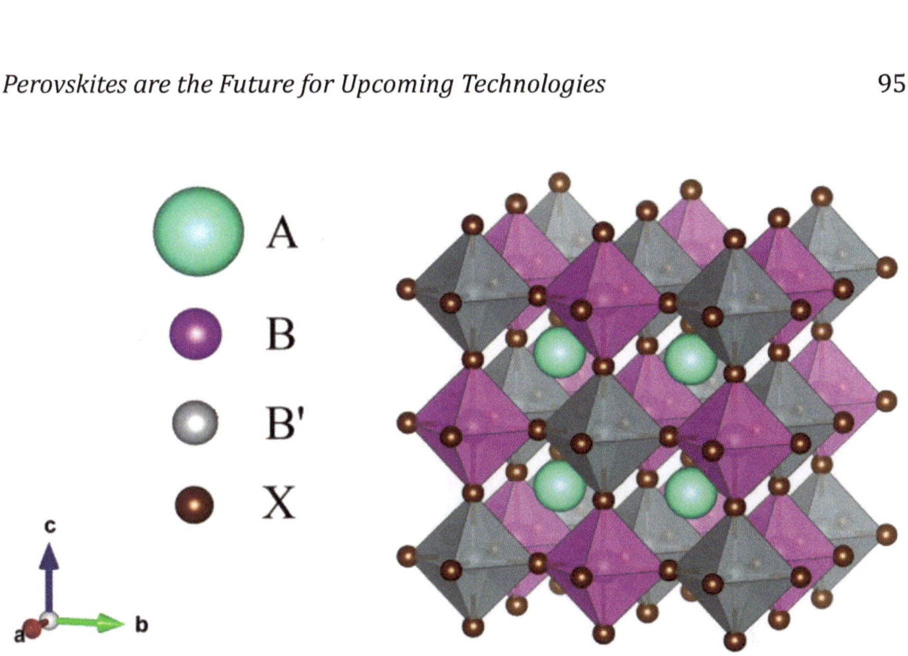

FIGURE 3.3 Crystal structure of $A_2BB'X_6$ double perovskite.

Perovskites have recently been used in a plethora of fascinating applications. Although perovskites have primarily been employed in photovoltaics in recent years, these materials offer unique properties that make them extremely useful in a variety of other applications. Intriguingly, the benefits of perovskites transcend far beyond photovoltaics expanding into the areas of light-emitting diodes, X-ray detection and imaging, thermoelectricity, forming a network of knowledge centered on a single material. Here, in the following sections, we will introduce a myriad of perovskite-related applications:

i) **Photovoltaics**

The significance of perovskites as a potential candidate for solar power conversion is undeniable. Perovskite has gained unprecedented attention in the community of solar cell research, owing to its meritorious properties such as large carrier diffusion length, tunable band gap, high absorption coefficient, and high charge carriers' mobility.[9–11] The HHP-based solar cells are now regarded as the most promising contender for the next generation of photovoltaics, capable of replacing commercially available wafer-based silicon solar cell technology. With the advancement of such unique material, we have been able to create highly efficient photovoltaic devices at a rate that has never been achieved in the history of solar cells. For instance, it took only a decade to fabricate a PSC with a high PCE of 25.7%,[12] whereas silicon solar cell technology took almost a half-century to reach a certified PCE of 26.7%.[13] The evolution of PSCs has accomplished the most significant advances in the field of photovoltaics. The device architecture of PSCs is shown in Figure 3.4.

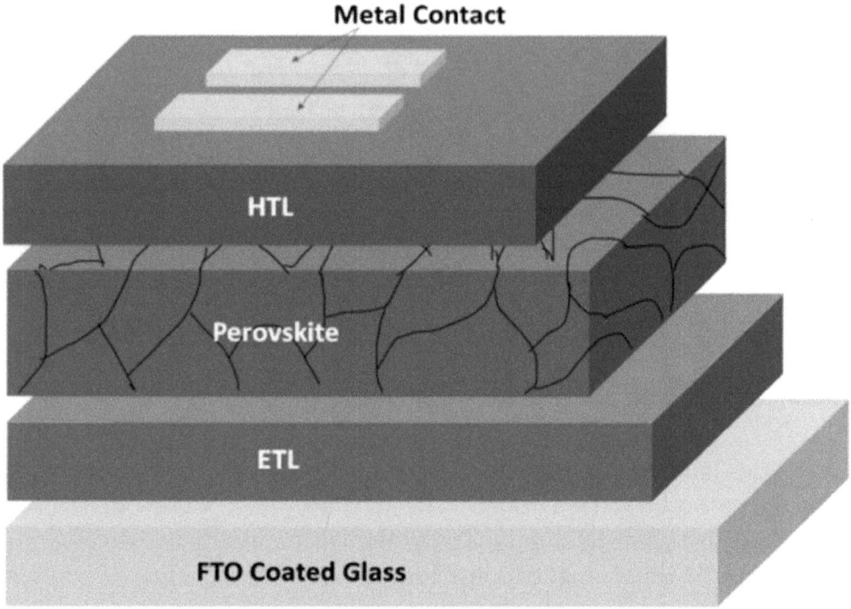

FIGURE 3.4 Device architecture of a perovskite solar cell.

Although the terminology "perovskite solar cell" may sound familiar to most people today, it was new in 2005 when the quest to investigate photovoltaic applications of halide perovskites started in the Miyasaka laboratory at Toin University in Yokohama, Japan. The first study to examine the photo power conversion of PSCs was reported by Miyasaka and coworkers in 2009.[14] In accordance with the DSSC framework, the cell presented a PCE of a mere 3.8%. However, the solar cell stability was fragile and received little attention as a result of the use of liquid electrolyte in the hole-transporting material (HTM). A similar trial was carried out by Park and his group[5] in 2011, yielding a PCE of 6.5%. The instability of HTMs brought on by the liquid medium was still the main issue with PSC's stability. Over the bygone few years, efficiency has advanced due to the use of solid-state HTM (2,20,7,70-tetrakis(N,N-di-pmethoxyphenamine)-9,90-spirobifluorene, also known as Spiro-OMeTAD) rather than liquid electrolyte on the highly crystalline absorbing layer.

Interestingly, a breakthrough PCE of 10.9% with an open-circuit voltage (V_{oc}) higher than 1.1 V was certified with single-crystalline silicon solar cells. Not only the high performance but also the facile, low-cost solution-based fabrication processes showcase the immense potential of PSCs as a future PV technology. Though, there has been extensive research work after its inception in the year 2009, and it has reached almost its theoretical limit (Shockley Queisser limit), still it is far away from commercialization due to issues like toxicity, instability, and low life expectancy. Further, scientists are making significant efforts in reducing the Pb toxicity and improving the stability of perovskite-based solar cells. However, there is a lot of scope for further improvement and realization of high-performance HHP solar modules with enhanced stability and flexibility.

ii) Opto-electronic devices

With their superior optoelectronic properties, hybrid perovskite materials offer a wide range of fascinating applications beyond photovoltaics such as a light-emitting diode (LED). One of the impressive aspects of halide perovskites is customizable photoluminescence property, particularly the tunable optical bandgap which makes them potential materials for LEDs.[17–19] Perovskite-based LEDs (PeLEDs) provide a number of benefits including high photoluminescence quantum yield (PLQY), optimal color quality, low operational voltage, fast response time, good brightness, and conformational versatility, which contribute to high color purity characteristics and low cost of PeLEDs.[20] In addition, PeLEDs can be designed in a way that permits high-luminance devices to be driven by low driving voltage, providing a significant advantage over conventional LEDs.[21,22] Several milestones are noteworthy in the PeLEDs roadmap: In 2012, Miyasaka and colleagues reported the high photoluminescence of $MAPbBr_3$ nanoparticles that can be readily synthesized using porous alumina media.[23] The resulting halide perovskite nanoparticles revealed narrow bandwidth and produced intense green emission. Significantly, in 2014, Friend and his coworkers put forward the first report on PeLEDs that can operate at room temperature, having a 20-nm-thick layer of $MAPbI_{3-x}Cl_x$ perovskite material.[24] With a peak emission wavelength of 754 nm, the device demonstrated an external quantum efficiency (EQE) of 0.76%. Furthermore, in 2016, Ling et al.[25] fabricated the highly intense and efficient green PeLEDs employing solution-processed $MAPbBr_3$ halide perovskite nanoplatelets as light emitters. These employed perovskite nanoplatelets displayed intense photoluminescence at 529 nm having a narrow spectral band. Also, these nanoplatelets demonstrated a substantially higher quantum yield of 85% when compared to polycrystalline bulk perovskites, which enabled the PeLEDs to be driven brightly. The device architecture for a PeLED based on $MAPbBr_3$ perovskite nanoplatelets and the electroluminescence spectrum at different applied voltages are shown in Figure 3.5.

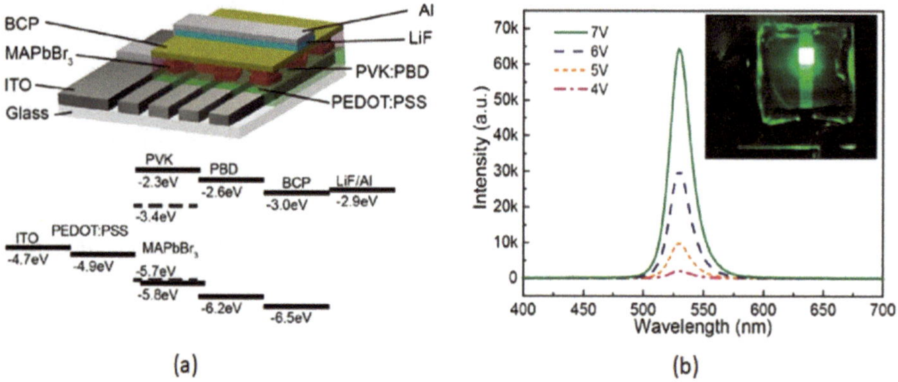

FIGURE 3.5 (a) Device architecture for a PeLED based on $MAPbBr_3$perovskite nanoplatelets along with the energy level diagram of different used materials with reference to vacuum. (b) Electroluminescence spectrum from a standard device with PVK:PBD weight ratio of 64/36 at an applied voltage of 4–7 V. Inset: Image of a working device at 7 V.

Source: Reproduced with permission from Ref. [25] Copyright 2015. John Wiley and Sons.

Immense research has already been done on PeLEDs. The latest evolution of PeLEDs is driven by two significant advancements: (i) methods to attain the maximum EQE and (ii) approaches to boost long-term stability. The proper selection of perovskite material and its consequent defect passivation play a pivotal role in enhancing the EQE of PeLEDs. The PeLEDs have made significant progress as their EQE has surpassed 20%, upgraded by a combination of surface/grain-boundary defect passivation and optimized creation of continuous pin-hole free perovskite layers by utilizing appropriate additives.[26,27] In the nutshell, PeLEDs hold great promise for display applications and beyond in the future. Extensive study in enhancing the optical quality of perovskite material resulted in photoluminescence quantum yield close to unity. This, along with the efficient device structure, exhibited the EQE in PeLEDs at par with quantum-dot and organic

LEDs. Nonetheless, PeLEDs continue to confront issues owing to their inferior lifetime, especially at stable emission color and high brightness conditions and toxicity. For instance, the most successful perovskite materials used in PeLEDs include lead, which introduces toxicity concerns. In order to extend the application of PeLEDs beyond display devices, in-depth research is required to develop novel perovskite material for NIR and UV (deep UV) emission.

iii) X-ray detection and imaging

X-ray has sparked unprecedented interests in the scientific community owing to its strong penetration capability. Just a year later after its discovery in 1895 by German physicist Wilhelm Röntgen, X-ray radiation was utilized many times in medical domain with excellent characteristics.[28,29] Notably, it has been widely used in many aspects of daily life such as security checking,[30] medical research,[31] and materials characterization.[32] Nowadays, X-ray detection and imaging technology have advanced rapidly for a variety of applications, providing enormous prospects for the industrial and scientific communities. The commercially available direct X-ray conversion detectors are primarily based on inorganic semiconductors such as Si, Ge, and CdTe, which necessitate high-cost and sophisticated fabrication techniques.[33] Intriguingly, amorphous selenium (a-Se) is one of the materials for direct X-ray detectors that could achieve high resolution imaging in mammography.[34] This, a-Se direct X-ray imager, on the other hand, has limited application due to its low intrinsic X-ray sensitivity, insufficient charge transport ability, and poor absorptivity of hard X-rays.[35] Therefore, there is a need to find promising materials for direct X-ray detection technology. In this regard, halide perovskites are recently emerged as potential materials for direct X-ray detection, owing to their low-cost fabrication processes, strong X-ray stopping power, high sensitivity, and excellent carrier transport performance.[36–38] Meanwhile, direct X-ray imagers based on perovskite have been successfully demonstrated.[39,40] In 2015, Yakunin et al.[39] utilized a $MAPbI_3$-based PSC as a single-pixel X-ray detector, performed scanning of the object, and also showed a simple X-ray image of a leaf fragment as depicted in Figure 3.6. Despite the ineffective imaging method and poor image quality, this work has undoubtedly inspired other researchers. Thereafter, Park and colleagues[40] reported the most reliable results of actual X-ray image detection utilizing a 2D patterned image sensor. The detector was a solution-processed perovskite device with X-ray absorber layer of thickness 0.83 mm. This device demonstrated good X-ray detection sensitivity, outperforming the sensitivity of amorphous Se-based direct detectors and Ti-doped CsI-based indirect detectors by an order of magnitude.

FIGURE 3.6 (a) Photograph and (b) corresponding X-ray image of a leaf.
Source: Reproduced with permission from Ref. [39] Copyright 2015. Springer Nature.

Furthermore, Zhang et al.[41] in 2020 fabricated an ultra-sensitive X-ray detector using all-inorganic lead-free $Cs_2AgBiBr_6$ perovskite film. The resultant device was encapsulated in metal shell with Be window and revealed high X-ray detection ability. The minimum detection dose rate was found to be 145.2 $nGy_{air}s^{-1}$, and the sensitivity obtained was 1.8×10^4 $\mu CGy_{air}^{-1}cm^{-2}$, making it 1000 times more sensitive than the commercially available a-Se based X-ray detector. It is noteworthy that even after 2 months of storage, the device maintained good detection performance. Due to the excellent properties of perovskite, direct perovskite-based X-ray detection and imaging have shown tremendous potential for futuristic applications. However, there are still some issues that are needed to be addressed in order to overcome the commercialization barrier. To begin with, more physical insights are urged in order to comprehend direct material optimization, device design to meet low noise and high sensitivity requirements. Similar to PSCs, long-term stability is a major problem that may be solved by developing chemically stable materials and device structural design. In the nutshell, proper understanding is still needed on properties, technological pathways, and bottlenecks to enhance the performance of perovskite direct X-ray detectors.

iv) Thermoelectrics

With the continuous rise in global energy demand and the limited reserves of fossil fuels, improving energy conversion efficiency is highly significant nowadays. About two-third of total energy, however, is lost as the waste heat. In this regard, the direct harvesting of this waste heat via thermoelectric materials (TE) has attracted a worldwide attention in the scientific community.[42–44] Using the Seeback effect, TE materials can convert waste heat from furnaces, industrial processes and engine exhaust streams into usable. The most commonly studied TE material, Bi_2Te_3, has a ZT value ~ 1 at near room temperature (RT) for both p- and n-types. Notably, alloying this TE material with other elements can further enhance its ZT value.[45] In addition, group-IV tellurides including GeTe, PbTe, and SnTe are mid-temperature TE materials (500 – 900 K)[46–49] whereas group-IV chalcogenides like SnSe and SiGe-based alloys are commonly regarded as high-temperature TE materials (> 900 K).[50,51] For these materials, the maximum ZT values are typically in the range of 1–3. In recent years, halide perovskites have come to the forefront of thermoelectric investigation owing to their high Seebeck coefficient, ultra-low thermal conductivity, and high charge carrier mobility along with inexpensive and facile solution-based fabrication processes. The observations of diffusive phonon transports and ballistic charge transport have also led to their designation as "phonon glass electron crystals" (PGEC).[52] The A-site cation is the cause of PGEC behavior of halide perovskite owing to its weak bond to the lattice and ease movement in the perovskite structure. In halide perovskites, both phonon and electron transport occur on the metal halide lattice, but anharmonicity caused by the motion of the A-site cation may scatter the phonon movements and lower the thermal conductivity into the ultralow zone ($0.5\ W\ m^{-1}\ K^{-1}$).[53] Some recent theoretical prediction indicates that perovskite TE materials may have ZT ~ 2 with charge density ~$10^{18-19}\ cm^{-3}$.[54–56] Lee et al.[57] synthesized the single-crystalline all-inorganic hybrid perovskite (AIHP) nanowires by low-temperature, facile solution-based fabrication processes. They extensively investigated the TE properties of all synthesized AIHP nanowires including $CsPbI_3$, $CsPbBr_3$, and $CsSnI_3$ and reported their ultralow thermal conductivity (~ 0.4 $Wm^{-1}K^{-1}$). Among them, $CsSnI_3$ uniquely showed simultaneously high-electrical conductivity, high hole mobility, ultra-low thermal conductivity, a power factor of 186 $\mu Wm^{-1}K^{-2}$ and ZT of 0.11 at 320 K. Moreover, halide perovskites have potential to show an ultra-high Seeback coefficient. Ye et al.[58] measured the Seeback coefficient and thermal conductivity at different temperatures for $MAPbI_3$ single crystal grown by modified fast and inverse temperature crystallization technique. It was observed that initially Seeback coefficient was around 920 $\mu V K^{-1}$ at RT and does not fluctuate very much in the temperature range of 298–330 K. Intriguingly, at 351 K, it demonstrated an ultra-high value ~ 1693 $\mu V K^{-1}$. Such type of Seeback coefficient behavior and thermal conductivity change with temperature variation as shown in Figure 3.7.

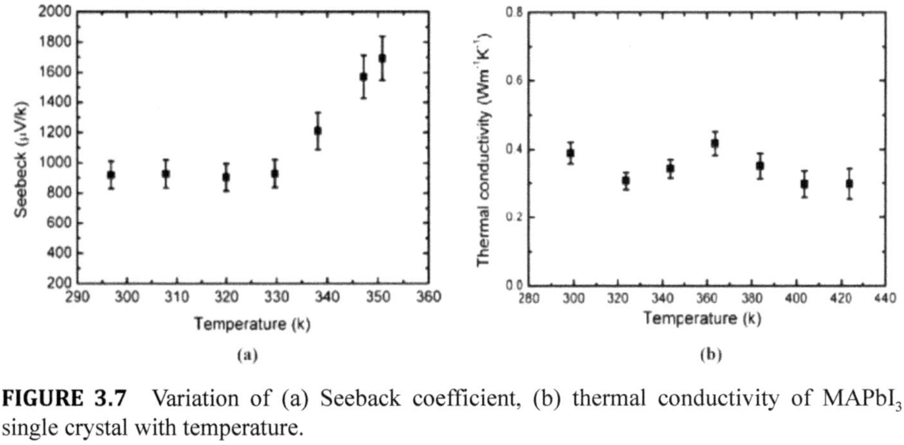

FIGURE 3.7 Variation of (a) Seeback coefficient, (b) thermal conductivity of $MAPbI_3$ single crystal with temperature.

Source: Reproduced with the Ref. [58] https://creativecommons.org/licenses/by/3.0/

Combining both large Seeback coefficient and ultra-low thermal conductivity, hybrid perovskites are expected to accomplish a ZT value higher than 1. Despite this, hybrid perovskites typically have low ZT and power factor owing to their electrically resistive nature. The current TE performance of halide perovskites is markedly lower than the state-of-the-art inorganic TE materials showing high ZT value such as Bi_2Te_3 and SnSe.[59] To the best of our knowledge, the highest ZT value for halide perovskites is 0.19.[60] Thus extensive efforts are required to enhance their TE performance in order to compete with conventional TE materials. The research in perovskite TE materials is still in its infancy and has only recently accelerated. The main advantage of perovskite as TE material is its low cost, which makes them a viable option to investigate. Furthermore, the decent Seeback coefficient and the intrinsic ultra-low thermal conductivity at RT make them appealing for further research. Perovskite materials with decent performance and improved stability can be utilized in low-grade heat recovery and electronic devices which require low power to begin with.

3.4 FUTURE TECHNOLOGY WITH PEROVSKITES

Metal halide perovskites can have additional futuristic applications in a variety of fields including space applications, thermochromic devices, data storage, and colloidal systems. The PSCs have been reported to have a maximum lifetime of less than 2 years.[61] The primary issue is the inherent instability of materials on exposure to oxygen and moisture. However, the prospect of using this material in space becomes more appealing, especially since PSCs would operate in no or very low levels of water and oxygen environmental conditions. In fact, recent studies revealed remarkable advantages due to the light weight of PSCs and their surprisingly high tolerance nature against high-energy particle irradiation.[62,63] Zhu and coworkers examined the stability of PSCs attached to a balloon raised to a height of 35 km in near space. The mixed cation PSC maintained 95% of their initial efficiency (~6.3%).[64] Further test with highly efficient devices will be required to demonstrate the enormous potential of perovskite for space solar cell applications. As stated earlier, perovskite can be utilized for data storage devices. Flash memory storage devices, such as SD cards, USB sticks, use nonvolatile memory which performs functions by erasing data units and rewriting them at the byte level. In 2020, Nazeeruddin and colleagues[65] fabricated a flash memory device having a transistor channel made from a single crystal of triple-cation perovskite. They created a single layer of nano floating gate with 10 nm thickness using organic semiconductors: fullerene for n-type and polyfluorene for p-type, which act as the charge storage unit. The resultant device demonstrated a memory window of 30 V, write/erase times of 50 ms, an on/off ratio of 9×10^7, and a retention time of 10^6 s. Moreover, thermochromic devices can make use of perovskite materials. The primary use of thermochromic material is to calculate the temperature variation and heat in chemical reaction. Their properties, on the other hand, are more widely applicable for energy-saving applications such as smart windows that can switch the incident light's transmittance between transparent and opaque states.[66] Bakr and coworkers[67] developed thermochromic perovskite inks for reversible smart windows applications by investigating opto-thermal properties of perovskites. By measuring the absorption spectra at different temperatures, 597 nm for 60°C, 615 nm for 90°C, and 651 nm for 120°C, respectively, they were able to confirm the chromatic variation of the perovskite materials. All of their works govern the potential of perovskite materials for future technology.

3.5 CONCLUSION

Halide perovskites have emerged as the most promising and exotic material for future technology. The viability of low-temperature, facile solution-based fabrication and a unique combination of electronic and optical properties indicate its potential for commercialization. This chapter makes an effort to highlight the advancements made in perovskite materials for various applications including photovoltaics, LEDs, X-ray detection, imaging, and thermoelectrics owing to their superior properties. The most recent research and techniques for advancing the use of these materials in these technologies have also been discussed. In addition, this chapter emphasizes the upcoming technology using these enigmatic perovskite materials. The main commercialization barrier for all these technologies is the instability of the perovskite materials. To sum up, the knowledge of properties, technological pathways, and bottlenecks are still required to enhance the performance and stability of perovskite materials which can pave a path toward their commercialization.

KEYWORDS

- **perovskite**
- **photovoltaics**
- **renewable energy**
- **optoelectronics**
- **thermoelectrics**

REFERENCES

1. Balamurugan, M.; Sahoo, S. K.; Sukchai, S. Application of Soft Computing Methods for Grid Connected PV System: A Technological and Status Review. *Renew. Sustain. Energy Rev.* **2017,** *75*, 1493–1508.
2. Kirchartz, T.; Bisquert, J.; Mora-Sero, I.; Garcia-Belmonte, G. Classification of Solar Cells According to Mechanisms of Charge Separation and Charge Collection. *Phys. Chem. Chem. Phys.* **2015,** *17* (6), 4007–4014.
3. Mayer, A. C.; Scully, S. R.; Hardin, B. E.; Rowell, M. W.; McGehee, M. D. Polymer-Based Solar Cells. *Mater. Today* **2007,** *10* (11), 28–33.
4. https://www.nrel.gov/pv/assets/pdfs/pv-efficiencies-20–03–2023.pdf
5. Im, J. H.; Lee, C. R.; Lee, J. W.; Park, S. W.; Park, N. G. 6.5% Efficient Perovskite Quantum-Dot-Sensitized Solar Cell. *Nanoscale* **2011,** *3* (10), 4088–4093.
6. Petrović, M.; Chellappan, V.; Ramakrishna, S. Perovskites: Solar Cells & Engineering Applications–Materials and Device Developments. *Solar Energy* **2015,** *122*, 678–699.
7. Christians, J. A.; Manser, J. S.; Kamat, P. V. Best Practices in Perovskite Solar Cell Efficiency Measurements: Avoiding the Error of Making Bad Cells Look Good. *J. Phys. Chem. Lett.* **2015,** *6* (5), 852–857.
8. Vasala, S.; Karppinen, M. A2B′ B ″O6 Perovskites: A Review. *Progress Solid State Chem.* **2015,** *43* (1–2), 1–36.
9. Kim, Y. H.; Cho, H.; Heo, J. H.; Kim, T. S.; Myoung, N.; Lee, C. L. et al. Multicolored Organic/Inorganic Hybrid Perovskite Light-Emitting Diodes. *Adv. Mater.* **2015,** *27* (7), 1248–1254.
10. Castelli, I. E.; García-Lastra, J. M.; Thygesen, K. S.; Jacobsen, K. W. Bandgap Calculations and Trends of Organometal Halide Perovskites. *APL Mater.* **2014,** *2* (8), 081514.
11. Eperon, G. E.; Stranks, S. D.; Menelaou, C.; Johnston, M. B.; Herz, L. M.; Snaith, H. J. Formamidinium Lead Trihalide: A Broadly Tunable Perovskite for Efficient Planar Heterojunction Solar Cells. *Energy Environ. Sci.* **2014,** *7* (3), 982–988.
12. National Renewable Energy Laboratory (NREL), Champion Photovoltaic Module Efficiency Chart. https://www.nrel.gov/pv/module-efficiency.html (accessed on July 2022).
13. Yoshikawa, K.; Kawasaki, H.; Yoshida, W.; Irie, T.; Konishi, K.; Nakano, K. et al. Silicon Heterojunction Solar Cell with Interdigitated Back Contacts for a Photoconversion Efficiency Over 26%. *Nat. Energy* **2017,** *2* (5), 1–8.
14. Kojima, A.; Teshima, K.; Shirai, Y.; Miyasaka, T. Organometal Halide Perovskites as Visible-Light Sensitizers for Photovoltaic Cells. *J. Am. Chem. Soc.* **2009,** *131* (17), 6050–6051.
15. Lee, M. M.; Teuscher, J.; Miyasaka, T.; Murakami, T. N.; Snaith, H. J. Efficient Hybrid Solar Cells Based on Meso-Super Structured Organometal Halide Perovskites. *Science* **2012,** *338* (6107), 643–647.
16. Saliba, M.; Matsui, T.; Domanski, K.; Seo, J. Y.; Ummadisingu, A.; Zakeeruddin, S. M. et al. Incorporation of Rubidium Cations into Perovskite Solar Cells Improves Photovoltaic Performance. *Science* **2016,** *354* (6309), 206–209.
17. Tan, Z. K.; Moghaddam, R. S.; Lai, M. L.; Docampo, P.; Higler, R.; Deschler, F. et al. Bright Light-Emitting Diodes Based on Organometal Halide Perovskite. *Nat. Nanotechnol.* **2014,** *9* (9), 687–692.

18. Yantara, N.; Bhaumik, S.; Yan, F.; Sabba, D.; Dewi, H. A.; Mathews, N. et al. Inorganic Halide Perovskites for Efficient Light-Emitting Diodes. *J. Phys. Chem. Lett.* **2015,** *6* (21), 4360–4364.
19. Cho, H.; Jeong, S. H.; Park, M. H.; Kim, Y. H.; Wolf, C.; Lee, C. L. et al. Overcoming the Electroluminescence Efficiency Limitations of Perovskite Light-Emitting Diodes. *Science* **2015,** *350* (6265), 1222–1225.
20. Alonso, C.; Byranvand, M. M.; Essig, S.; Saliba, M. Perovskites: Weaving a Network of Knowledge Beyond Photovoltaics. *J. Mater. Chem. A* **2022.**
21. Yang, X.; Zhang, X.; Deng, J.; Chu, Z.; Jiang, Q.; Meng, J. et al. Efficient Green Light-Emitting Diodes Based on Quasi-Two-Dimensional Composition and Phase Engineered Perovskite with Surface Passivation. *Nat. Commun.* **2018,** *9* (1), 570.
22. Wang, J.; Wang, N.; Jin, Y.; Si, J.; Tan, Z. K.; Du, H. et al. Interfacial Control Toward Efficient and Low-Voltage Perovskite Light-Emitting Diodes. *Adv. Mater.* **2015,** *27* (14), 2311–2316.
23. Kojima, A.; Ikegami, M.; Teshima, K.; Miyasaka, T. Highly Luminescent Lead Bromide Perovskite Nanoparticles Synthesized with Porous Alumina Media. *Chem. Lett.* **2012,** *41* (4), 397–399.
24. Tan, Z. K.; Moghaddam, R. S.; Lai, M. L.; Docampo, P.; Higler, R.; Deschler, F. et al. Bright Light-Emitting Diodes Based on Organometal Halide Perovskite. *Nat. Nanotechnol.* **2014,** *9* (9), 687–692.
25. Ling, Y.; Yuan, Z.; Tian, Y.; Wang, X.; Wang, J. C.; Xin, Y. et al. Bright Light-Emitting Diodes Based on Organometal Halide Perovskite Nanoplatelets. *Adv. Mater.* **2016,** *28* (2), 305–311.
26. Lin, K.; Xing, J.; Quan, L. N.; de Arquer, F. P. G.; Gong, X.; Lu, J. et al. Perovskite Light-Emitting Diodes with External Quantum Efficiency Exceeding 20 Per Cent. *Nature* **2018,** *562* (7726), 245–248.
27. Xu, W.; Hu, Q.; Bai, S.; Bao, C.; Miao, Y.; Yuan, Z. et al. *Nat. Photonics* **2019,** *13*, 418–424.
28. Cattell, H. W. Application of the X-rays to Surgery. *Science* **1896,** *3* (62), 344–346.
29. Clark, C. F. Location of Fragment of Steel in Eye, by X-rays. *Trans. Am. Ophthalmol. Soc.* **1896,** *7*, 711.
30. Wells, K.; Bradley, D. A. A Review of X-ray Explosives Detection Techniques for Checked Baggage. *Appl. Radiat. Isotopes* **2012,** *70* (8), 1729–1746..
31. Bernheim, A.; Mei, X.; Huang, M.; Yang, Y.; Fayad, Z. A.; Zhang, N. et al. Chest CT Findings in Coronavirus Disease-19 (COVID-19): Relationship to Duration of Infection. *Radiology* **2020,** *295* (3), 685–691.
32. Fadley, C. S.; Shirley, D. A. Electronic Densities of States from X-ray Photoelectron Spectroscopy. *J. Res. Natl. Bureau Standards: Sect. A, Phys. Chem.* **1970,** *74* (4), 543.
33. Li, W.; Xu, Y.; Peng, J.; Li, R.; Song, J.; Huang, H. et al. Evaporated Perovskite Thick Junctions for X-ray Detection. *ACS Appl. Mater. Interfaces* **2021,** *13* (2), 2971–2978.
34. Kasap, S.; Frey, J. B.; Belev, G.; Tousignant, O.; Mani, H.; Greenspan, J. et al. Amorphous and Polycrystalline Photoconductors for Direct Conversion Flat Panel X-ray Image Sensors. *Sensors* **2011,** *11*, 5112–5157.
35. Su, Y.; Ma, W.; Yang, Y. M. Perovskite Semiconductors for Direct X-ray Detection and Imaging. *J. Semiconductors* **2020,** *41* (5), 051204.
36. Wei, H.; Fang, Y.; Mulligan, P.; Chuirazzi, W.; Fang, H. H.; Wang, C. et al. Sensitive X-ray Detectors Made of Methylammonium Lead Tribromide Perovskite Single Crystals. *Nat. Photonics* **2016,** *10* (5), 333–339.

37. Pan, W.; Wu, H.; Luo, J.; Deng, Z.; Ge, C.; Chen, C. et al. Cs2AgBiBr6 Single-Crystal X-ray Detectors with a Low Detection Limit. *Nat. Photonics* **2017,** *11* (11), 726–732.
38. Zhuang, R.; Wang, X.; Ma, W.; Wu, Y.; Chen, X.; Tang, L. et al. Highly Sensitive X-ray Detector Made of Layered Perovskite-Like $(NH_4)_3Bi_2I_9$ Single Crystal with Anisotropic Response. *Nat. Photonics* **2019,** *13* (9), 602–608.
39. Yakunin, S.; Sytnyk, M.; Kriegner, D.; Shrestha, S.; Richter, M.; Matt, G. J. et al. Detection of X-ray Photons by Solution-Processed Lead Halide Perovskites. *Nature Photonics* **2015,** *9* (7), 444–449.
40. Kim, Y. C.; Kim, K. H.; Son, D. Y.; Jeong, D. N.; Seo, J. Y.; Choi, Y. S. et al. Printable Organometallic Perovskite Enables Large-Area, Low-Dose X-ray Imaging. *Nature* **2017,** *550* (7674), 87–91.
41. Zhang, H.; Dun, G.; Feng, Q.; Zhao, R.; Liang, R.; Gao, Z. et al. Encapsulated X-ray Detector Enabled by All-Inorganic Lead-Free Perovskite Film with High Sensitivity and Low Detection Limit. *IEEE Trans. Electron Devices* **2020,** *67* (8), 3191–3198.
42. Bell, L. E. Cooling, Heating, Generating Power, and Recovering Waste Heat with Thermoelectric Systems. *Science* **2008,** *321* (5895), 1457–1461.
43. Majumdar, A. Thermoelectricity in Semiconductor Nanostructures. *Science* **2004,** *303* (5659), 777–778.
44. Zhang, X.; Zhao, L. D. Thermoelectric Materials: Energy Conversion Between Heat and Electricity. *J. Materiomics* **2015,** *1* (2), 92–105.
45. Kim, S. I.; Lee, K. H.; Mun, H. A.; Kim, H. S.; Hwang, S. W.; Roh, J. W. et al. Dense Dislocation Arrays Embedded in Grain Boundaries for High-Performance Bulk Thermoelectrics. *Science* **2015,** *348* (6230), 109–114.
46. Pei, Y.; Shi, X.; LaLonde, A.; Wang, H.; Chen, L.; Snyder, G. J. Convergence of Electronic Bands for High Performance Bulk Thermoelectrics. *Nature* **2011,** *473* (7345), 66–69.
47. Heremans, J. P.; Jovovic, V.; Toberer, E. S.; Saramat, A.; Kurosaki, K.; Charoenphakdee, A. et al. Enhancement of Thermoelectric Efficiency in PbTe by Distortion of the Electronic Density of States. *Science*, *321* (5888), 554–557.
48. Perumal, S.; Roychowdhury, S.; Negi, D. S.; Datta, R.; Biswas, K. High Thermoelectric Performance and Enhanced Mechanical Stability of p-Type Ge1–x Sb x Te. *Chem. Mater.* **2015,** *27* (20), 7171–7178.
49. Moshwan, R.; Yang, L.; Zou, J.; Chen, Z. G. Eco-Friendly SnTe Thermoelectric Materials: Progress and Future Challenges. *Adv. Funct. Mater.* **2017,** *27* (43), 1703278.
50. Basu, R.; Bhattacharya, S.; Bhatt, R.; Roy, M.; Ahmad, S.; Singh, A. et al. Improved Thermoelectric Performance of Hot Pressed Nanostructured n-type SiGe Bulk Alloys. *J. Mater. Chem. A* **2014,** *2* (19), 6922–6930.
51. Zhao, L. D.; Lo, S. H.; Zhang, Y.; Sun, H.; Tan, G.; Uher, C. et al. Ultralow Thermal Conductivity and High Thermoelectric Figure of Merit in SnSe Crystals. *Nature* **2014,** *508* (7496), 373–377.
52. Miyata, K.; Atallah, T. L.; Zhu, X. Y. Lead Halide Perovskites: Crystal-Liquid Duality, Phonon Glass Electron Crystals, and Large Polaron Formation. *Sci. Adv.* **2017,** *3* (10), e1701469.
53. Liu, T.; Yue, S. Y.; Ratnasingham, S.; Degousée, T.; Varsini, P.; Briscoe, J. et al. Unusual Thermal Boundary Resistance in Halide Perovskites: A Way to Tune Ultralow Thermal Conductivity for Thermoelectrics. *ACS Appl. Mater. Interfaces* **2019,** *11* (50), 47507–47515.

54. Lee, C.; Hong, J.; Stroppa, A.; Whangbo, M. H.; Shim, J. H. Organic–Inorganic Hybrid Perovskites ABI_3 (A= CH_3NH_3, NH_2CHNH_2; B= Sn, Pb) as Potential Thermoelectric Materials: A Density Functional Evaluation. *RSC Adv.* **2015,** *5* (96), 78701–78707.
55. Filippetti, A.; Caddeo, C.; Delugas, P.; Mattoni, A. Appealing Perspectives of Hybrid Lead–Iodide Perovskites as Thermoelectric Materials. *J. Phys. Chem. C* **2016,** *120* (50), 28472–28479.
56. Zhao, T.; Wang, D.; Shuai, Z. Doping Optimization of Organic-Inorganic Hybrid Perovskite $CH_3NH_3PbI_3$ for High Thermoelectric Efficiency. *Synthetic Metals* **2017,** *225*, 108–114.
57. Lee, W.; Li, H.; Wong, A. B.; Zhang, D.; Lai, M.; Yu, Y. et al. Ultralow Thermal Conductivity in All-Inorganic Halide Perovskites. *Proc. Natl. Acad. Sci.* **2017,** *114* (33), 8693–8697.
58. Ye, T.; Wang, X.; Li, X.; Yan, A. Q.; Ramakrishna, S.; Xu, J. Ultra-High Seebeck Coefficient and Low Thermal Conductivity of a Centimeter-Sized Perovskite Single Crystal Acquired by a Modified Fast Growth Method. *J. Mater. Chem. C* **2017,** 5 (5), 1255–1260.
59. Zhao, L. D.; Lo, S. H.; Zhang, Y.; Sun, H.; Tan, G.; Uher, C. et al. Ultralow Thermal Conductivity and High Thermoelectric Figure of Merit in SnSe Crystals. *Nature* **2014,** *508* (7496), 373–377.
60. Zheng, L.; Zhu, T.; Li, Y.; Wu, H.; Yi, C.; Zhu, J.; Gong, X. Enhanced Thermoelectric Performance of F4-TCNQ Doped $FASnI_3$ Thin Films. *J. Mater. Chem. A* **2020,** *8* (47), 25431–25442.
61. Khenkin, M. V.; Katz, E. A.; Abate, A.; Bardizza, G.; Berry, J. J.; Brabec, C. et al. Consensus Statement for Stability Assessment and Reporting for Perovskite Photovoltaics Based on ISOS Procedures. *Nat. Energy* **2020,** *5* (1), 35–49.
62. Kang, S.; Jeong, J.; Cho, S.; Yoon, Y. J.; Park, S.; Lim, S. et al. Ultrathin, Lightweight and Flexible Perovskite Solar Cells with an Excellent Power-Per-Weight Performance. *J. Mater. Chem. A* **2019,** *7* (3), 1107–1114.
63. Klein-Kedem, N.; Cahen, D.; Hodes, G. Effects of Light and Electron Beam Irradiation on Halide Perovskites and Their Solar Cells. *Acc. Chem. Res.* **2016,** *49* (2), 347–354.
64. Tu, Y.; Xu, G.; Yang, X.; Zhang, Y.; Li, Z.; Su, R. et al. Mixed-Cation Perovskite Solar Cells in Space. *Sci. China Phys. Mech. Astronomy* **2019,** *62*, 1–4.
65. Vasilopoulou, M.; Kim, B. S.; Kim, H. P.; da Silva, W. J.; Schneider, F. K.; Mat Teridi, M. A. et al. Perovskite Flash Memory with a Single-Layer Nanofloating Gate. *Nano Lett.* **2020,** *20* (7), 5081–5089.
66. Wang, Y.; Runnerstrom, E. L.; Milliron, D. J. *Annu. Rev. Chem. Biomol. Eng.* **2016,** *7*, 283–304.
67. De Bastiani, M.; Saidaminov, M. I.; Dursun, I.; Sinatra, L.; Peng, W.; Buttner, U. et al. Thermochromic Perovskite Inks for Reversible Smart Window Applications. *Chem. Mater.* **2017,** *29* (8), 3367–3370.

CHAPTER 4

Perovskite Materials for Photovoltaic Applications

AMIT KUMAR SHARMA[1,2], SANGEETA PRASHER[3], and MUKESH KUMAR[1]

[1]*Department of Physics, Lovely Professional University, Phagwara, India*

[2]*Department of Physics, Maharana Pratap Govt. College Amb, Himachal Pradesh, India*

[3]*Department of Physics, Kanya Maha Vidhayalya, Jalandar, Punjab, India*

ABSTRACT

Natural abundant perovskite material has centered material scientist's attention on future technologies. The properties of these materials aligned and advanced research in superconductivity, energy storage, photovoltaics, lasers, sensors, and many other fields. This chapter is focussed on the photovoltaic applications of perovskite materials due to their efficient light-absorbing power. Perovskites are cheap and simple to synthesize in comparison to traditional photovoltaic materials available in the market. Perovskite materials based solar cells are advancing very rapidly and approaching power conversion efficiency more than 25% due to remarkable intrinsic electrical and optical properties like long charge carrier lifetimes, energy band gap tunability, remarkable charge carrier mobilities, tolerance against high defects, high absorption coefficient, small binding energy of excitons, and large diffusion lengths of charge carriers. Despite all of these materials' favorable characteristics for future solar technology, instability and degradation pose the biggest obstacles to their commercialization. Moreover, perovskite-based photovoltaic technology is advancing steadily toward commercialization.

Advanced Materials for Next-Generation Technologies: Challenges and New Prospects. Shrikaant Kulkarni, Vipul Srivastava, Rabah Khenata, & Yarub Al-Douri (Eds.)

4.1 INTRODUCTION

Renewable energy sources are the only options to overcome the problem of global warming and environmental issues. Sun, being the primary origin of renewable energy, is continuously supplying light and heat since the origin of life on earth. Proper and efficient harvesting as well as storage of the energy from the sun is the prime need of researchers. One of the easy and popular technologies of converting sun energy into electrical energy is photovoltaics (PV), which can solve the problem of harvesting the energy. The harvested energy could be used through power storage technology in the future, even in the absence of natural energy source. In the present scenario, single-junction crystalline silicon-based photovoltaics are delivering more than 25% power conversion efficiency[1,2] and commercially ruling 90% of the PV market.[3] But silicon-based PV technology is limiting its power conversion efficiencies to a maximum theoretical value of 29.4% for further enhancement due to its electronic nontunability.[4] Since the beginning of the 21st century many alternate materials for silicon were focussed and explored for light absorbance in solar cells for better power conversion efficiencies. Perovskite solar cells (PSCs),[10,11] dye sensitized solar cells (DSSCs),[5] quantum dot solar cells (QDSCs),[8,9] and organic solar cells (OSCs)[6,7] are the advance generation solar cells competent to first-generation silicon-based solar cell technology. One more prime factor limiting the use of silicon-based solar cell technology is "high cost."[12–14] The complete PV technology is tabulated generation wise in the diagram below.[15]

4.2 CLASSIFICATION

4.2.1 PEROVSKITE MATERIAL

The name "Perovskite" was derived from its inventor's name Lev Perovski (1792–1856). The material was originated by Gustav Rose in Russian mountain range[16] and was first reported in the literature in 1945.[17] The fundamental anatomy of perovskite material is analogous to lattice structure of calcium titanium oxide (CaTio2),[18,19] whose natural look is shown in Figure 4.2(a).

4.2.2 STRUCTURE

All perovskite materials are represented in the form of general abbreviation ABX_3, A, B are metallic cation (cation A is larger in size than cation B), X is metal anion routinely halide and oxide. B is octahedrally connected to X making BX_6, forming fundamental block of perovskite material (Figure 4.3), BX_6 so formed is coordinated to metal cation A in three-dimensional cubic structure.[20,21] Few oxides of these materials are also known, called as chalcogenide perovskite having exceptional superconducting and ferroelectric properties for various applications.[22] Oxides perovskites (chalcogenide perovskites) have different properties and applications[23] in comparison with halide perovskites.[24]

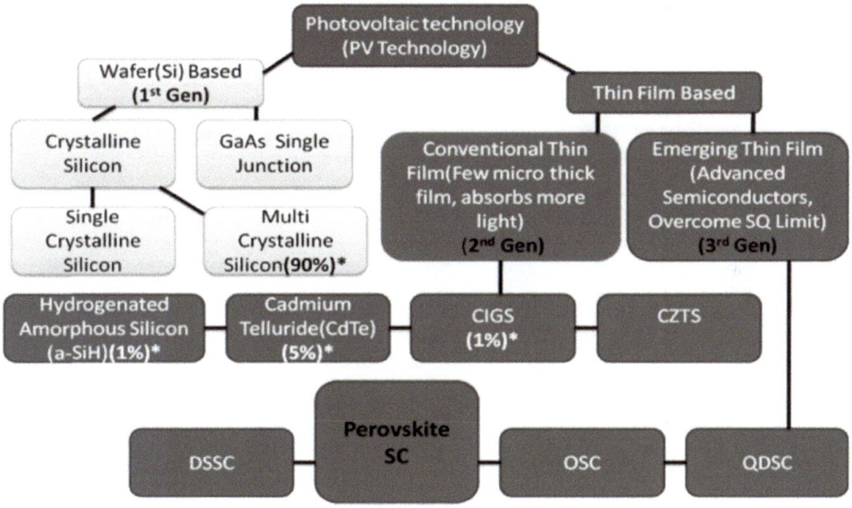

FIGURE 4.1 Generation of PV technology.

Source: Adapted from Ref. [15]

FIGURE 4.2 Naturally occurred perovskite material. Image source: www.perovskite-info.com.

Source: Rob Lavinsky, iRocks.com. https://creativecommons.org/licenses/by-sa/3.0/

Due to flexible options available, the combinations of different cations and anions can form many types of perovskite materials. The cation pairing in the perovskite materials is possible in four general ways, that are $A^{+}B^{2+}X^{-}_{3}$(1:2 perovskite), $A^{2+}B^{4+}X^{-}_{3}$(2:4 perovskite), $A^{3+}B^{3+}X^{-}_{3}$(3:3 perovskite), and $A^{+}B^{5+}X^{-}_{3}$(1:5 perovskite).

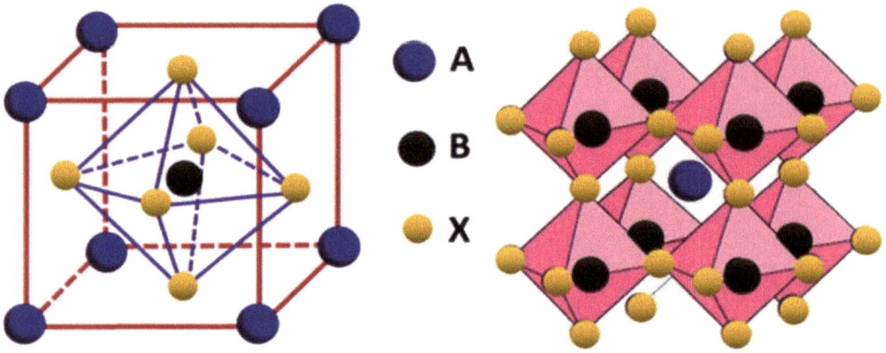

FIGURE 4.3 Building block structure of perovskite material.

Source: Adapted from Ref. [84]

Primarily the perovskites are divided into three main categories as given below (Figure 4.4).

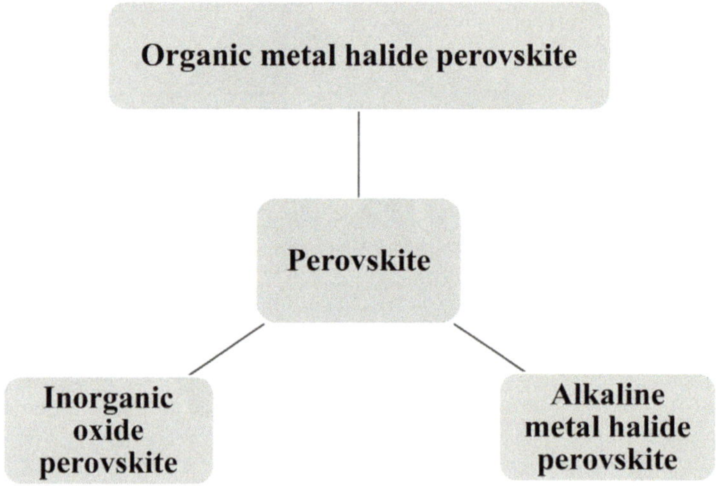

FIGURE 4.4 Classification of perovskite material.

In this article only perovskite materials based on halides are on our focus highlighting their photovoltaic applications. The hybrid mixing of organic and inorganic compounds developed the remarkable physical and optoelectrical properties in organometallic halide perovskites. Organometallic halide perovskites (ABX_3) have the same structure as $CaTio_3$ with different oxidation states with popular combination of cations and anions as given (for A, B and X):

A= (Methyl ammonium cation), (Formamidine cation) alternatively
A (Inorganic monovalent cation) =Cesium (Cs^{2+}), rubidium (Rb^{2+})
B- inorganic cation (Pb^{2+}, Sn^{2+}, Ge^{2}, Mg^{2+}, Ca^{2+})
X- chlorine, bromine, iodine anions

As from these combinations of anions and cations the most common and earlier reported combination is mixed halide perovskite material ($CH_3NH_3PbI_3$).[25,26] In addition to this famous combination, some more combinations like triple cation perovskite and quadruple cation perovskites showed better performance concerning efficiency and stability for photovoltaic applications.[27,28] The current generation in the perovskite materials is basically due to the electrons which rest or are trapped in the shallow energy states near or in the bands (conduction or valence).[29] This property of perovskite materials results in long carrier diffusion lengths of order of micrometers which highlighted them for photovoltaic applications.[30] Organometallic halide perovskite exists and shows both p- and n-type conducting properties for charge transport even for thick films.[31]

4.2.3 PRESENCE OF LEAD IN PEROVSKITE MATERIALS—TOXICITY

Lead is still imperative among all perovskite materials for excellent photovoltaic applications. But toxicity of lead in these materials is still a challenge and a huge hurdle affecting commercialization of these materials in photovoltaic technology.[32] Less toxic material is required for the purpose, Tin (Sn) is the best option among the available elements from the same group as lead (IVA of periodic table) of elements. In addition to this, tin has energy band gap of 1.2–1.4 eV[33] near to band gap 1.38 eV (Shockley-Queissner) under standard sun illumination limits.[34] Field-effect transistors based on tin were studied and reported more than twenty years ago for their conducting properties and as conducting channels in thin films.[35,36] In the year 2014, the PCE of these materials touched 6%,[37,38] which inspired the group of researchers to look more in tin-based perovskite materials for their excellent photovoltaic applications. Further, 9% PCE is listed in the literature for low-dimensional tin-based perovskite solar cell.[39] The main issue with tin-based PSCs is necessitous performance in the surrounding atmosphere ascribed to facile oxidation from Sn^{2+} to Sn^{4+}. To overcome the problem many syntheses method and technologies were developed; moreover, a PCE near to13% with a better stability is achieved.[40] The issue of the stability and performance of the tin-based PSCs could be addressed through the main emphasis on the following three main factors.[41]

a) Suitable reducing additive (to minimize oxidation of Sn^{2+} to Sn^{4+}).
b) Suitable solvent for enhancing superficial morphology.
c) Modifying compositional strategies for better performance.

4.3 APPLICATIONS OF PEROVSKITE MATERIALS

Different properties related to different perovskite materials opened the research in multifield applications. Applications-based classification of these materials to be used in oxides or halides form is given as follows.

4.3.1 *PEROVSKITE OXIDES*

Perovskite oxides for their effective chemical and physical properties are explored in many fields like piezoelectricity, dielectricity, ferroelectricity, energy conversion, energy storage, and many more.[42] These properties of perovskite oxides advance the research for the following applications[43]:

a) Solid-state batteries
b) Solid oxide fuel cells
c) Oxygen evolution and reduction reactions
d) Electrochemical water splitting
e) Metal-air batteries etc.

4.3.2 *PEROVSKITE HALIDES*

Perovskite halides are the potential candidates for excellent optoelectronic properties. Tunable band gap property allows the researchers to explore for photovoltaic and optoelectrical applications. Single-crystal perovskite in comparison of all other available perovskite halides shows better performance.[44] Hence, these materials in their halide form (single crystal) are the excellent and promising candidates for optoelectronic and photovoltaic applications.[45–53] Single-crystal mixed (organic–inorganic) halide perovskites also show better photovoltaic performance in contrasted to inorganic perovskites. The intrinsic properties of the halide perovskite materials responsible for photovoltaic activities are summarized as follows:

a) High carrier mobility[48,50]
b) Long balanced electron-hole diffusion lengths[49,50,53–55]
c) Long carrier life time[50,55,56–59]
d) High absorption coefficient in visible region[57,60]

These properties make perovskite halides a potential contestant for photovoltaic and optoelectrical applications in the following devices:

a) Photodetectors
b) Light-emitting diodes
c) Lasers
d) Solar cells
e) Field-effect transistors

and many more.

4.4 PHOTOVOLTAIC TECHNOLOGY (PV TECHNOLOGY)

Becquerel in the eighteenth century noted and explored the photoelectric phenomena on electrodes immersed in electrolyte; the phenomena were also explored for solids in the same century. Without using outer source of voltage, electromotive force may be generated. But in the nineteenth century the phenomena were extensively highlighted and deeply researched. The photovoltaic (PV) word is made of two distinguished technologies, that is, photo and voltaic, which shows photons (light) and voltage (electric current). In simple words in photovoltaic technology the photons (light) generate voltage (electric currents). The production of electric current by the application of light is called photovoltaic phenomena. This phenomenon was first observed in 1950 at Bell Laboratory. The photovoltaic devices are generally known as solar cells, which lean on the principal of photoelectric effect. As and when the light strikes on the solar cells, the basic electronic properties of the cell change. The working of a solar cell comprises three main processes listed below:

a) Incident photon absorption
b) Photogenerated charge carrier separation (electrons and holes)
c) Charge carrier collection and transportation to external circuit

a) **Incident photon absorption**
Not every incident photon will generate charge carriers (electrons and holes), but photons having energy beyond valence and conduction band gap (Eg) will generate them. The material which absorbs the photons is semiconductor or the combination of the semiconductors. So, in a p–n junction solar cell, the photons will hit depletion region, got absorbed, and further supply their energy to the valence electrons resulting in the generation of charge carrier pairs.
b) **Photogenerated charge carrier separation (electrons and holes)**
Separated electrons and holes after photogeneration are swept outside the depletion region by induced electric field (voltage), which prevents the diffusion of electrons to p-side and holes to n-side. These charge carriers remained separated until they reunite to form electron–hole pair.
c) **Charge carrier collection and transportation to external circuit**
The third and very important part of the photovoltaic process is the transportation of charge carriers to external circuit after collection for current generation. The movement of electrons from n-side to p-side of the cell device generates current. Electrons that left the cell in the form of currents supply their energy to the external connected source and after that re-entre the cell for repeating the whole process to produce more output (solar energy).

Figure 4.5 depicts the complete process.

4.4.1 PV TECHNOLOGY GENERATIONS

With improved efficiencies of the solar cells, the researchers and scientists explored many materials and technologies. Based upon the crystallinity, traditional and commercial solar cells (first generation) based on silicon are mainly divided into four types:

a) Amorphous silicon cells
b) Monocrystalline silicon cells
c) Polycrystalline silicon cells
d) Hybrid solar cells

Monocrystalline silicon is the pure and basic form of the material in comparison of other two, so it gives better efficiency, but manufacturing/processing cost is high.

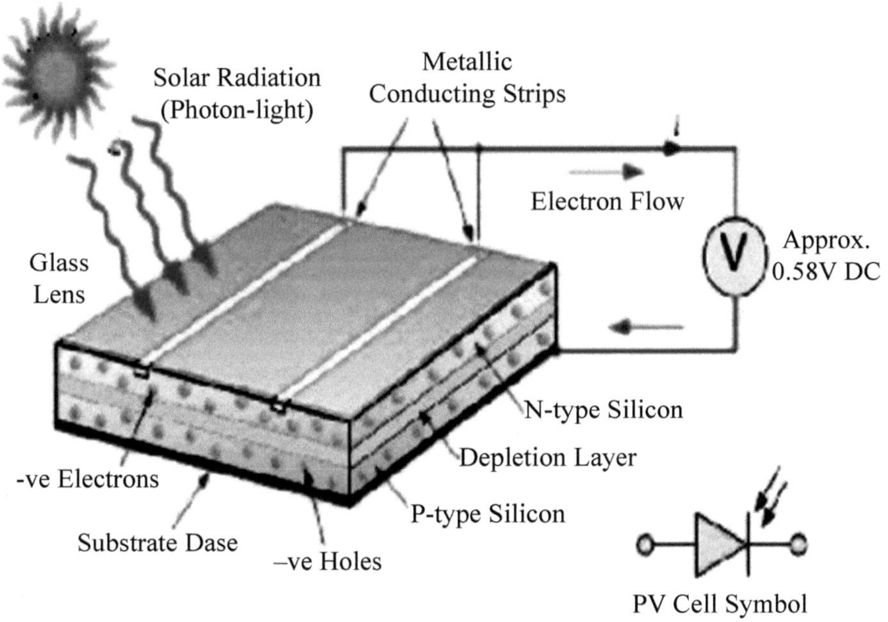

FIGURE 4.5 Basic solar cell showing complete PV process.
Source: Adapted from Ref. [61]

Also, on the basis of the technology the solar cells are divided into three categories

a) Discrete cell
b) Integrated thin-film cell
c) Multi-crystalline silicon cells.

In a solar cell, for excellent power output (efficiency), selection of absorbing material as well as technology is very important. The complete overview of the emerging technologies generation wise is shown in Figure 4.1.

Thin-film-based solar cells having a few microns thickness, that is, conventional thin-film technology is the second generation of cells (Figure 4.6).

Thin-film solar cells are made using different layers intercede between two metal contact layers. CdTe (cadmium telluride) and CIGS (copper-indium-gallium-selenide) based thin films are the common example. These thin-film cells are more efficient and can harvest more photon from the sun[62,63] compared to traditional cells. But synthesis cost, nonabundance, and toxic nature of these material hinder their use toward commercialization.[64,65] So, there was the dire need of the absorber material for solar cell that can overcome these obstacles.

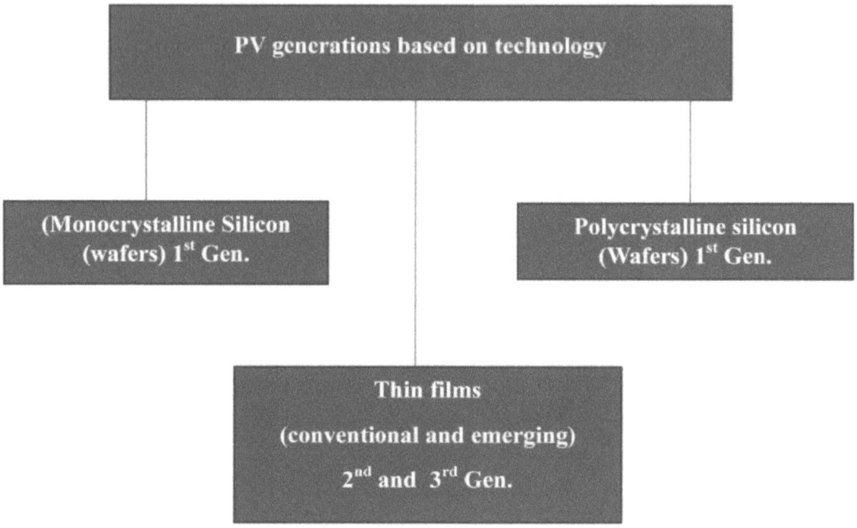

FIGURE 4.6 PV generation technology.

Further, a new emerging generation was researched, that is, third generation of PV technology. This generation of solar cell does not really depend upon typical p–n junction techniques of separating charge carriers. But this generation has multiple advantages over other two, such as tunable band gap, collecting photoenergy from whole sun spectra, synthesis on flexible substrate etc.[66–68] Quantum dot solar cells (QDSCs), dye sensitized solar cells (DSSCs), organic solar cells (OSCs), and perovskite solar cells (PSCs) are included in the third generation. Moreover, Tandem (mixed) solar cells (TSCs) may be abbreviated as fourth generation.

4.5 SYNTHESIS METHODS

Natural abundance of perovskite materials makes them potential candidate for commercialization of the PV technology in suppression of traditional silicon-based technology. The two basic configurations for the layered structure of solar cells based on perovskite material (PSCs) are given below

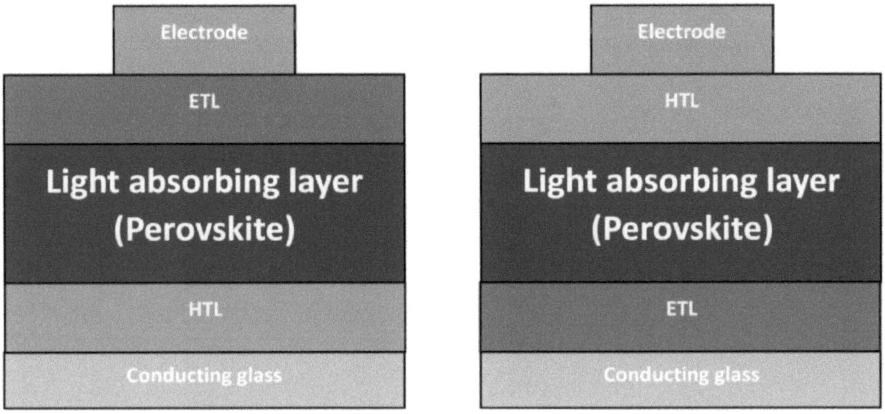

(a) (b)

FIGURE 4.7 (a) Regular planer configuration, (b) inverted planer configuration.

The basic point of difference between regular planer and inverted planer configuration is, interchanging of HTL (hole transport layer) and ETL (electron transport layer) over the conducting glass. The light-harvesting or light-absorbing layer, that is, perovskite layer intercedes between ETL and HTL.

Miyasaka and coworkers (2009) explored organic–inorganic perovskite halides for photoelectric and photoconductive properties in the electrochemical cells. The properties of perovskite materials had been explored since a few decades, which are highly dependent on synthesis techniques adopted for making crystalline thin films. The thin films so synthesized or grown may be of single crystals, nanocrystals, or polycrystals.[69–72] However so far single-crystal perovskites, due to their monocrystal nitty showed size and shape dependent electrical and optical properties.[73]

4.5.1 SINGLE-CRYSTAL GROWTH METHODS

For commercialization of technology of solar cells, quality and size of synthesized perovskite material is of utmost importance. Here, a few single-crystal synthesis techniques are given below[74]:

1. Layered solution growth (LSG) method
2. Inverse temperature crystallization (ITC) method
3. Anti-solvent vapor-assisted crystallization (AVC) method
4. Droplet pinned crystallization method (DPC)
5. Solution temperature lowering (STL) method

6. Bridgman (BG) method
7. Modified inverse temperature crystallization (MITC) method
8. Solvothermal growth (SG) method
9. Slow evaporation (SE) method

The chemical, physical, and optoelectrical characteristics of perovskite materials are mainly dependent upon blend processes and techniques. To have a need of desired property from the synthesized material, synthesis method selection is a very important part. Morphology of the synthesized material and its crystal structure may be affected by the synthesis technique.[75] In general, the synthesis of these material is done on the basis of the following synthesis techniques:[76]

a) Solid-state synthesis, for making bulk materials
b) Liquid-state synthesis, for making nanomaterials
c) Gaseous-state synthesis, for making thin films

4.5.2 SOLID-STATE SYNTHESIS (BULK MATERIALS)

This method is also called ceramic method due to synthesizing polycrystalline ceramics. Heating at high temperatures is the main art of this technique for maximizing the chemical rates through the following processes.

a) Mechanical ball-milling method
b) High-energy ball-milling method

4.5.3 LIQUID-STATE SYNTHESIS (NANOMATERIALS)

The materials synthesized by this method are nanomaterials of oxides, nitrates, acetates, and oxalates at room temperatures. The synthesis is done through the following processes:

a) Sol-gel method
b) Auto combustion method
c) Co-precipitation method

4.5.4 GASEOUS STATE SYNTHESIS (THIN FILMS)

A variety of nanomaterials as well as thin films are synthesized using this method. There are many processes in this type of synthesis technique which enable the perovskite materials excellent candidate for photovoltaic applications especially solar cell applications.[77–83]

a) DC sputtering
b) Electron beam evaporation
c) Molecular beam epitaxy
d) Chemical vapor deposition
e) Magnetron sputtering
f) Thermal evaporation
g) Laser ablation

4.6 FUTURE PROSPECTIVE

In a short span of time, the perovskite-based solar cell has shown the power conversion efficiency above 25% in comparison to the traditional silicon-based solar cells. Moreover, the environment concerns are linked with lead in these materials due to its toxicity. However, the replacement of lead with other elements like tin is the bright option. Stability is still hindering the commercialization of these material in photovoltaic technology. But, in the future, improved engineering and encapsulation processes will definitely overcome these obstacles.

KEYWORDS

- **perovskite materials**
- **photovoltaic technology**
- **charge carriers**
- **lead toxicity**
- **stability**
- **optical properties**

REFERENCES

1. Battaglia, C.; Cuevas, A.; De Wolf, S. High-Efficiency Crystalline Silicon Solar Cells: Status and Perspectives. *Energy Environ. Sci.* **2016,** *9* (5), 1552–1576.

2. Yoshikawa, K.; Kawasaki, H.; Yoshida, W.; Irie, T.; Konishi, K.; Nakano, K. et al. Silicon Heterojunction Solar Cell with Interdigitated Back Contacts for a Photoconversion Efficiency Over 26%. *Nat. Energy* **2017,** *2* (5), 1–8.
3. Green, M.; Dunlop, E.; Hohl-Ebinger, J.; Yoshita, M.; Kopidakis, N.; Hao, X. Solar Cell Efficiency Tables (Version 57). *Progress Photovoltaics: Res. App.* **2021,** *29* (1), 3–15.
4. Richter, A.; Hermle, M.; Glunz, S. W. Reassessment of the Limiting Efficiency for Crystalline Silicon Solar Cells. *IEEE J. Photovoltaics* **2013,** *3* (4), 1184–1191.
5. O'regan, B.; Grätzel, M. A Low-Cost, High-Efficiency Solar Cell Based on Dye-Sensitized Colloidal TiO_2 Films. *Nature* **1991,** *353* (6346), 737–740.
6. Zhang, J.; Zhu, L.; Wei, Z. Toward Over 15% Power Conversion Efficiency for Organic Solar Cells: Current Status and Perspectives. *Small Methods* **2017,** *1* (12), 1700258.
7. Lin, Y.; Firdaus, Y.; Nugraha, M. I.; Liu, F.; Karuthedath, S.; Emwas, A. H. et al. 17.1% Efficient Single-Junction Organic Solar Cells Enabled by n-Type Doping of the Bulk-Heterojunction. *Adv. Sci.* **2020,** *7* (7), 1903419.
8. Tavakoli, M. M.; Nasilowski, M.; Zhao, J.; Bawendi, M. G.; Kong, J. Efficient Semitransparent CsPbI3 Quantum Dots Photovoltaics Using a Graphene Electrode. *Small Methods* **2019,** *3* (12), 1900449.
9. Semonin, O. E.; Luther, J. M.; Choi, S.; Chen, H. Y.; Gao, J.; Nozik, A. J.; Beard, M. C. Peak External Photocurrent Quantum Efficiency Exceeding 100% via MEG in a Quantum Dot Solar Cell. *Science* **2011,** *334* (6062), 1530–1533.
10. Macdonald, T. J.; Batmunkh, M.; Lin, C.; Kim, J.; Tune, D. D.; Ambroz, F.; Li, X.; Xu, S.; Sol, C. Papakonstantinou, I.; McLachlan, M. A.; Parkin, I. P.; Shapter, J. G.; Durrant, J. R. *Small Methods* **2019,** *3*, 1900164.
11. Lee, M. M.; Teuscher, J.; Miyasaka, T.; Murakami, T. N.; Snaith, H. J. Efficient Hybrid Solar Cells Based on Meso-Superstructured Organometal Halide Perovskites. *Science* **2012,** *338* (6107), 643–647.
12. Noel, N. K.; Stranks, S. D.; Abate, A.; Wehrenfennig, C.; Guarnera, S.; Haghighirad, A. A.; Sadhanala, A.; Eperon, G. E.; Pathak, S. K.; Johnston, M. B.; Petrozza, A.; Herz, L. M.; Snaith, H. J. Lead-free Organic−Inorganic Tin Halide Perovskites for Photovoltaic Applications. *Energy Environ. Sci.* **2014,** *7*, 3061−3068.
13. Tai, Q. D.; Tang, K. C.; Yan, F. Recent Progress of Inorganic Perovskite Solar Cells. *Energy Environ. Sci.* **2019,** *12*, 2375−2405.
14. Wang, R.; Mujahid, M.; Duan, Y.; Wang, Z. K.; Xue, J. J.; Yang, Y. A Review of Perovskites Solar Cell Stability. *Adv. Funct. Mater.* **2019,** *29*, 1808843.
15. Sharma, A. K.; Kumar, D. Sn-Based Perovskites for Photovoltaic Applications. In: *Advances in Functional and Smart Materials: Select Proceedings of ICFMMP 2021*; Springer Nature Singapore: Singapore, 2022; pp 303–310.
16. Cheng, Z.; Lin, J. Layered Organic–Inorganic Hybrid Perovskites: Structure, Optical Properties, Film Preparation, Patterning and Templating Engineering. *Cryst. Eng. Comm.* **2010,** *12*, 2646–2662.
17. Szuromi, P.; Grocholski, B. Natural and Engineered Perovskites. *Science* **2017,** *358* (6364), 732–733.
18. Kathryn, B. *Crystal Structure Prediction for Complex Modular Materials*; Doctor in Philosophy, University of Liverpool, 2017; p 135.
19. Jeon, N. J.; Noh, J. H.; Yang, W. S. Compositional Engineering of Perovskite Materials for High-Performance Solar Cells. *Nature* **2015,** *517* (7535), 476–480.

20. Atta, N. F.; Ahmed, G.; El-Ads Ekram, H. Perovskite Nanomaterials—Synthesis, Characterization, and Applications. *Tech. Chapter.* **2016,** *4*, 108–151.
21. Schaak, R. E.; Mallouk, T. E. Prying Apart Ruddlesden-Popper Phases: Exfoliation into Sheets and Nanotubes for Assembly of Perovskite Thin Films. *Chem. Mater.* **2000,** *12*, 3427–3434.
22. Tsunoda, Y.; Sugimoto, W.; Sugahara, Y. Intercalation Behavior of n-Alkylamines into a Protonated Form of a Layered Perovskite Derived from Aurivillius Phase $Bi_2SrTa_2O_9$. *Chem. Mater.* **2003,** *15*, 632–635.
23. Kreisel, J.; Glazer, A.; Jones, G.; Thomas, P.; Abello, L.; Lucazeau, G. An X-ray Diffraction and Raman Spectroscopy Investigation of A-Site Substituted Perovskites Compounds: The $(Na_{1-x}K_x)_{0.5}Bi_{0.5}TiO_3$ (0_x1) solid solution. *J. Phys.: Condens. Matter.* **2000,** *12*, 3267.
24. Fu, P.; Qingsong, S.; Yuequn, S.; Jizhong, S.; Haibo, Z.; Zhijun, N.; Jinkang, G. Perovskite Nanocrystals: Synthesis, Properties and Applications. *Sci. Bull.* **2017,** *62*, 369–380.
25. Ono, L. K.; Juarez-Perez, E. J.; Qi, Y. Progress on Perovskite Materials and Solar Cells with Mixed Cations and Halide Anions. *ACS Appl. Mater. Interfaces* **2017,** *9* (36), 30197–30246.
26. Kojima, A.; Teshima, K.; Shirai, Y.; Miyasaka, T. Organometal Halide Perovskites as Visible-Light Sensitizers for Photovoltaic Cells. *J. Am. Chem. Soc.* **2009,** *131* (17), 6050–6051.
27. Saliba, M.; Matsui, T.; Domanski, K.; Seo, J. Y.; Ummadisingu, A.; Zakeeruddin, S. M. et al. Incorporation of Rubidium Cations into Perovskite Solar Cells Improves Photovoltaic Performance. *Science* **2016,** *354* (6309), 206–209.
28. Saliba, M.; Matsui, T.; Seo, J. Y.; Domanski, K.; Correa-Baena, J. P.; Nazeeruddin, M. K. et al. Cesium-Containing Triple Cation Perovskite Solar Cells: Improved Stability, Reproducibility and High Efficiency. *Energy Environ. Sci.* **2016,** *9* (6), 1989–1997.
29. Jena, A. K.; Kulkarni, A.; Miyasaka, T. Halide Perovskite Photovoltaics: Background, Status, and Future Prospects. *Chem. Rev.* **2019,** *119* (5), 3036–3103.
30. Kojima, A.; Ikegami, M.; Teshima, K.; Miyasaka, T. Highly Luminescent Lead Bromide Perovskite Nanoparticles Synthesized with Porous Alumina Media. *Chem. Lett.* **2012,** *41* (4), 397–399.
31. Giorgi, G.; Yamashita, K. Organic–Inorganic Halide Perovskites: An Ambipolar Class of Materials with Enhanced Photovoltaic Performances. *J. Mater. Chem. A* **2015,** *3* (17), 8981–8991.
32. Chang, X.; Yang, G.; Zhong, J. X.; Tan, Y.; Wu, W. Q. New Pathways Toward Sustainable Sn-Related Perovskite Solar Cells. *Adv. Energy Sustain. Res.* **2023,** 2200175.
33. Stoumpos, C. C.; Malliakas, C. D.; Kanatzidis, M. G. Semiconducting Tin and Lead Iodide Perovskites with Organic Cations: Phase Transitions, High Mobilities, and Near-Infrared Photoluminescent Properties. *Inorg. Chem.* **2013,** *52* (15), 9019–9038.
34. Shockley, W.; Queisser, H. J. Detailed Balance Limit of Efficiency of p–n Junction Solar Cells. *J. Appl. Phys.*, *32* (3), 510–519.
35. Kagan, C. R.; Mitzi, D. B.; Dimitrakopoulos, C. D. Organic-Inorganic Hybrid Materials as Semiconducting Channels in Thin-Film Field-Effect Transistors. *Science* **1999,** *286* (5441), 945–947.
36. Mitzi, D. B.; Dimitrakopoulos, C. D.; Kosbar, L. L. Structurally Tailored Organic–Inorganic Perovskites: Optical Properties and Solution-Processed Channel Materials for Thin-Film Transistors. *Chem. Mater.* **2001,** *13* (10), 3728–3740.

37. Hao, F.; Stoumpos, C. C.; Cao, D. H.; Chang, R. P.; Kanatzidis, M. G. Lead-Free Solid-State Organic–Inorganic Halide Perovskite Solar Cells. *Nat. Photon.* **2014,** *8* (6), 489–494.
38. Noel, N. K.; Stranks, S. D.; Abate, A.; Wehrenfennig, C.; Guarnera, S.; Haghighirad, A. A. et al. Lead-Free Organic–Inorganic Tin Halide Perovskites for Photovoltaic Applications. *Energy Environ. Sci.* **2014,** *7* (9), 3061–3068.
39. Shao, S.; Liu, J.; Portale, G.; Fang, H. H.; Blake, G. R.; ten Brink, G. H. et al. Highly Reproducible Sn-Based Hybrid Perovskite Solar Cells with 9% Efficiency. *Adv. Energy Mater.* **2018,** *8* (4), 1702019.
40. Eperon, G. E.; Stranks, S. D.; Menelaou, C.; Johnston, M. B.; Herz, L. M.; Snaith, H. J. Formamidinium Lead Trihalide: A Broadly Tunable Perovskite for Efficient Planar Heterojunction Solar Cells. *Energy Environ. Sci.* **2014,** *7* (3), 982–988.
41. Jeon, N. J.; Noh, J. H.; Yang, W. S.; Kim, Y. C.; Ryu, S.; Seo, J.; Seok, S. I. Compositional Engineering of Perovskite Materials for High-Performance Solar Cells. *Nature* **2015,** *517* (7535), 476–480.
42. Bhalla, A. S.; Guo, R.; Roy, R. The Perovskite Structure—A Review of Its Role in Ceramic Science and Technology. *Mater. Res. Innov.* **2000,** *4* (1), 3–26.
43. Sun, C.; Alonso, J. A.; Bian, J. Recent Advances in Perovskite-Type Oxides for Energy Conversion and Storage Applications. *Adv. Energy Mater.* **2021,** *11* (2), 2000459.
44. Liu, Y.; Yang, Z.; Liu, S. Recent Progress in Single-Crystalline Perovskite Research Including Crystal Preparation, Property Evaluation, and Applications. *Adv. Sci.* **2018,** *5* (1), 1700471.
45. Murali, B.; Kolli, H. K.; Yin, J.; Ketavath, R.; Bakr, O. M.; Mohammed, O. F. Single Crystals: The Next Big Wave of Perovskite Optoelectronics. *ACS Mater. Lett.* **2019,** *2* (2), 184–214.
46. Chen, Y.; He, M.; Peng, J.; Sun, Y.; Liang, Z. Structure and Growth Control of Organic–Inorganic Halide Perovskites for Optoelectronics: From Polycrystalline Films to Single Crystals. *Adv. Sci.* **2016,** *3* (4), 1500392.
47. Saidaminov, M. I.; Adinolfi, V.; Comin, R.; Abdelhady, A. L.; Peng, W.; Dursun, I. et al. Planar-Integrated Single-Crystalline Perovskite Photodetectors. *Nat. Commun.* **2015,** *6* (1), 8724.
48. Wehrenfennig, C.; Eperon, G. E.; Johnston, M. B.; Snaith, H. J.; Herz, L. M. High Charge Carrier Mobilities and Lifetimes in Organolead Trihalide Perovskites. *Adv. Mater.* **2014,** *26* (10), 1584–1589.
49. Xing, G.; Mathews, N.; Sun, S.; Lim, S. S.; Lam, Y. M.; Grätzel, M. et al. Long-Range Balanced Electron-and Hole-Transport Lengths in Organic-Inorganic CH3NH3PbI3. *Science* **2013,** *342* (6156), 344–347.
50. Dong, Q.; Fang, Y.; Shao, Y.; Mulligan, P.; Qiu, J.; Cao, L.; Huang, J. Electron-Hole Diffusion Lengths> 175 μm in Solution-Grown CH3NH3PbI3 Single Crystals. *Science* **2015,** *347* (6225), 967–970.
51. Liu, Y.; Zhang, Y.; Yang, Z.; Yang, D.; Ren, X.; Pang, L.; Liu, S. Thinness-and Shape-Controlled Growth for Ultrathin Single-Crystalline Perovskite Wafers for Mass Production of Superior Photoelectronic Devices. *Adv. Mater.* **2016,** *28* (41), 9204–9209.
52. Song, Y.; Bi, W.; Wang, A.; Liu, X.; Kang, Y.; Dong, Q. Efficient Lateral-Structure Perovskite Single Crystal Solar Cells with High Operational Stability. *Nat. Commun.* **2020,** *11* (1), 274.

53. Johnston, M. B.; Herz, L. M. Hybrid Perovskites for Photovoltaics: Charge-Carrier Recombination, Diffusion, and Radiative Efficiencies. *Acc. Chem. Res.* **2016,** *49* (1), 146–154.
54. Stranks, S. D.; Eperon, G. E.; Grancini, G.; Menelaou, C.; Alcocer, M. J.; Leijtens, T. et al. Electron-Hole Diffusion Lengths Exceeding 1 Micrometer in an Organometal Trihalide Perovskite Absorber. *Science* **2013,** *342* (6156), 341–344.
55. Williams, S. T.; Zuo, F.; Chueh, C. C.; Liao, C. Y.; Liang, P. W.; Jen, A. K. Y. Role of Chloride in the Morphological Evolution of Organo-Lead Halide Perovskite Thin Films. *ACS Nano* **2014,** *8* (10), 10640–10654.
56. Saidaminov, M. I.; Abdelhady, A. L.; Murali, B.; Alarousu, E.; Burlakov, V. M.; Peng, W. et al. High-Quality Bulk Hybrid Perovskite Single Crystals Within Minutes by Inverse Temperature Crystallization. *Nat. Commun.* **2015,** *6* (1), 7586.
57. Green, M. A.; Ho-Baillie, A.; Snaith, H. J. The Emergence of Perovskite Solar Cells. *Nat. Photonics* **2014,** *8* (7), 506–514.
58. Bi, Y.; Hutter, E. M.; Fang, Y.; Dong, Q.; Huang, J.; Savenije, T. J. Charge Carrier Lifetimes Exceeding 15 μs in Methylammonium Lead Iodide Single Crystals. J. Phys. *Chem. Lett.* **2016,** *7* (5), 923–928.
59. Shi, D.; Adinolfi, V.; Comin, R.; Yuan, M.; Alarousu, E.; Buin, A. et al. Low Trap-State Density and Long Carrier Diffusion in Organolead Trihalide Perovskite Single Crystals. *Science* **2015,** *347* (6221), 519–522.
60. Jiang, Y.; Green, M. A.; Sheng, R.; Ho-Baillie, A. Room Temperature Optical Properties of Organic–Inorganic Lead Halide Perovskites. *Solar Energy Mater. Solar Cells* **2015,** *137,* 253–257.
61. Solar Cell Construction Diagram and Working Principle (electricalengineering123.com).
62. Hossain, M. I. Prospects of CZTS Solar Cells from the Perspective of Material Properties, Fabrication Methods and Current Challenges. *Chalcogenide Lett.* **2012,** *9* (6).
63. Guo, L.; Zhu, Y.; Gunawan, O.; Gokmen, T.; Deline, V. R.; Ahmed, S. et al. Electrodeposited $Cu_2ZnSnSe_4$ Thin Film Solar Cell with 7% Power Conversion Efficiency. *Progress Photovoltaics: Res. App.* **2014,** *22* (1), 58–68.
64. Siebentritt, S. Why Are Kesterite Solar Cells Not 20% Efficient? *Thin Solid Films* **2013,** *535,* 1–4.
65. Bag, S.; Gunawan, O.; Gokmen, T.; Zhu, Y.; Todorov, T. K.; Mitzi, D. B. Low Band Gap Liquid-Processed CZTSe Solar Cell with 10.1% Efficiency. *Energy Environ. Sci.* **2012,** *5* (5), 7060–7065.
66. https://energyprofessionalsymposium.com/?p 5 15173i
67. https://www.iitk.ac.in/ispst/?page 5 solar_cells
68. https://www.growthconsulting.frost.com/web/images.nsf/0/5C8C038AEE4690D96525 744E001A9C9B/$File/TI.html
69. Ye, T.; Wang, X.; Li, X.; Yan, A. Q.; Ramakrishna, S.; Xu, J. Ultra-High Seebeck Coefficient and Low Thermal Conductivity of a Centimeter-Sized Perovskite Single Crystal Acquired by a Modified Fast Growth Method. *J. Mater. Chem. C* **2017,** *5* (5), 1255–1260.
70. Stoumpos, C. C.; Kanatzidis, M. G. Halide Perovskites: Poor Man's High-Performance Semiconductors. Adv. Mater. 2016, 28 (28), 5778–5793.
71. Cheng, X.; Jing, L.; Zhao, Y.; Du, S.; Ding, J.; Zhou, T. Crystal Orientation-Dependent Optoelectronic Properties of $MAPbCl_3$ Single Crystals. *J. Mater. Chem.C* **2018,** *6* (6), 1579–1586.

72. Ding, J.; Lian, Z.; Li, Y.; Wang, S.; Yan, Q. The Role of Surface Defects in Photoluminescence and Decay Dynamics of High-Quality Perovskite $MAPbI_3$ Single Crystals. *J. Phys. Chem. Lett.* **2018,** *9* (15), 4221–4226.
73. Chen, Y.; He, M.; Peng, J.; Sun, Y.; Liang, Z. Structure and Growth Control of Organic–Inorganic Halide Perovskites for Optoelectronics: From Polycrystalline Films to Single Crystals. **Adv. Sci.** *2016,* 3 (4), 1500392.
74. Arya, S.; Mahajan, P. N. k. Tailor, Satapathi, S.; RR Sumathi, Datt, R.; and, Gupta, V. A Comprehensive Review on Synthesis and Applications of Single Crystal Perovskite Halides. *Prog. Solid State Chem.* **2020,** *60,* 100286.
75. Ring, T. A. *Fundamentals of Ceramic Powder Processing and Synthesis;* Elsevier, 1996.
76. Kumar, D.; Yadav, R. S.; Singh, A. K.; Rai, S. B. Synthesis Techniques and Applications of Perovskite Materials. *Perovskite Materials, Devices and Integration,* 2020.
77. Kwak, B. S.; Zhang, K.; Boyd, E. P.; Erbil, A.; Wilkens, B. J. Metalorganic Chemical Vapor Deposition of $BaTiO_3$ Thin Films. *J. Appl. Phys.* **1991,** *69* (2), 767–772.
78. Yu, Z.; Ramdani, J.; Curless, J. A.; Finder, J. M.; Overgaard, C. D.; Droopad, R. et al. Epitaxial Perovskite Thin Films Grown on Silicon by Molecular Beam Epitaxy. J. *Vacuum Sci. Technol. B* **2000,** *18* (3), 1653–1657.
79. Matsumuro, Y.; Oishi, Y.; Wu, W. B.; Okuyama, M. Preparation of PMN-PT Films by Laser Ablation Method. *Integr. Ferroelectrics* **1998,** *20* (1–4), 95–105.
80. Yang, B.; Wang, J. Y.; Jia, Y. M.; Huang, Y. Preparation of PbTiO3 Thin Film by DC Single-Target Magnetron Sputtering. In: *International Conference on Thin Film Physics and Applications,* Vol. 1519; SPIE, 1991; pp 725–728.
81. Lu, C. J.; Shen, H. M. Preparation and Crystallization of Pb (Zr0.95Ti0.05) O3 Thin Films Deposited by Radio-Frequency Magnetron Sputtering with a Stoichiometric Ceramic Target. *Appl. Phys. A* **1998,** *67,* 253–258.
82. Li, Y.; Xu, X.; Wang, C.; Wang, C.; Xie, F.; Yang, J.; Gao, Y. Investigation on Thermal Evaporated CH3NH3PbI3 Thin Films. *AIP Adv.* **2015,** *5* (9), 097111.
83. Pae, S. R.; Byun, S.; Kim, J.; Kim, M.; Gereige, I.; Shin, B. Improving Uniformity and Reproducibility of Hybrid Perovskite Solar Cells via a Low-Temperature Vacuum Deposition Process for NiOx Hole Transport Layers. *ACS Appl. Mater. Interfaces* **2018,** *10* (1), 534–540.
84. Perovskite-info. An introduction to Perovskites. https://www.perovskite-info.com/introduction

PART II

Structure, Synthesis, and Processing of the Materials

CHAPTER 5

Structural Analysis of Spinel Material

FARZANA FAROOQ, PRABHPREET KAUR, and SUMAN RANI

Department of Physics, School of Chemical Engineering & Physical Sciences, Lovely Professional University, Punjab, India

ABSTRACT

The future of emerging technology lies with spinel. As a possible material for various optoelectronic applications in the visible and ultraviolet (UV) ranges, they have drawn a lot of interest. Their distinct structure (normal and inverse) and bandgap area provide for their adaptability. Almost all of the transition metals and main group metals are included in the large family of metal elements known as spinels (TM). The spinel structures generally consist of AB_2O_4 in their elemental composition (A and B are metal ions). When the metal cations are different, spinel compounds form AB_2O_4, while when the metal cations are the same, spinel compounds form AB_2O_4. The spinel structure (occasionally called garnet structure) and general composition is AB_2O_4. The unit cell length of MAS, a face-centered cubic material with the Fd_3m space group, is 0.8084 nm. The anions of the typical spinel structure are arranged in a dense array. Half of the octahedral holes are occupied by the B-site cations, while one-eighth of the tetrahedral holes are occupied by the A-site cations. Each unit cell has a composition of $A_8B_{16}O_{32}$ and eight formula units. The A-site ions and half of the B-site ions exchange positions in inverse spinels, which have a structurally similar design and share the same big unit cell. Thus, inverse spinels are represented by the formula $B(AB)O_4$, where the parenthesized AB ions fill up octahedral positions while the other B ions fill up tetrahedral sites. Among these spinels, magnesium aluminate ($MgAl_2O_4$) spinel (MAS) and zinc aluminate ($ZnAl_2O_4$) have received great attention.

Advanced Materials for Next-Generation Technologies: Challenges and New Prospects. Shrikaant Kulkarni, Vipul Srivastava, Rabah Khenata, & Yarub Al-Douri (Eds.)

5.1 INTRODUCTION

Spinels form a big family of metal elements that contains almost all the transition metals (TMs) and main group metals. The term "spinels" refers to a group of minerals with the basic formula AB_2X_4 that crystallize in the cubic crystal system, with the cations A and B filling up some or all of the octahedral and tetrahedral sites in the lattice and the X anions sorted in a cubic close-filled lattice. To form a spinel, only those mixed oxides will be used in which there are two metal ions, one with a +2 oxidation state (OS) and the other with a +3 OS. There is no condition on metals; both metal cations (A and B) can be the same or different.[1] If the metal cations are different, then the spinel compound takes the form of AB_2O_4, and if the metal cations are the same, then the spinel compounds have a formation like A_3O_4.[2,3] Generally, tetrahedral sites are smaller than octahedral sites. The tetrahedral site is preferred by cations with smaller ionic radii, whereas the octahedral site is preferred by those with greater ionic radii. The spinel structure's (occasionally called garnet structure) general composition is AB_2O_4. Table 5.1 shows details of atoms that can be placed at the A site, B site, and X site, respectively. Multiple factors induce the allotment of cations in spinels such as (i) the crystal field effects of the octahedral site preference energy (OSPE) of cations, (ii) cation radius, and (iii) the crystal field stabilization energy (CFSE). According to crystal field theory (CFT), the OSPE is explained as the stable crystal field energy difference between the tetrahedral and octahedral fields. The larger the value of OSPE, the greater the chances for the cations to preferably fill up the octahedral interstices.[4] When various chalcogenides make up the anion sublattice in addition to oxygen, the structure is known as a thiospinel.

The most prevalent mineral in the spinel group, magnetite, Fe_3O_4 (as $Fe^{2+}Fe^{3+}{}_2O^{2}{}_4$), is an example of a material with both A and B as the same metal but with distinct valences. Spinel ($MgAl_2O_4$), formerly known as spinel ruby, is the source of the name for the group.[5] Among spinel metals, zinc aluminate ($ZnAl_2O_4$) is a very familiar spinel semiconductor material with an energy gap of 3.5 to 3.9 eV.[6] The unfamiliar mineral $ZnAl_2O_4$, which has a normal spinel structure, exhibits excellent electrical, magnetic, and optical behavior. Zinc aluminate is used as a catalyst, optical material, and high-temperature material.[6] Some of the reported spinel materials, along with their crystallite size and particle size, are shown in Table 5.2.

In recent years, a remarkable increase in the number of publications on spinels has been reported, with a record of 18,700 papers between 2001 and 2010 (as per Scopus). To demonstrate interest in this field, the total number of published papers on spinels has been displayed in Figure 5.1.

TABLE 5.1 Valence of Atoms Occupying A, B, and X Sites.[4]

A site		B site		X site	
Atom	**Oxidation state**	**Atom**	**Oxidation state**	**Atom**	**Oxidation state**
Mg	+2	Zn	+3	F	−1
Ca	+2	Al	+3	O	−2
Ba	+2	Cr	+3	S	−2
Mn	+2	Mn	+3	Se	−2
Fe	+2	Fe	+3	Te	−2
Co	+2	Co	+3	N	−3
Ni	+2	Ni	+3		
Cu	+2				
Zn	+2				
Cd	+2				

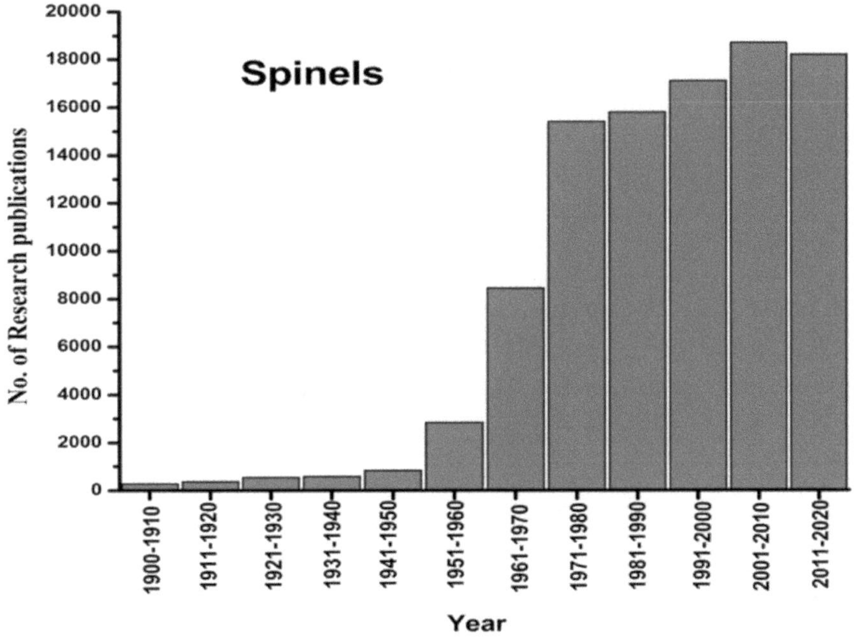

FIGURE 5.1 Number of research publications on spinels.

5.2 DIFFERENT SPINEL TYPES

There are three types of spinels obtained from close-packed lattices. They are especially important because of the material properties they possess in their compounds. The three types are normal spinels, inverse spinels, and mixed spinels.

Normal Spinels: Normal spinels are represented as $(A^{+2})^{tet}\ (B^{+3})_2^{Oct}O_4$, as shown in Figure 5.2. In normal spinels, the tetrahedral voids are filled up by divalent A^{+3} ions, whereas the octahedral voids are filled up by trivalent B^{+3} ions in the close-packed arrangement of oxide ions. Examples of normal spinels are $FeCr_2O_4$, $MgAl_2O_4$, $ZnFe_2O_4$, and Mn_3O_4.

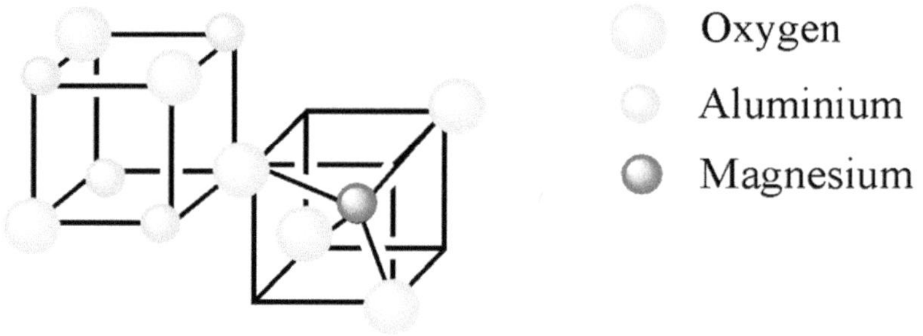

FIGURE 5.2 Normal spinel structure.

Source: Adapted from Ref. [7]

Inverse Spinels: Inverse spinels are indicated as $(B^{+3})^{tet}\ (A^{+2}\ B^{+3})^{oct}\ O_4$, as shown in Figure 5.3. In inverse spinel, the octahedral voids are filled up by A^{+2} ions, whereas the tetrahedral voids are filled up by half of B^{+3} ions. Examples of inverse spinels are $NiFe_2O_4$, $CoFe_2O_4$, and Fe_3O_4 (ferrite).

Mixed Spinels: Mixed spinels are interposed between the normal and the inverse spinel structures. Eg: $MnFe_2O_4 = MgFe_2O_4$

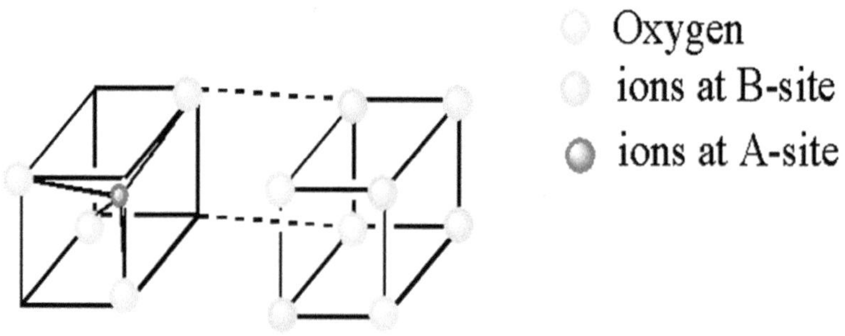

FIGURE 5.3 Inverse spinel structure.

Source: Adapted from Ref. [8]

TABLE 5.2 Crystallite Size and Particle Size of Reported Spinels.

S no.	Spinels	Crystallite Size (nm)	Particle Size (nm)	References
1	$ZnAl_2O_4$	23 nm	>1 nm	[9]
2	$ZnFe_2O_4$	43–49 nm	42–67 nm	[10]
3	(Co, Mg, Mn, Ni)Fe_2O_4	12–21 nm	28.70 nm	[11]
4	$MgAl_2O_4$	14 nm	20 nm	[12]
5	$NiGa_2O_4$	8 nm	5 nm	[13]
6	$ZnCo_2O_4$	0.2333 nm	15 nm	[14]
7	$CoAl_2O_4$	20–30 nm	<30 and 75 nm	[15]
8	$ZnGa_2O_4$	5 nm	>5 nm	[16]
9	$MnCo_2O_4$	<50 nm	<0.2 nm	[17]
10	$NiCo_2O_4$	~10.8 nm	13 nm	[18]
11	$MgCr_2O_4$	10–40 nm	<20 nm	[19]
12	$CoFe_2O_4$	8 nm	4–10 nm	[20]
13	$MgGa_2O_4$	6.7–11.2 nm	>5 nm	[21]

5.3 DIFFERENT SYNTHESIS TECHNIQUES

Spinels can be prepared using a variety of processes and procedures. The preferred synthesis technique will depend on the application because it will have an impact on the material's crystallinity and shape. Due to several significant advancements in numerous fields, nanoparticle preparation is in high demand. Sol-gel preparation is preferred over other methods such as autocombustion, coprecipitation, Pechini, and others for the formation of nanoparticles. The synthesis details of the sol-gel method are shown in Figure 5.4. The sol-gel method is more effective than other kinds of material synthesis because it is able to synthesize a broad variety of materials, including composites, ceramics, and glasses, with high purity and good compositional and structural control. It is also a method that is attractive for industrial-scale production because it is straightforward and inexpensive.[22]

The prepared material followed several characterizations for the structural investigation, as shown in Figure 5.5. Using a straightforward sol-gel process in conjunction with a hydrothermal synthesis technique, zinc oxide-based combined nanomaterials were flourishingly created. Results from the X-ray powder diffraction and the Fourier transform infrared spectroscopy showed that the samples of the zinc oxide-based combined nanomaterials solely include the matching metal oxides and do not contain any other contaminants. The $ZnO/ZnAl_2O_4$ complex materials are made up of rhombic fine particles and tiny nanoparticles according to scanning electron microscopic pictures. $ZnO/ZnAl_2O_4$ complex materials have a broad light absorption volume than $ZnAl_2O_4$ and ZnO, which may indicate that they have high antibacterial activity. The $ZnO/ZnAl_2O_4$ complex materials show better separation efficiency and charge transfer in the visible spectrum, which is an indication of their strong antibacterial activity according to fluorescence spectrum analysis. The $ZnO/ZnAl_2O_4$ composite materials exhibit three visible emission peaks at 430, 420, and 372 nm, with an excitation wavelength of 250 nm. Staphylococcus aureus and Escherichia coli were chosen as examples of gram-negative and gram-positive bacteria, respectively, and the antibacterial qualities of complex materials made of $ZnO/ZnAl_2O_4$ were assessed. The ratio of the mass of ZnO to $ZnAl_2O_4$, production process, light absorption capacity, separation efficiency, specific surface area, and charge transfer work together synergistically to give the $ZnO/ZnAl_2O_4$ complex materials a greater photocatalytic deactivation rate than earlier recorded rate. Using theoretical and experimental research, a potential photocatalytic deactivation mechanism of $ZnAl_2O_4$ was recommended.

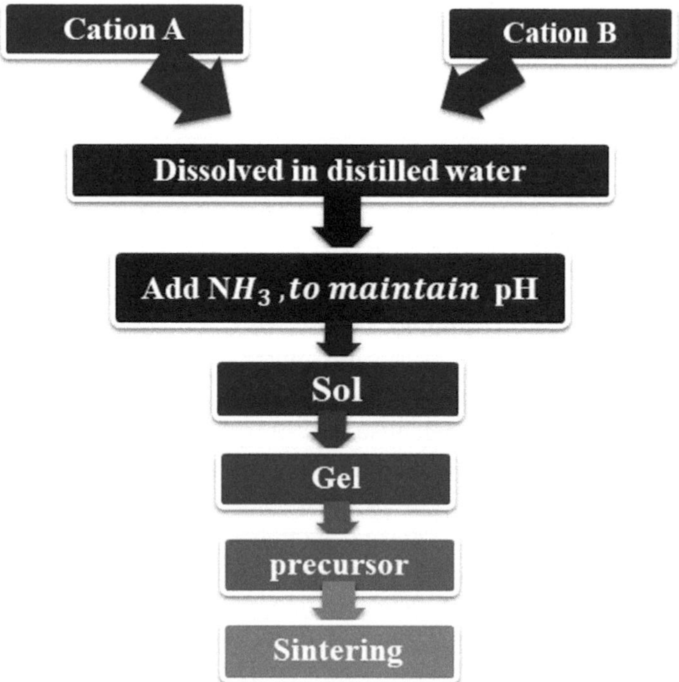

FIGURE 5.4 Flow chart of a sol-gel technique for the construction of nanoparticles.

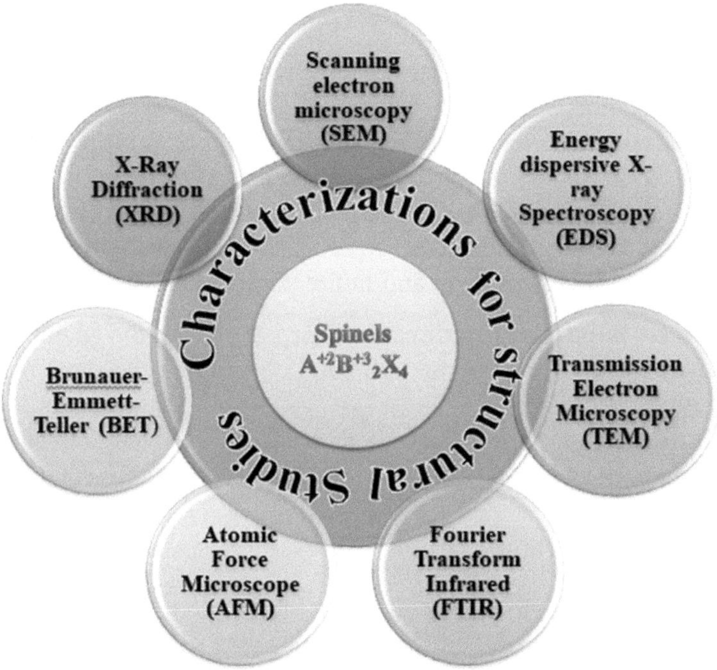

FIGURE 5.5 Systematic view for the structural characterization of spinel.

Three distinct processes—solvothermal, citrate precursor, and hydrothermal—were used to make $ZnAl_2O_4$ nanoparticles. By using energy-dispersive X-ray spectra (EDX), X-ray powder diffraction (XRD), Brunauer–Emmett–Teller (BET), scanning electronic microscopy (SEM), Fourier transform infrared spectroscopy (FT-IR), and UV-vis diffuse reflectance spectroscopy (DRS) techniques, their structural properties were systematically examined. To illustrate the photoinduced charge separation in the samples, surface photovoltage (SPV) measurements were used. To compare the photocatalytic performances of nanostructured TiO_2 samples and $ZnAl_2O_4$ samples, the degradation of gaseous toluene under UV lamp irradiation in an in situ FTIR reactor was used. The outcomes showed that the sample produced using a simple solvothermal approach had a toluene photocatalytic efficiency of nearly 90%. Toluene was clarified into mostly water and carbon dioxide. A promising method for air purification is the photocatalytic decomposition of gaseous pollutants over UV-illuminated zinc aluminate ($ZnAl_2O_4$).

5.4 BANDGAP

A bandgap, also known as an energy gap, can be explained as the minimum energy variation between the bottom of the conduction band and the top of the valence band in the insulators and semiconductors. It is the energy needed to be given to a valence electron, which is bound to an atom to become a conduction electron, which is then independent to move in the crystal lattice and behaves as a charge transporter to conduct the electric current. Two types of bandgaps are direct bandgap and indirect bandgap.

a) **Direct bandgap:** In a direct bandgap, the top of the valence band and the bottom of the conduction band coincide at the same momentum value.
b) **Indirect bandgap:** In the indirect bandgap, the maximum energy in the valence band is higher than the minimum energy in the conduction band.

5.5 DEFECT IN SPINEL STRUCTURE

A^{3+} ions fill up octahedral 16d sites, and B^{2+} ions fill up tetrahedral 8a sites in a perfectly typical AB_2O_4 spinel. In truth, certain cations are switched between the cation sublattices, resulting in pairs of antisite defects and a certain amount of "inversion." Here, the effect of antisite defects on populations of other intrinsic defects, including those connected to the Schottky and Frenkel processes, is being studied using atomic simulation. One effect is an increase in the total magnesium interstitial concentration relative to the aluminum interstitial concentration that is significantly greater and a lesser rise in the magnesium vacancy concentration relative to the aluminum vacancy concentration.[23]

In spinel crystals, point defects, like lattice flaws and impurity ions or their complexes, create optically active centers. It is feasible to gather important knowledge about the nature and electronic states of these flaws using various optical spectroscopies. A greater knowledge of the nature of color centers and the connections between many of them and defects in zinc aluminate spinels has come about over the course of the previous several decades as a result of research on color centers formed by various treatments (irradiation, annealing, etc.). However, due to the complexity of the spinel structure, the wide range of potential compositions, and the wide range of impurity atoms, it can be difficult to identify faults. Furthermore, growth and processing circumstances affect both ion and vacancy distributions.[24]

5.6 APPLICATION AREAS OF SPINELS

5.6.1 PHOTOCATALYTIC

Due to their high mechanical resistance, great chemical and thermal stability, high quantum yields with hydrophobic properties, low surface activity, and wide surface area, metal aluminate spinels with the formula AB_2O_4 have attracted substantial attention for numerous applications. Such Fd_3m space group-belonging metal aluminate spinels could also facilitate photocatalytic degradation processes under visible light and UV with bandgaps in the region of 2–3 eV. Such metal aluminates have been produced using a variety of methods in the literature, including the complexation method, coprecipitation, combustion method, solid-state reaction, hot pressing, etc. Each of these methods had unique equipment requirements, high equipment costs, and additional limitations such as lack of homogeneity and low product surface area. The method used in this work has the built-in benefits of being simple to set up, affordable, and a highly chemically homogeneous nanopowder.[25] The significance of spinel ferrite nanocomposites (SFNCs) and spinel ferrite nanoparticles (SFNPs) in the treatment of water and wastewater has been demonstrated in several research studies. Particularly, spinel ferrite nanocomposites (SFNCs) used for wastewater and water treatments are easily regenerable, very stable, and without losing any of their qualities are being reused for many cycles, resulting in lower treatment costs. This review critically examines the most recent developments and possible applications of spinel ferrite nanoparticles and composites for the adsorption-based removal of inorganic and organic pollutants. Additionally, a full discussion of the use of spinel ferrite nanoparticles (SFNPs) for the removal of several contaminants existing in water for quantitative evaluation at mark levels is provided. Finally, this review covered potential methods of reuse and recovery, knowledge gaps, toxicity, and the outlook for SFNPs in the future. Depending on this analysis, it is feasible to conclude that the composites and SFNPs that are derived from them have a limitless ability to address a variety of issues in wastewater and water treatment. However, further research and analysis must be done on the largescale practical implementation of SFNPs.[26]

In the past ten years, the management of hazardous waste has become a sensitive issue and a major priority on a global scale. Most industrialized nations are dealing with major environmental problems related to groundwater contamination, toxic wastes, soil diseases, and dangerous air pollutants in their defense, civic, and commercial sectors. Crushing these toxic blots from the environment is a basic requirement of humans to lessen their detrimental effects on human life. Numerous solutions to this problem have been proposed, but each one has flaws. The development of semiconductor photocatalysis is of great interest to researchers to eliminate and transform these dangerous elements into toxin-free nature. Heterogeneous photodegradation has numerous uses in environmental problems such as disinfectant remediation, soil cleaning, and water and air purification. Large amounts of UV light are absorbed by the solid particles during the photocatalytic process. The result is the production of oxidants. Titania was the first material ever utilized to study photocatalytic action since it is affordable and sturdy. The rejoining of holes and electrons produced by the incident photoreaction lowers a photocatalyst's efficiency. This element reduces and prevents the technique's usage in the treatment of wastewater and the elimination of toxins from the air and water. A photocatalyst's performance can be adjusted by enhancing its stability, bandgap, and capacity to regulate electron/hole rejoining. To completely utilize solar energy, a photocatalyst needs to be exceptionally efficient at absorbing light. To create novel materials with the desired properties, many schemes are used. These approaches entail creating the material, doping the relevant metals, and creating material composites.[27]

The technology used for the treatment of water waste is known as photocatalysis. We can also say that an effective process for completely degrading organic molecules is known as photocatalysis. It is crucial for the breakdown of harmful chemical molecules by converting air and water into secure components. A semiconductor is commonly used in a photocatalytic process because it is available at a reasonable cost, chemically stable, and nontoxic. In semiconductor materials, the photocatalytic process is operated at room temperature and is used for the removal of harmful organic pollutants like phenol and the purification of contaminated water.

The rate at which a photochemical reaction is increased by the substance without being consumed itself is known as photocatalysis rate. ZnO as a catalyst was used to study the photocatalytic degradation process of 2-chlorophenol (2-cp) in water. The different forms of ZnO were characterized by UV-visible spectra and photoluminescence. Apart from this, various applications have been explored by spinel materials using the sol-gel method, which is tabulated in Table 5.3.

TABLE 5.3 Target Application Areas of Spinels Synthesized Using the Sol-Gel Technique.

S No.	Material	Sintering temperature (°C)	Crystallite size (nm)	Target application area	Publication year	References
1	$CuFe_2O_4$	Room temperature	17.75	Microwave circuits	2023	[28]
2	$MgFe_2O_4$	Room temperature	85	–	2023	[29]
3	$Ni_xCo_yO_4$	450	250–300µm	Oxidation of CO and toluene	2023	[30]
4	$MgAl_2O_4$	1100	49	Optoelectronics	2023	[22]
5	$CoAl_2O_4$	1200	88	Ceramic ink	2023	[31]
6	$(Cu, Mn, Co)_3O_4$	800	26	Solid-oxide fuel cell	2023	[32]
7	$ZnCr_2O_4$	500	45–80	Nanomedicines	2023	[33]
8	$SrFe_2O_4$	800	30	Photocatalytic applications	2022	[34]
9	$MgAl_2O_4$	1600	170	Nanofibers	2022	[35]
10	$CuCr_2O_4$	700	46	Photodegradation	2022	[36]
11	$(Cu, Zn, Mn)Fe_2O_4$	700	20–30	Antibacterial	2022	[37]
12	$ZnAl_2O_4$	450		Optoelectronics	2022	[38]
13	$CuFe_2O_4$	600	37	Photocatalytic	2022	[39]
14	Mg_2MnO_4	700	30–50	Li-ion batteries	2021	[40]
15	$ZnAl_2O_4$	800	35	Photoelectric devices	2021	[41]
16	$ZnFe_2O_4$	950	30	Ni-MH batteries	2021	[42]
17	$MnCo_2O_4$	750	70	Electronic applications	2021	[43]
18	$LiMn_2O_4$	750	17–36	Li-ion batteries	2021	[44]
19	$NiMn_2O_4$	500	10	Capacitors	2021	[45]
20	$ZnAl_2O_4$	700	25	Photocatalytic	2021	[46]

KEYWORDS

- **spinel**
- **synthesis**
- **defects**
- **oxides**
- **bandgap**
- **structure**

REFERENCES

1. Sickafus, K. E.; Wills, J. M.; Grimes, N. W. Structure of Spinel. *J. Am. Ceram. Soc.* **1999,** *82* (12), 3279–3292. DOI: 10.1111/j.1151-2916.1999.tb02241.x.
2. Goodenough, J. B.; Loeb, A. L. Theory of Ionic Ordering, Crystal Distortion, and Magnetic Exchange Due to Covalent Forces in Spinels. *Phys. Rev.* **1955,** *98* (2), 391–408. DOI: 10.1103/PhysRev.98.391.
3. Kaur, P.; Rani, S. Defect Induced Emission of $MgAl_2O_4$ Sintered at High Temperature. *J. Phys. Conf. Ser.* **2022,** *2267* (1). DOI: 10.1088/1742-6596/2267/1/012100.
4. Zhao, Q.; Yan, Z.; Chen, C.; Chen, J. Spinels: Controlled Preparation, Oxygen Reduction/Evolution Reaction Application, and Beyond. *Chem. Rev.* **2017,** *117* (15), 10121–10211. DOI: 10.1021/acs.chemrev.7b00051.
5. Ayres, F.; De Souza, S. S.; Blak, A. R. Dipole Defects in $MgAl_2O_4$ Spinel Crystals. *Radiat. Eff. Defects Solids* **2001,** *156* (1), 305–309. DOI: 10.1080/10420150108216910.
6. Sofia Nirmala, T.; Iyandurai, N.; Yuvaraj, S.; Sundararajan, M. Effect of Cu_2+ Ions on Structural, Morphological, Optical and Magnetic Behaviors of $ZnAl_2$O4 Spinel. *Mater. Res. Express* **2020,** *7* (4). DOI: 10.1088/2053-1591/ab7a7a.
7. Tsurkan, V.; H. A. Krug von Nidda, Deisenhofer, J.; Lunkenheimer, P.; Loidl, A. On the Complexity of Spinels: Magnetic, Electronic, and Polar Ground States. *Phys. Rep.* **2021,** *926*, 1–86. DOI: 10.1016/j.physrep.2021.04.002.
8. Jain, R. A Review on the Development of XRD in Ferrite Nanoparticles. *J. Supercond. Nov. Magn.* **2022,** *35* (5), 1033–1047. DOI: 10.1007/s10948-022-06213-9.
9. Ianoş, R.; Borcănescu, S.; Lazău, R. Large Surface Area $ZnAl_2O_4$ Powders Prepared by a Modified Combustion Technique. *Chem. Eng. J.* **2014,** *240* (3), 260–263. DOI: 10.1016/j.cej.2013.11.082.
10. Manikandan, A.; Kennedy, L. J.; Bououdina, M.; Vijaya, J. J. Synthesis, Optical and Magnetic Properties of Pure and Co-Doped $ZnFe_2O_4$ Nanoparticles by Microwave Combustion Method. *J. Magn. Magn. Mater.* **2014,** *349*, 249–258. DOI: 10.1016/j.jmmm.2013.09.013.
11. Gherca, D.; Pui, A.; Cornei, N.; Cojocariu, A.; Nica, V.; Caltun, O. Synthesis, Characterization and Magnetic Properties of MFe_2O_4 (M=Co, Mg, Mn, Ni) Nanoparticles Using Ricin Oil as Capping Agent. *J. Magn. Magn. Mater.* **2012,** *324* (22), 3906–3911. DOI: 10.1016/j.jmmm.2012.06.027.

12. Pati, R. K.; Pramanik, P. Low Temp Chemical Synthesis of Nano $Mgal_2O_4$.Pdf. **2000,** *24* (189117), 1999–2001.
13. Yan, S. C. et al. A Room-Temperature Reactive-Template Route to Mesoporous $ZnGa_2O_4$ with Improved Photocatalytic Activity in Reduction of CO_2. *Angew. Chemie* **2010,** *122* (36), 6544–6548. DOI: 10.1002/ange.201003270.
14. Wei, X.; Chen, D.; Tang, W. Preparation and Characterization of the Spinel Oxide $ZnCo_2O_4$ Obtained by Sol-Gel Method. *Mater. Chem. Phys.* **2007,** *103* (1), 54–58. DOI: 10.1016/j.matchemphys.2007.01.006.
15. Salem, S.; Jazayeri, S. H.; Bondioli, F.; Allahverdi, A.; Shirvani, M.; Ferrari, A. M. $CoAl_2O_4$ Nano Pigment Obtained by Combustion Synthesis. *Int. J. Appl. Ceram. Technol.* **2012,** 9 (5), 968–978. DOI: 10.1111/j.1744-7402.2011.02704.x.
16. Fernández-Osorio, A.; Tapia, M.; Vázquez-Olmos, A. R.; Chávez, J. Enhanced Luminescence Properties of $ZnGa2O_4$:Cr_3+ Nanoparticles with an Average Crystallite Size of 5 nm. *J. Solid State Chem.* **2019,** *269,* 328–335. DOI: 10.1016/j.jssc.2018.10.003.
17. Borges, F. M. M. et al. Magnetic Behavior of Nanocrystalline $MnCo_2O_4$ Spinels. *J. Magn. Magn. Mater.* **2006,** *302* (2), 273–277. DOI: 10.1016/j.jmmm.2005.09.017.
18. Rahulan, K. M.; Padmanathan, N.; Philip, R.; Balamurugan, S.; Kanakam, C. C. Structural Evaluation and Nonlinear Optical Properties of Ni/NiO, Ni/$NiCo_2O_4$ and Co/Co_3O_4 Nanocomposites. *Appl. Surf. Sci.* **2013,** *282,* 656–661. DOI: 10.1016/j.apsusc.2013.06.028.
19. Mykhailovych, V. et al. Structural, Optical, and Catalytic Properties of $MgCr_2O_4$ Spinel-Type Nanostructures Synthesized by Sol–Gel Auto-Combustion Method. *Catalysts* **2021,** *11* (12), 1–10. DOI: 10.3390/catal11121476.
20. Kumar, V.; Rana, A.; Yadav, M. S.; Pant, R. P. Size-Induced Effect on Nano-Crystalline $CoFe_2O_4$. *J. Magn. Magn. Mater.* **2008,** 320 (11), 1729–1734. DOI: 10.1016/j.jmmm.2008.01.021.
21. Sharma, G.; Castro, R. H. R. Synthesis and Surface Enthalpy of $MgGa_2O_4$ Spinel. *Thermochim. Acta* **2017,** *655,* 326–330. DOI: 10.1016/j.tca.2017.07.018.
22. Kaur, P.; Rani, S. Effect of Sintering Temperature on Structural and Optical Properties of Magnesium Aluminate Spinel. *J. Opt.* **2023,** 0123456789. DOI: 10.1007/s12596-023-01167-0.
23. Ball, J. A. et al. Defect Processes in $MgAl_2O_4$ Spinel. *Solid State Sci.* **2008,** *10* (6), 717–724. DOI: 10.1016/j.solidstatesciences.2007.04.005.
24. Gritsyna, V. T.; Afanasyev-Charkin, I. V.; Kobyakov, V. A.; Sickafus, K. E. Structure and Electronic States of Defects in Spinel of Different Compositions $MgOnAl_2O_3$:Me. *J. Am. Ceram. Soc.* **1999,** *82* (12), 3365–3373. DOI: 10.1111/j.1151-2916.1999.tb02252.x.
25. Goswami, B.; Rani, N.; Vats, R.; Bhukkal, C.; Ahlawat, R. Highly Crystalline and Narrow Bandgap $MgAl_2O_4$: Synthesis and Characterization. *AIP Conf. Proc.* **2021,** *2352* (3), 1–6. DOI: 10.1063/5.0052473.
26. Kefeni, K. K.; Mamba, B. B.; Msagati, T. A. M. Application of Spinel Ferrite Nanoparticles in Water and Wastewater Treatment: A Review. *Sep. Purif. Technol.* **2017,** *188,* 399–422. DOI: 10.1016/j.seppur.2017.07.015.
27. Shaheen, N.; Yousuf, M. A.; Shakir, I.; Zulfiqar, S.; Agboola, P. O.; Warsi, M. F. Wet Chemical Route Synthesis of Spinel Oxide Nano-Catalysts for Photocatalytic Applications. *Phys. B Condens. Matter* **2020,** *580,* 411820. DOI: 10.1016/j.physb.2019.411820.
28. Samikannu, K.; Ahmad, A. F.; Ismail, I.; Aspanut, Z.; Majid, W. H. A. Sol–Gel Combustion Synthesis and Study the High Frequency Dielectric Properties of Copper Ferrite Nanoparticles. *J. Clust. Sci.* **2023,** 0123456789. DOI: 10.1007/s10876-023-02408-9.

29. Chihi, I.; Baazaoui, M.; Hamdaoui, N.; Oumezzine, M.; Farah, K. Role of Ag Doping on Structural, Electrical, and Dielectric Properties of $MgFe_2O_4$ Synthesized via the Sol–Gel Route. *J. Supercond. Nov. Magn.* **2023,** 36 (2), 759–768. DOI: 10.1007/s10948-023-06516-5.
30. Le, M. T.; Nguyen, P. A.; Tran, T. T. H., Chu, T. H. N., Wang, Y.; Arandiyan, H. Catalytic Performance of Spinel-Type Ni-Co Oxides for Oxidation of Carbon Monoxide and Toluene. *Top. Catal.* **2023,** 66 (1–4), 117–125. DOI: 10.1007/s11244-022-01676-5.
31. Podder, M.; Ahmed, M. F.; Moni, M. R.; Rahman, M. L.; Biswas, B.; Sharmin, N. Effect of Metal Ions on Structural, Morphological and Optical Properties of Nano-Crystallite Spinel Cobalt-Aluminate ($CoAl_2O_4$). *Arab. J. Chem.* **2023,** *16* (5), 104700. DOI: 10.1016/j.arabjc.2023.104700.
32. Mah, J. C. W., Aznam, I.; Muchtar, A.; Somalu, M. R.;; Raharjo, J. Synthesis of $(Cu,Mn,Co)_3O_4$ Spinel: Effects of Citrate-to-Nitrate Ratio on Its Homogeneity and Electrical Properties. *Energies* **2023,** *16* (3). DOI: 10.3390/en16031382.
33. Ghotekar, S. et al. Spinel ZnCr2O4 Nanorods Synthesized by Facile Sol-Gel Auto Combustion Method with Biomedical Properties. *J. Sol-Gel Sci. Technol.* **2023,** *105* (1), 176–185. DOI: 10.1007/s10971-022-05964-0.
34. Attia, S.; Helaili, N.; Rekhila, G.; Bessekhouad, Y.; Trari, M. Physical and Photo-Electrochemical Properties of the Spinel $SrFe_2O_4$: Application to Hydrogen Production Under Visible Light. *J. Mater. Sci. Mater. Electron.* **2022,** *33* (13), 9976–9987. DOI: 10.1007/s10854-022-07989-1.
35. Xu, K. et al. Preparation of Magnesium Aluminate Spinel Nanofibers with High Temperature Resistance by Electrospinning Process Based on Non-Hydrolytic Sol-Gel Method. *Russ. J. Phys. Chem. B* **2022,** *16* (4), 787–796. DOI: 10.1134/S1990793122040054.
36. Benrighi, Y. et al. Characterization and Application of the Spinel $CuCr_2O_4$ Synthesized by Sol–Gel Method for Sunset Yellow Photodegradation. *J. Sol-Gel Sci. Technol.* **2022,** *101* (2), 390–400. DOI: 10.1007/s10971-021-05697-6.
37. Dabagh, S.; Haris, S. A.; Ertas, Y. N. Synthesis, Characterization and Potent Antibacterial Activity of Metal-Substituted Spinel Ferrite Nanoparticles. *J. Clust. Sci.* **2022,** 0123456789. DOI: 10.1007/s10876-022-02373-9.
38. Albukhari, S. M.; Ismail, A. A.; Alsheheri, S. Z.; Mahmoud, M. H. H. Synthesis and Design of Spinel $ZnAl_2O_4$ Nanowires with PtO Nanoparticles Decoration for Enhanced Induced Visible Light Hg(II) Reduction. *Int. J. Environ. Sci. Technol.* **2022,** *19* (8), 7067–7078. DOI: 10.1007/s13762-021-03825-x.
39. Soufi, A.; Hajjaoui, H.; Elmoubarki, R.; Abdennouri, M.; Qourzal, S.; Barka, N. Heterogeneous Fenton-Like Degradation of Tartrazine Using $CuFe_2O_4$ Nanoparticles Synthesized by Sol-Gel Combustion. *Appl. Surf. Sci. Adv.* **2022,** *9*, 100251. DOI: 10.1016/j.apsadv.2022.100251.
40. Wang, Z. et al. Synthesis, Electronic Structure, and Electrochemical Properties of the Cubic Mg_2MnO_4 Spinel with Porous-Spongy Structure. *Nanomaterials* **2021,** *11* (5), 1–13. DOI: 10.3390/nano11051122.
41. Mohanty, P.; Mohapatro, S.; Mahapatra, R.; Mishra, D. K. Low Cost Synthesis Route of Spinel $ZnAl_2O_4$. *Mater. Today Proc.* **2021,** *35* (3), 130–132. DOI: 10.1016/j.matpr.2020.03.508.
42. Zayani, W. et al. Electrochemical Behavior of a Spinel Zinc Ferrite Alloy Obtained by a Simple Sol-Gel Route for Ni-MH Battery Applications. *Int. J. Energy Res.* **2021,** *45* (4), 5235–5247. DOI: 10.1002/er.6140.

43. Zaouali, A. et al. High Electrical Conductivity at Room Temperature of $MnCo_2O_4$ Cobaltite Spinel Prepared by Sol–Gel Method. *J. Mater. Sci. Mater. Electron.* **2021,** *32* (1), 1221–1232. DOI: 10.1007/s10854-020-04895-2.
44. Chand, P.; Bansal, V.; Lal, S.; Joshi, A.; Sukriti; Kumar, A. Effect of Annealing Temperatures on Optical and Electrochemical Behavior of Spinel $LiMn_2O_4$ as Cathode Materials for Lithium-Ion Batteries Applications. *J. Mater. Sci. Mater. Electron.* **2021,** *32* (5), 6648–6659. DOI: 10.1007/s10854-021-05380-0.
45. Dhas, S. D. et al. Fabrication of Efficient Electrochemical Capacitors Rooted in Sol-Gel Derived $NiMn_2O_4$ Nanoparticles. *J. Electroanal. Chem.* **2021,** *897*, 115548. DOI: 10.1016/j.jelechem.2021.115548.
46. Saffar, A.; Ahangar, H. A.; Salehi, S.; Fekri, M. H.; Rabbani, A. Synthesis of Novel $ZnAl_2O_4/Al_2O_3$ Nanocomposite by Sol–Gel Method and Its Application as Adsorbent. *J. Sol-Gel Sci. Technol.* **2021,** *99* (1), 158–168. DOI: 10.1007/s10971-021-05559-1.

CHAPTER 6

Green Synthesis of Metal Oxides for Grow Light

JASMEEN KAUR and SUMAN RANI

Department of Physics, School of Chemical Engineering & Physical Sciences, Lovely Professional University, Punjab, India

ABSTRACT

Houseplants possess a remarkable ability to bring life and tranquility to any space. They are highly favored by decorators and are essential for maintaining good health. Indoor plants play a vital role in our daily lives. Fortunately, the best indoor plants for your home are easy to care for. Adequate light is crucial for their growth as they cannot survive without a sufficient amount of it. Therefore, it is essential to provide the appropriate amount of light to ensure the healthy development of the plant. Plants require light within the range of 400–700, known as the Photosynthetically Active Radiation (PAR) region. The spectrum of grow lights consists of electromagnetic wavelengths that simulate the light produced by the sun, facilitating plant growth. During photosynthesis, plants utilize light within the PAR region. Chlorophyll, present in plants, converts light energy into chemical energy, primarily absorbing blue and red wavelengths in the photosynthetic spectrum. Light serves as a crucial component for plants, enabling photosynthesis through specific chemicals such as chlorophyll *a* and *b* found in the leaves. These chemicals have peaks of absorbance in the blue and red sections, indicating the significance of these wavelengths for photosynthesis. Blue light is essential for the morphologically healthy development of plants, while red light supports photosynthesis. This study focuses on highlighting the importance

Advanced Materials for Next-Generation Technologies: Challenges and New Prospects. Shrikaant Kulkarni, Vipul Srivastava, Rabah Khenata, & Yarub Al-Douri (Eds.)

of the light spectrum for the healthy growth of plants and explores the use of grow lights, which serve as artificial light sources for indoor cultivation.

6.1 INTRODUCTION

Since light is a plant's main source of energy, the spectrum of light it receives is essential to its healthy growth and development. Grow light is basically an electric light that assists the plant in growing. It is an attempt to provide the same light spectrum as that of the sun. As there are different stages of the plants, every stage needs a different spectrum of light such as the germination phase, flowering, etc. Basically, a specific range of spectrum temperature and luminous efficiency is required by a plant. Grow light is utilized in indoor gardens, propagation of plants, food production, aquatic plants, etc. Grow Light is suitable for both the households as well as for industrial purposes. The wavelengths of light that chlorophyll, the pigment responsible for photosynthesis in plants, absorbs most effectively are blue and red.[1] Artificial grow lights are a common option for growing plants inside because they offer a regulated source of light that can be adapted to the individual requirements of the plants being cultivated. This study will concentrate on the significance of the light spectrum, particularly the blue and red wavelengths, for plants' healthy growth through absorption of chlorophyll. It will also focus on how grow lights can be used as a synthetic light source for indoor growth. We may increase the effectiveness and success of indoor plant culture by comprehending how the light spectrum affects plant growth as well as the advantages and disadvantages of various grow light technologies. Chlorophyll is a pigment that is present in plants, algae, and some bacteria. It is in charge of absorbing light energy and using photosynthesis to change it into chemical energy. Chlorophyll comes in a variety of forms, including chlorophyll *a*, chlorophyll *b*, and chlorophyll *c*, and each one has certain spectral characteristics that affect how well it absorbs light at various wavelengths. Light absorption, or the pigment's capacity to take in photons of light energy, is one of the most significant spectrum characteristics of chlorophyll. Chlorophyll's ability to absorb light is crucial for photosynthesis because it enables plants to transform solar energy into chemical energy.[1]

Chlorophyll primarily absorbs light in the blue and red parts of the electromagnetic range, with peaks at 400–450 and 650–700 nm, respectively. When a photon of light with the right amount of energy is absorbed, it can excite an electron from its ground state to a higher energy level.[2] These wavelengths match the energy levels of the bonds holding the atoms together in the pigment molecule. Chemical reactions like the production of glucose from carbon dioxide and water can be powered by this excitation energy. However, chlorophyll does not absorb light equally over the entire electromagnetic spectrum. The reason leaves appear green to the human eye is that chlorophyll has a much lower absorption in the green portion of the spectrum.

This is due to the chlorophyll pigment, which reflects most of the green light but absorbs most of the blue and red light that strikes the leaf.[1–3] Plants also contain additional pigments, such as carotenoids and flavonoids, which have various spectral characteristics and may absorb light in various regions of the electromagnetic spectrum, in addition to chlorophyll. Though they are not as good at absorbing light as chlorophyll, some pigments can nonetheless contribute to photosynthesis.

Due to the potential of green synthesis to increase plant growth and productivity in an environmentally responsible and sustainable manner, it has attracted a lot of interest in the field of plant growth and development. The term "green synthesis" describes the process of creating chemicals, materials, and certain other substances using ecofriendly practices. Due to worries about the effects standard chemical synthesis processes, which frequently use dangerous chemicals and produce considerable amounts of waste, have on the environment and human health, green synthesis has drawn more and more attention in recent years.

The usage of artificial lighting systems providing red and blue light is one way that green synthesis has been implemented in this field. In controlled environments like greenhouses or indoor grow rooms, these systems serve the purpose of providing plants with the light they require to develop and thrive. They can be employed to supplement natural sunlight or to supply light in the absence of sunlight as they are able to mimic the wavelengths that are absorbed most efficiently by chlorophyll.[2] These systems can also be well-controlled by the researchers in terms of intensity and duration of exposure so as to study the effects of different light conditions on growth and development. In this way, it is possible to improve plant development and growth in a sustainable and ecologically friendly way by analyzing the light spectrum of plant growth and then using the artificial light systems accordingly. This, in turn, may have applications in agriculture and horticulture.

6.2 BACKGROUND

Gates et al. (1965) conducted a study on the spectral properties of plants, specifically the absorbance, transmittance, and reflectance of light by different plant tissues and pigments. Using a spectrophotometer, the authors measured the spectral responses of various plant tissues, including leaves, stems, and flowers, under different wavelengths of light. They found that different plant tissues exhibit distinct spectral characteristics and that certain wavelengths are more efficiently absorbed or reflected by certain tissues. The authors also studied the effect of pigments, such as chlorophyll and carotenoids, on the spectral properties of plants.[1] Dutta Gupta and Agarwal reviewed the history and development of artificial lighting systems for plant growth and development in agriculture. They compared the advantages and disadvantages of different artificial lighting systems and provided recommendations for selecting the most suitable lighting system.[2] Pritam Kumar Dikshit et al. studied the nanoparticles through the green synthesis approach because it is advantageous for our environment. In the review, the author highlights the green synthesis of metal oxides from plants and microorganisms. As compared to the chemical method, the green synthesis is ecofriendlier. It was determined from the literature that the various types of metal oxides are synthesized from plant material, consisting of extract of leaf, fruit, seed, etc., and also from microorganisms. Metal oxides have good potential, for example, Au, Ag, Pt, Pd, Ni, Si, CU, CuO, and TiO_2. Different characteristic techniques are used to study the structure and morphology of the sample material such as XRD, SEM, TEM, EDX, FTIR, etc.[3] Mohamed Abd Elkodous et al. (2021) studied the utilization of waste products for the synthesis of metal oxide nanoparticles for various applications in bioscience. Different types of waste are obtained in our daily lives and can be utilized for antimicrobial, anticancer, and antioxidant agents. This review paper emphasizes the importance of the waste product utilization because of its advantages for the environment.[4] Reena et al. (2021) prepared iron NP from the peel of Artocarpus heterophyllous (fruit). The peel of this fruit is antioxidant, so it can be used as a bioreduction and capping agent during the process of green synthesis. Iron NPs are characterized by FTIR, SEM, TEM, XRD, and EDX. The average size of the prepared material was 33 nm. It was determined that the sample material was suitable for catalytic activity. This study relates the applications of iron with heterogeneous Fenton catalysts for the degradation of Fuchsin dye, which follows a surface-controlled reaction. The author highlights the preparation of catalyst from the green synthesis due to the excellent yields of this method, which is quite suitable for the environment.[5] Siti Mafu 'a h et al. (2021) studied the ZnO nanoparticles by green synthesis using banana peel as a capping agent. From the XRD data, it was determined that ZnO NPs were prepared successfully. An agar diffusion test was conducted, which shows that the ZnO nanoparticles are suitable for antibacterial agents. The diameter of the clear zone was found to be 8–10 nm. In the study, three types of banana peels were taken, and the Tanduk banana peel gave the best result.[6] Fall et al. (2021) studied the synthesis of metal oxides by using the peel of citrus sinensis (orange peel). The method used to prepare it is green synthesis, which is an ecofriendly technique that acts as a reducing agent. The sample material was characterized by XRD, ART-FTIR, PL, and SEM. The temperature of annealing is 300°C to 500°C for 2 hours. Nickel oxide nanoparticles are successfully synthesized with green synthesis with high-purity crystalline.[7] Matthew Huston et al. (2021) studied the green synthesis of nanomaterials, which is a very innovative technique nowadays. In this review, the study highlighted a new research area by using a green approach, which is very beneficial for our environment.[8] Md. Shakhawant Hossan Bhiyan et al. (2020) stated that iron oxide was prepared by using papaya leaf by the green synthesis method. As it has a very ecofriendly and nontoxic behavior, different characterizations are used to study the sample material, which include FTIR, XRD, FESEM, EDX, and TGA. Photocatalytic degradation of yellow RR dye and antibacterial activity are used for different applications. It was found in the study that the sample material showed the intermediate antibacterial activity against Klebsiella spp., E. coli, pseudomonas spp., S. aureus, and bacterial strains. The nanoparticles show the highest toxicity at the highest doses and brilliant/excellent activity; around 95% of cells get damaged against BHK 21 and HELA.[9] Kaiyi Luo et al. (2020) studied CuO/Ceo_2/ZnO and ternary by using two-step sol-gel methods. Cuo–CeO_2 is effective for visible light. This study is basically a cost-effective method for the synthesis of a photocatalyst that contains multiple components with a clear division of work.[10]

6.3 MATERIALS AND METHODS

Green synthesis of metal oxides for grow lights is an innovative approach that combines sustainability and technology to enhance plant growth. Traditional methods of synthesizing metal oxides often involve harsh chemicals and high energy consumption, leading to negative environmental impacts. However, with the green synthesis approach, the focus is on using ecofriendly and renewable resources to produce metal oxides for grow lights.

One promising method of green synthesis involves the utilization of plant extracts or natural compounds as reducing and stabilizing agents. These plant extracts, rich in bioactive molecules, can effectively convert metal precursors into metal oxide nanoparticles. The choice of plants and extraction methods can be tailored to target specific metal oxides, ensuring optimal performance for grow lights.

The green synthesis of metal oxides offers several advantages. Firstly, it reduces the reliance on hazardous chemicals, minimizing potential risks to human health and the environment. Additionally, it promotes the utilization of natural resources, making it a sustainable alternative to conventional synthesis methods.

Furthermore, metal oxide nanoparticles synthesized through green methods have shown excellent properties for growing lights. These nanoparticles can be tuned to emit specific wavelengths of light, matching the requirements of plants at different growth stages. By controlling the composition, size, and shape of the nanoparticles, it is possible to optimize their light absorption and scattering properties, maximizing their efficiency in promoting plant growth.

The green synthesis of metal oxides for grow lights holds great promise in revolutionizing indoor plant cultivation. By harnessing the power of nature and sustainable practices, we can create a greener and more efficient approach to providing artificial light for plants. This not only benefits the environment but also contributes to healthier and more abundant plant growth, ultimately improving food production and the overall well-being of the society.

6.3.1 DIFFERENT APPROACHES FOR THE SYNTHESIS OF METAL OXIDES

The different approaches to synthesize the metal oxides are as follows:

Top-up: The top-up approach of green synthesis is an innovative strategy that aims to enhance the efficiency and sustainability of metal oxide synthesis. Traditional synthesis methods often involve large quantities of reagents and high energy consumption, leading to environmental concerns. In contrast, the top-up approach focuses on utilizing smaller amounts of green precursors and gradually adding them to the reaction mixture.

The top-up approach begins with the preparation of a reaction mixture containing a solvent, metal precursors, and green reducing agents or stabilizers. Rather than adding all the precursors at once, the top-up approach involves a stepwise addition of the green precursors during the reaction. This gradual addition allows for better control over the reaction kinetics and product formation, resulting in improved efficiency and higher-quality metal oxide nanoparticles.

One of the key advantages of the top-up approach is its ability to minimize the excess use of reagents. By adding the precursors in smaller quantities over time, the reaction can be fine-tuned to achieve optimal conversion and yield, reducing waste generation and improving resource utilization. This approach also enables better control over the size, shape, and composition of the synthesized metal oxide nanoparticles, leading to enhanced properties and performance.

Another benefit of the top-up approach is its compatibility with green precursors and reducing agents. These green precursors can be derived from renewable resources, such as plant extracts or bio-based compounds, minimizing the reliance on nonrenewable materials. The gradual addition of these green precursors promotes their effective utilization, maximizing their potential as reducing and stabilizing agents in the synthesis process.

The top-up approach of green synthesis has shown promising results in various applications, including the synthesis of metal oxide nanoparticles for catalysis, sensing, and energy storage. Its ability to reduce waste, enhance control, and utilize green precursors makes it a sustainable and efficient alternative to traditional synthesis methods.

By adopting the top-up approach in green synthesis, we can achieve greener and more sustainable production of metal oxide nanoparticles. This approach not only minimizes environmental impacts but also contributes to the development of advanced materials with improved properties and functionalities.[10]

Bottom-up: The bottom-up approach of green synthesis is a strategy that involves building complex structures or materials from individual components or molecules. This method focuses on utilizing environmentally friendly and sustainable resources to create new materials with desired properties and functionalities.

In the bottom-up approach, the synthesis process starts with small, well-defined building blocks such as molecules or nanoparticles. These building blocks serve as the foundation for the construction of larger structures or materials. By carefully controlling the reaction conditions and interactions between the building blocks, it is possible to assemble them into more complex and organized architectures.

One of the key advantages of the bottom-up approach is its ability to achieve precise control over the final product's characteristics. By manipulating the size, shape, composition, and arrangement of the building blocks, researchers can tailor the properties of the synthesized materials to meet specific requirements. This level of control enables the design of materials with enhanced performance and functionality for various applications.

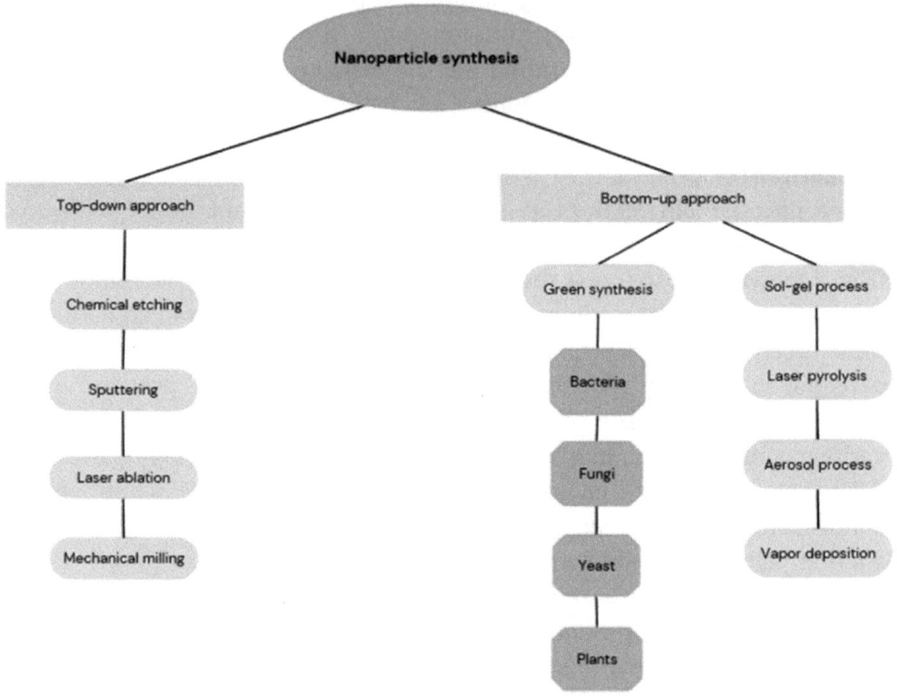

FIGURE 6.1 Flowchart representing the different techniques to prepare metal oxides.

The bottom-up approach also promotes sustainability by utilizing green precursors and ecofriendly synthesis methods. Green precursors, derived from renewable resources, are often employed in this approach to minimize the environmental impact associated with traditional synthesis routes. Furthermore, the use of environmentally friendly solvents and reducing agents further contributes to the sustainability of the process.

Moreover, the bottom-up approach allows for the incorporation of functional groups or biomolecules into the synthesized materials. This opens up possibilities for creating materials with unique properties, such as responsiveness to stimuli, biocompatibility, or self-healing capabilities. These functionalized materials find applications in fields such as biomedicine, energy storage, and environmental remediation.

The bottom-up approach of green synthesis has shown significant potential in various research areas. It enables the production of advanced materials with tailored properties while also promoting sustainability and reducing environmental impact. By harnessing the principles of self-assembly and utilizing green precursors, this approach offers a promising avenue for the development of innovative and environmentally friendly materials.

In conclusion, the bottom-up approach of green synthesis provides a sustainable and versatile method for constructing complex materials. By starting from individual building blocks and carefully controlling their assembly, researchers can create materials with precisely engineered properties and functionalities, contributing to advancements in various scientific and technological fields.

Green synthesis is a nonhazardous and environmentally friendly method, which is beneficial for other methods to prepare metal oxides.

It is a new technique to explore the growth of the nanomaterial. The synthesis of metal oxide through green synthesis is very advantageous for environmental sustainability and also removes the limitations of the sample material that was synthesized from chemicals. This new technique will solve many environmental problems as it is ecofriendly. Green synthesis can be done from various parts of the plants as the researchers can see that the metal oxides can be synthesized from the different parts of the plants such as the roots, seeds, bark, flowers, tubers, etc., but they also notice that the leaf extracts of the plants are more effective than others. Also, the reaction conditions utilized during the synthesis are mild, so green synthesis is a simple and nonrisky, cheap, and efficient way for the production of metal oxides. Plants are used for the reduction of metallic nanoparticles, which consist of biomolecules such as carbohydrates, proteins, and coenzymes that help to reduce the metal salt into nanoparticles.

Basically, green synthesis consists of two major routes.

1. Where water is utilized as a solvent
2. Secondly, natural source/plant extract is used as a main component

Nonetheless, the precise fundamental mechanism for the preparation of metal oxide nanoparticles using plant extracts is still unknown. In general, there are three stages in the synthesis of metallic nanoparticles from plant extracts:

1. Activation (bioreduction of metal ions/salts and nucleation of the reduced metal ions),
2. Growth (a spontaneous combination of tiny particles with larger ones) via a process known as Ostwald ripening, and
3. Termination (defining the final shape of the nanoparticles)

The preparation of metals and metal oxides using green synthesis methods has gained significant attention due to its potential for sustainable and environmentally friendly production. Green synthesis approaches utilize ecofriendly precursors, renewable resources, and energy-efficient techniques to minimize the environmental impact associated with traditional synthesis methods. Here, we discuss some common green synthesis methods for the preparation of metals and metal oxides, along with relevant references.

Plant-mediated synthesis: Various plant extracts have been used as reducing agents and stabilizers in the synthesis of metal nanoparticles and metal oxides. These plant extracts contain bioactive compounds such as phenols, flavonoids, and terpenoids, which can effectively reduce metal ions and promote the formation of nanoparticles. For example, Kumar et al. (2019) utilized aloe vera extract to synthesize silver nanoparticles with potential antibacterial activity.[11] Similarly, Mohanpuria et al. (2008) employed neem leaf extract for the green synthesis of zinc oxide nanoparticles.[12]

Microbial-assisted synthesis: Microorganisms, including bacteria, fungi, and algae, have been utilized for the synthesis of metal nanoparticles and metal oxides. These microorganisms can serve as both reducing agents and templates for nanoparticle formation. For instance, Sanghi et al. (2009) reported the biosynthesis of gold nanoparticles using the bacterium *Escherichia coli.*[13] In another study, Mukherjee et al. (2001) demonstrated the synthesis of silver nanoparticles using the fungus Fusarium oxysporum.[14]

Sol-gel synthesis: Sol-gel synthesis is a versatile method for the preparation of metal oxides. It involves the hydrolysis and condensation of metal alkoxides or metal salts in a solution, followed by gelation and drying to obtain the desired metal oxide materials. Green sol-gel synthesis techniques utilize environmentally friendly solvents and renewable precursors. For example, Chaudhary et al. (2016) reported the green synthesis of zinc oxide nanoparticles using neem leaf extract and a sol-gel method.[15] Li et al. (2017) developed a green sol-gel method for the synthesis of titanium dioxide nanoparticles using a plant-based reducing agent.[16]

These are just a few examples of the green synthesis methods employed for the preparation of metals and metal oxides. Green synthesis approaches offer numerous advantages, such as sustainability, low toxicity, and energy efficiency, making them highly desirable for various applications.

In order to evaluate the differences between analyses in both structural and optical terms, different characterization methods are used. XRD is a technique used to examine the structural characteristics of a processed material. To identify the crystalline formations, XRD is utilized. Using

Bruker equipment, a dry sample is photographed with a scanning range of 10–90° and a wavelength of 1.5406A. Utilizing FTIR, it was possible to examine the numerous functional groups and chemical bonds that were present in the samples of both organic and inorganic substances. To examine the optical properties of materials, absorption spectroscopy, which makes use of the ultraviolet and visible regions of the electromagnetic field, is used. The use of a UV spectrometer is one of the simplest techniques for evaluating interactions between molecules with variations in wavelength and absorbance. Fluorescence spectroscopy is used to investigate the optical properties of the material. A Perkin Elmer spectrometer was used to look at the luminescent spectrum of the sample material. The wavelengths of the lamp's excitation as well as emission range from 200 to 900 nm.[17–19]

6.4 STRUCTURAL AND OPTICAL PROPERTIES OF METAL OXIDES PREPARED BY GREEN SYNTHESIS

6.4.1 STRUCTURAL PROPERTIES OF METAL OXIDES

Figure 6.2 represents the XRD spectra of pure CuO, NiO, and CuO/NiO nanocomposites. CuO nanoparticles' monoclinic structure was discovered and indexed as a JCPDS card number 80–1268. The average crystallite size of CuO, which was calculated using Scherer's formula, is 18.89 nm. The FCC structure of NiO is shown by the XRD pattern data and is referenced by JCPDS card no. 78–0643.[20] The NiO-prepared material has an average crystal size of 11.27 nm. CuO and NiO mixed nanocomposite phase concentrations were measured, and the presence of both CuO and NiO diffraction peals in the prepared sample was studied, which supports the development of nanocomposites.

6.4.2 FESEM

The image of the materials' microstructure is captured using an innovative method known as field emission scanning electron microscopy (FESEM). As would be expected, the average grain size is slightly greater than the crystallite size, demonstrating the strong magnetic interaction. The CuO/NiO nanocomposite's structure and morphology were examined using FESEM (field emission scanning electron microscopy), as shown in Figure 6.3. The aggregated hierarchical three-dimensional (3D) nanostructures are made up of parallel self-assembled nanosheets, which are clearly visible in the FESEM image. Metal oxides are made up of spherical grains. As would be expected, the average grain size is slightly greater than the crystallite size, demonstrating the strong magnetic interaction.[20–21]

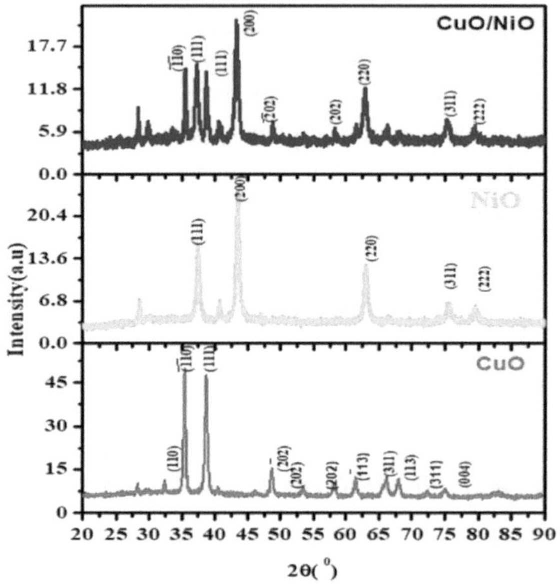

FIGURE 6.2 Representation of CuO, NiO, and CuO/NiO XRD spectra synthesized from banana peels.

Source: Adapted from Ref. [20]

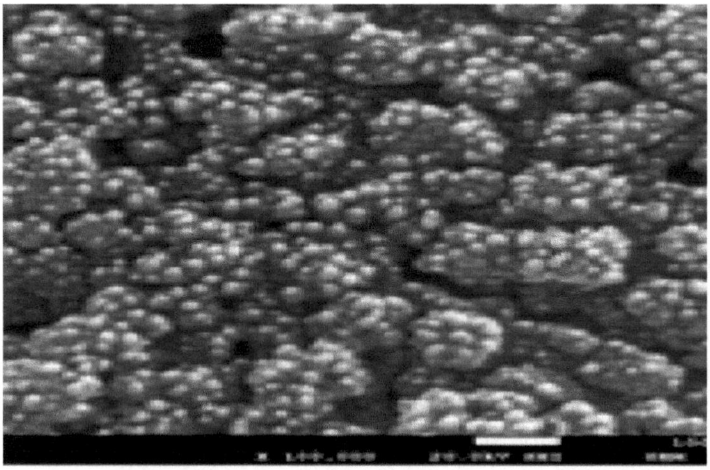

FIGURE 6.3 FESEM micrograph of CuO/NiO nanocomposites prepared from banana peels.

Source: Adapted from Ref. [20]

6.4.3 OPTICAL BAND GAP OF METAL OXIDES PREPARED BY GREEN SYNTHESIS

6.4.3.1 OPTICAL ENERGY BAND GAP

The UV-Vis absorption spectrum for the sample material made from the solid banana peel waste of CuO, NiO, and a nanocomposite of CuO/NiO is shown in Figure 6.4. The CuO, NiO, and CuO/NiO exhibit a high absorption band and are excited at 310 nm, as seen in the UV-Vis spectra. For NiO, CuO, and CuO/NiO nanocomposites, distinct optical band gaps were also computed using the Tauc relation. The absorption was used to calculate the prepared material's optical band gap. CuO, NiO, and NiO/CuO have computed band gaps of 2.9 eV, 2.8 eV, and 2.6 eV, respectively.

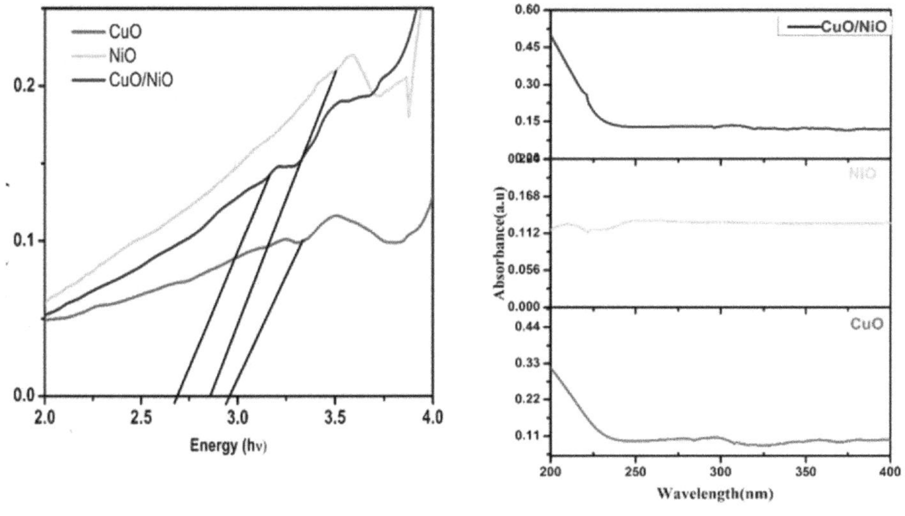

FIGURE 6.4 (a) UV Spectrum of Pure CuO, NiO, and CuO/NiO. (b) Tauc plot calculation of the optical band gap of CuO, NiO, and CuO/NiO sintered at 400°C.

Source: Adapted from Ref. [20]

6.4.3.2 EMISSION SPECTRA OF METAL OXIDES PREPARED BY GREEN SYNTHESIS

The crystal amount and the purity of the substance are evaluated using the luminescence characteristics. By using 310 nm excitation, broad emission spectra between 300 and 700 nm were achieved. CuO/NiO's visible emission was mostly caused by the defect center in the energy band gap, whereas metal oxides' UV emission is caused by the band gap edge. Defect-induced emission in metal oxides refers to the phenomenon where the presence of structural defects within the crystal lattice of metal oxide materials leads to the emission of light. Metal oxides are compounds composed of metallic elements bonded with oxygen atoms, such as titanium dioxide (TiO_2), zinc oxide (ZnO), copper oxide (CuO), nickel oxide (NiO), and tungsten trioxide (WO_3).

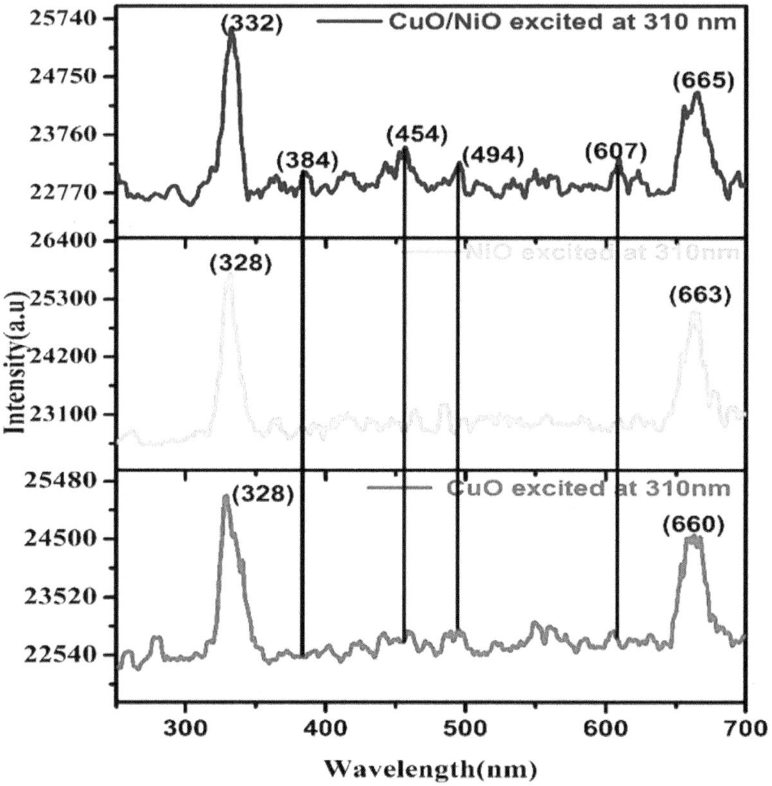

FIGURE 6.5 Fluorescence spectroscopy of banana peel-derived pure CuO, NiO, and CuO/NiO layers sintered at 400°C.

Source: Adapted from Ref. [20]

In a perfect crystal lattice, all atoms are arranged in an ordered and symmetric manner. However, during the synthesis or processing of metal oxide materials, various defects can occur. These defects can be classified into two main categories: intrinsic and extrinsic defects.

Intrinsic defects arise from imperfections within the crystal structure itself, such as vacancies (missing atoms), interstitials (extra atoms), or dislocations (structural irregularities). These defects can introduce energy levels within the band gap of the metal oxide, leading to the generation of electron-hole pairs upon excitation. When these pairs recombine, excess energy is released in the form of light, resulting in defect-induced emissions.

Extrinsic defects, on the other hand, result from the intentional introduction of impurities or dopants into the metal oxide material. These impurities can disrupt the crystal lattice and create additional energy levels. The recombination of electron-hole pairs at these energy levels can also lead to defect-induced emission.

Defect-induced emission in metal oxides has garnered significant interest due to its potential applications in various fields. It has been explored for the development of optoelectronic devices, including light-emitting diodes (LEDs), lasers, and sensors. The emission properties can be tuned by controlling the type and concentration of defects, allowing for customized light emission in specific wavelength ranges.

Furthermore, defect-induced emission has provided insights into the fundamental properties of metal oxides and their defect chemistry. Researchers continue to investigate the underlying mechanisms and explore novel strategies to enhance and control defect-induced emission in metal oxide materials.

In summary, defect-induced emission in metal oxides arises from structural defects within the crystal lattice, either intrinsic or extrinsic in nature. This phenomenon has both fundamental significance and practical implications, paving the way for advancements in optoelectronics and material science.

Plants are particularly sensitive to changes in light intensity spectrum (300–700 nm) and time; they can detect even little variations. The wavelength and spectrum of light, which are sensed by photoreceptors, affect how a plant grows. The three main categories of photoreceptors are as follows:

1. Cryptochromes
2. Phytotropins
3. Phytochromes

Under UV-stimulated conditions, pure NiO and CuO respond well to ultraviolet and red light. The CuO/NiO composite exhibits an improvement in the intensity of the green and blue colors coupled with UV blue light when used as a grow light. Fluorescence spectroscopy shows that the emission spectrum of a green synthetic material made from solid banana peel residue has the potential to be further explored since it has the potential for usage as a grow light.[14]

6.5 CONCLUSIONS

The green synthesis technique is used for the production of metal oxides used in grow lights; NiO and CuO/NiO are all examples of oxides. The structural and optical analyses were carried out using a variety of characterization techniques. Fluorescence spectroscopy shows that CuO, NiO, and CuO/NiO emission spectra indicate that green synthetic material has promise to be further investigated. It also provides grow light spectra that are advantageous for indoor planting. An electric light that assists the plant in growing provides the same light spectrum as that of the sun, provides abundant and sufficient light throughout the year, and is suitable for both the households as well as for industrial purposes. In addition to that, there are various applications of metal oxides and their composites in electrochemistry as catalysts, high-temperature conductors, insulators, and super magnetic materials.

KEYWORDS

- **health**
- **green**
- **space**
- **light**
- **spectrum**

REFERENCES

1. Gates, D. M.; Keegan, H. J.; Schleter, J. C.; Weidner, V. R. Spectral Properties of Plants. *Appl. Opt.* **1965,** *4* (1), 11–20.
2. Dutta Gupta, S.; Agarwal, A. Artificial Lighting System for Plant Growth and Development: Chronological Advancement, Working Principles, and Comparative Assessment. In: *Light Emitting Diodes for Agriculture*; Singapore: Singapore, 2017; pp 1–25.
3. Dikshit, P. K.; Kumar, J.; Das, A. K.; Sadhu, S.; Sharma, S.; Singh, S.; Gupta, P. K.; Kim, B. S. Green Synthesis of Metallic Nanoparticles: Applications and Limitations. *Catalysts* **2021,** *11* (8), 902.
4. Abd Elkodous, M.; El-Husseiny, H. M.; El-Sayyad, G. S.; Hashem, A. H.; Doghish, A. S.; Elfadil, D.; Radwan, Y.; El-Zeiny, H. M.; Bedair, H.; Ikhdair, O. A.; Hashim, H. Recent Advances in Waste-Recycled Nanomaterials for Biomedical Applications: Waste-to-Wealth. *Nanotechnol. Rev.* **2021,** *10* (1), 1662–1739.

5. Jain, R.; Mendiratta, S.; Kumar, L.; Srivastava, A. Green Synthesis of Iron Nanoparticles Using *Artocarpus heterophyllus* Peel Extract and Their Application as a Heterogeneous Fenton-Like Catalyst for the Degradation of Fuchsin Basic Dye. *Curr. Res. Green Sustain. Chem.* **2021,** *4*, 100086.
6. Marfu'ah, S.; Rohma, S. M.; Fanani, F.; Hidayati, E. N.; Nitasari, D. W.; Primadi, T. R.; Ciptawati, E.; Sumari, S.; Fajaroh, F. Green Synthesis of ZnO Nanoparticles by Using Banana Peel Extract as Capping Agent and Its Bacterial Activity. In: *IOP Conference Series: Materials Science and Engineering*, Vol. 833, No. 1; IOP Publishing, 2020; p 012076.
7. Fall, A.; Mayedwa, N.; Bucher, R.; Ngom, B. D.; Maaza, M. Bio-inspired Nickel Oxides Nanoscale Synthesis by using Peel of Citrus Sinensis. *MRS Adv.* **2020,** *5* (21–22), 1157–1166.
8. Huston, M.; DeBella, M.; DiBella, M.; Gupta, A. Green Synthesis of Nanomaterials. *Nanomaterials* **2021,** *11* (8), 2130.
9. Luo, K.; Li, J.; Hu, W.; Li, H.; Zhang, Q.; Yuan, H.; Yu, F.; Xu, M.; Xu, S. Synthesizing $CuO/CeO_2/ZnO$ Ternary Nano-Photocatalyst with Highly Effective Utilization of Photo-Excited Carriers Under Sunlight. *Nanomaterials* **2020,** *10* (10), 1946.
10. Kumar, V.; Yadav, S. K. Plant-Mediated Synthesis of Silver and Gold Nanoparticles and Their Applications. *J. Chem. Technol. Biotechnol.* **2019,** *94*, 286–296).
11. Mohanpuria, P.; Rana, N. K.; Yadav, S. K. Biosynthesis of Nanoparticles: Technological Concepts and Future Applications. *J. Nanopart. Res.* **2008,** *10*, 507–517.
12. Sanghi, R.; Verma, P.; Puri, S. Enzymatic Formation of Gold Nanoparticles Using *Phanerochaete chrysosporium*. *Adv. Chem. Eng. Sci.* **2009,** *9*, 1–8.
13. Mukherjee, P.; Ahmad, A.; Mandal, D. et al. Fungus-Mediated Synthesis of Silver Nanoparticles and Their Immobilization in the Mycelial Matrix: A Novel Biological Approach to Nanoparticle Synthesis. *Nano Lett.* **2001,** *1*, 515–519.
14. Chaudhary, G. R.; Satsangee, S. P.; Rao, B. S. et al. Neem mediated Synthesis of ZnO Nanoparticles: Characterization and Its Evaluation for Thermal Stability. *Adv. Sci. Lett.* **2019,** *22*, 3693–3697
15. Meyers, A.; Mishra, A.; Benson, D. J. Mechanical properties of nanocrystalline materials. *Progress Mater. Sci.* **2005,** *51* (4), 427–556. https://doi.org/10.1016/j.pmatsci.2005.08.003.
16. Ighalo, J. O.; Sagboye, P. A.; Umenweke, G.; Ajala, O. J.; Omoarukhe, F. O.; Adeyanju, C. A. et al. CuO Nanoparticles (CuO NPs) for Water Treatment: A Review of Recent Advances. *Environ. Nanotechnol., Monitor. Manag.* **2021,** *15*, 100443.
17. Rahdar, A.; Aliahmad, M.; Azizi, Y.; Keikha, N.; Moudi, M.; Keshavarzi, F. CuO-NiO Nano Composites: Synthesis, Characterization, and Cytotoxicity Evaluation. *Nanomed. Res. J.* **2017,** *2* (2), 78–86.
18. Arun, L.; Karthikeyan, C.; Philip, D.; Unni, C. Optical, Magnetic, Electrical, and Chemo-Catalytic Properties of Bio-Synthesized CuO/NiO Nanocomposites. *J. Phys. Chem. Solids* **2020,** *136*, 109155.
19. Kaur, J.; Rani, S. CuO/NiO Nano-Composite Synthesized from Banana Peels for Grow Light. *Mater. Today: Proc.* **2023.**
20. Chatterjee, S.; Ray, A.; Mandal, M.; Das, S.; Bhattacharya, S. K. Synthesis and Characterization of CuO-NiO Nanocomposites for Electrochemical Supercapacitors. *J. Mater. Eng. Perform.* **2020,** *29*, 8036–8048.

CHAPTER 7

Materials Processing of Metals, Ceramics, and Polymers for Industrial Products

RIYADH A. AL-SAMARAI[1] and Y. AL-DOURI[2,3,4]

[1]*Electromechanical Engineering Department, Engineering College, University of Samarra, Samarra, Iraq*

[2]*College of Health and Medical Techniques, Al-Bayan University, Baghdad, Iraq*

[3]*Department of Mechanical Engineering, Faculty of Engineering, Piri Reis University, EflatunSk, No: 8, 34940 Tuzla, Istanbul, Turkey*

[4]*Nanotechnology and Catalysis Research Centre, University of Malaya, 50603 Kuala Lumpur, Malaysia*

ABSTRACT

In this chapter, we will go through items to consider while employing typical fabrication processes. Once we understand the structure and properties of engineered materials, we can discuss how these materials can be developed or manufactured. There is little to no variation in the descriptions of the most essential processing between the procedures that manufacture integrated circuits and components at the macro and micro scales. This chapter's conclusion should be useful because it lists and describes the many techniques of working with metal, calculates the force and depth of penetration of induction-heated metals operating on a workpiece, determines the best grinding procedure for creating the optimum-sized metal and ceramic powders, calculates the pressure to determine the rolling neutral point, and describes the various methods of ceramic sludge processing. There is a separation between kinetic and transport speed constraints in

Advanced Materials for Next-Generation Technologies: Challenges and New Prospects. Shrikaant Kulkarni, Vipul Srivastava, Rabah Khenata, & Yarub Al-Douri (Eds.)

the manufacturing of vapor-phase ceramics. Several batch and continuous polymer processing methods are listed and described. Change equations can be used to construct simple design equations for flow and pressure decrease in polymer processing technologies. The advantages of plasma therapy in improving biological product biocompatibility are described. In addition to the measurement technique, this enhances the accuracy of coefficient, electrical resistivity, and thermal conductivity measurement data (based on diffusion and heat capacity). Potentially inaccurate data is detected. The calculations of the concentration and mobility of charge carriers, as well as the measurement of the Hall coefficient, are included due to their usefulness in comprehending material handling operations.

7.1 INTRODUCTION

We may talk about how engineered materials can be created or transformed into practical goods and components if we have a firm grasp of their structure and properties. With little to no distinction made between microscale and macroscale processes, the majority of significant processing techniques are discussed here. For instance, technologies that produce both integrated circuits and bridge components for automobiles are included in this article.[1–3] The fact that all the chemical and physical phenomena required for the application of these processing methods have previously been covered in earlier chapters is something that all of these phenomena have in common. This chapter's conclusion should enable you to do the following:

- List and explain the numerous techniques for working with metal.
- Determine the force that is being applied to the workpiece during forging.
- Calculate the induction heating power and depth of penetration of metals. Identify the proper milling method for creating size-appropriate metal and ceramic powders.
- Determine the neutral point in rolling operations by computing the pressure.
- List and describe the various ceramic slurry treatment techniques.
- In the production of ceramics in the vapor phase, there is a distinction between kinetic and transport speed limitations.
- List and describe the various batch- and continuous-polymer-processing techniques.
- Simple design equations for flow and pressure drop in polymer processing technologies can be derived using change equations.

- To differentiate between the manufacturing of fiber-reinforced composite materials, comparisons were made between the pultrusion, resin injection molding, and filament winding processes.

7.2 PROCESSING OF METALS AND ALLOYS

Metals continue to remain the predominant group of structural materials for many applications, including the automotive, aerospace, and construction sectors, despite the relatively recent emergence of polymers and composites as structural materials. This is partly a result of the components' formability or ease of manufacture. The three most popular and commonly utilized metal processing techniques—casting, forging, and powder metallurgy—will be discussed in this section. Although significant, the subject of connecting metals (such as welding) is outside the purview of this article.[4]

7.2.1 CASTING

Casting is the process of melting metal and solidifying it in a hollow, known as a mold, to create an item whose shape is determined by the mold's design. It is used in the context of metalworking. Almost all metals undergo casting at some point during production. They can be cast in two different ways: directly into the shape of the component (option A) or as ingots (option B), which can then be molded into the necessary shape using the procedures outlined later in this section. Compared to other metal-forming techniques, casting has a number of advantages, including the ability to produce large, complicated components in large quantities and to give parts the same mechanical and physical qualities. From a financial perspective, since additional processes like shaping, pressing, annealing, and joining raise the cost of the metal components, it would be preferable to produce the majority of the metal components directly from the casting. The microstructure of cast materials is substantially coarser than that of heat-treated materials or annealed metals due to the fact that the rate of growth of crystalline phases in the liquid state is much higher than in the solid state. As a result, cast pieces often have lower strength than forged or rolled parts. When structural components will not be subjected to heavy loads while in use, when high strength can be gained through subsequent heat treatment, when the component is too large to be easily molded, or when the shape calls for it, casting is often employed as the primary production procedure if the component is too difficult to manage. Let us take a closer look at the crucial casting process phases to better understand how some of the qualities might be connected to the crucial processing variables at each stage.[5–7]

7.2.1.1 MELTING

Melting is the first step in the casting process. There are many well-known ways to melt metals, such as resistance heating components and open flames, but there are a few other techniques that have particular advantages when melting some metals. For instance, arc melting, in which a direct or alternating current is supplied to the electrode and the substance of the crucible, is used to melt metals like zirconium, titanium, and molybdenum in water-cooled copper crucibles. If the electrode is composed of tungsten or carbon, it may not be consumable, but if it is formed of molten metal, it may be. The electrode and the charge form an arc, and the energy from the arc causes the charge to melt. This method is performed in a vacuum or in an inert environment, such as helium or argon, to avoid contamination with oxygen or nitrogen. The metal is placed in a container inside an electric coil that generates a high-frequency alternating magnetic field to initiate the induction melting process (see Figure 7.1). An alternating magnetic field produced by an alternating current passing via an induction coil induces a current in the metal known as the workpiece. Due to the passage of these secondary currents that electromagnetic induction induces in the coils, heat is produced in the metal. As a result, the electrical energy that passes through the coil is transformed into magnetic energy, which is then transformed once more into heat parts as a result of eddy currents, which are induced secondary electrical currents. The skin effect, in which the majority of the eddy currents occur on the surface of the workpiece, determines the current density in the workpiece. The current density diminishes exponentially as one gets closer to the center. The current is virtually nil in the workpiece's middle. There is no need for a transmission channel like air, water, or a mechanical connection because heat is released within the workpiece itself. Heat conduction eventually causes the entire workpiece to melt as the metal's outer layers start to melt. The diameter of the melt or billet must, of course, be bigger than the depth of penetration. Another common environment for the induction process is a vacuum or an inert gas atmosphere. Induction melting has the benefit of having a quick, clean melt. Since the crucible can also be employed as a current collector, it can be used for practically any metal or alloy, even if it is not inductively susceptible to the process. Additionally, secondary current induction lines stir the molten charge and increase the homogeneity of the melt when the metal is in the molten stage. For highly refractory metals like tantalum, other techniques like flame plasma melting and electron beam melting are used. They are frequently carried out in a vacuum as well.[8]

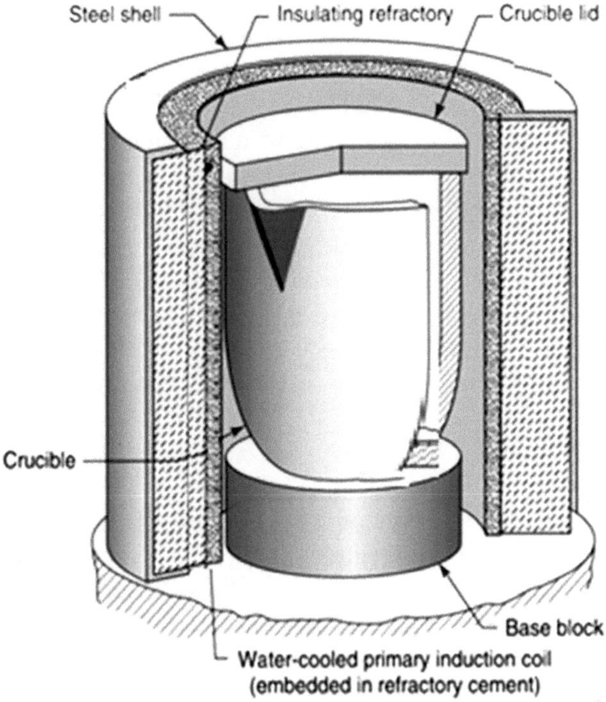

FIGURE 7.1 Diagram of the crucible and induction furnace used in the melting of copper.[1]
Source: Adapted from Ref. [1]

The vacuum is used in smelting processes to stop reactions like oxidation as well as to stop and even eliminate dissolved gases from metals. In their liquid state, the majority of metals disperse a sizable amount of gas. As the metal solidifies, gases can be released from the melt, creating holes and pinholes in the metal ingot. These gases are much less soluble in the solid state than in the liquid state. Porosity results from this, which can harm the characteristics of the metal or alloy. While melting, oxidation should not always be prevented. This crucial phase in the manufacturing of steel involves melting pig iron, the byproduct of pig iron production, in a stream of oxygen. About 4 weight percent of cast iron is made up of carbon, which, when exposed to oxygen, produces carbon monoxide. Because of the oxygen explosion it provides, the furnace used to smelt iron and create steel is known as a blast furnace. In order to create steel with the correct carbon content, the amount of oxygen is carefully controlled so that only the necessary amount of carbon is removed. Additionally, cast iron's highly reactive trace impurities, like silicon, aluminum, and manganese, react with oxygen to generate oxides. Due to their lower density than the molten metal, these oxides float to the melt's surface where they combine with other materials to produce slag, which can be skimmed off. The fact that these interactions with carbon and trace impurities generate a significant amount of energy that may be used to melt iron is another benefit of oxygen blasting. Numerous changes and alterations have since been made to this procedure, and there may be variants, such as the creation of pig iron from iron ore (Fe_2O_3). The chemistry involved in the steel-making process is rather complicated, so readers interested in learning more are directed to the metallurgical texts mentioned at the end of this chapter.[9]

7.2.1.2 FORMING OPERATIONS

The molten metal could then theoretically be put into a mold with the desired shape and hardened there to create the part. The most frequent process after melting is the creation of a sizable metal "blank" called an ingot because, as was previously said and will be further explored in the following section, solidification from casting procedures typically results in undesirable crystal formations. Thus, ingot casting and mold casting are the two primary categories of metal casting operations. The majority of metal castings are ingot castings, which can be further broken down into three types: static casting, semicontinuous or direct cooling casting, and continuous casting. Simply pouring molten metal into a permanent refractory mold with a plug or removable bottom is called static ingot casting.[10]

As graphically displayed in Figure 7.2, the ingots feature a little cone-shaped bottom that helps with extraction and speeds up the solidification process. If required, the metal can be drained from the mold during remelting thanks to a cork at the bottom of the mold. Aluminum castings are often produced by semicontinuous ingot casting. In this procedure, the molten metal is transferred into a mold with a moveable base and a lengthy piston that is water-cooled (see Figure 7.3a).

The piston descends, and the reservoir continues to fill with metal as the metal solidifies (see Figure 7.3b). The piston can move its entire length before coming to a stop. Both the steel and copper industries employ continuous casting. Similar to a semicontinuous technique, molten metal is fed into a

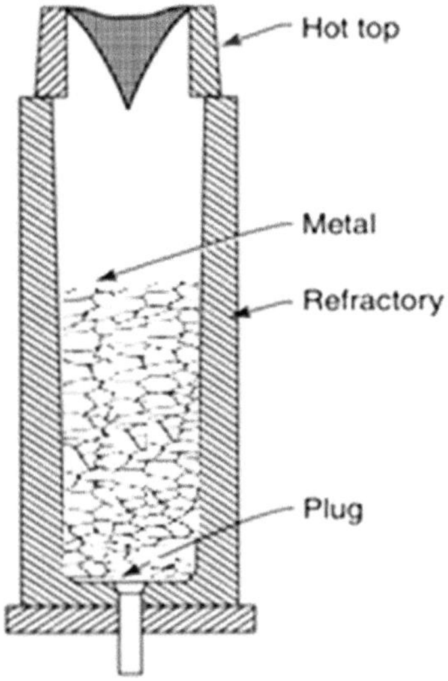

FIGURE 7.2 An ingot in a refractory mold is shown schematically (From the second edition of The Nature and Properties of Engineering Materials).

Source: Adapted from Ref. [2]

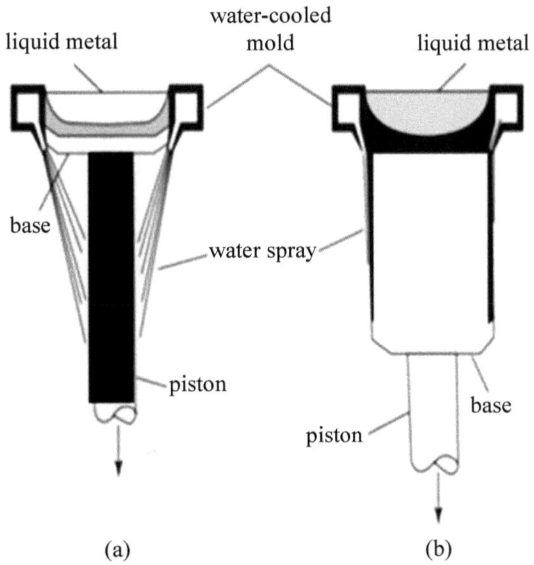

FIGURE 7.3 Schematic representation of a semicontinuous casting process in which (a) molten metal cools in a mold with a detachable base and (b) the detachable base is lowered downward to allow for the addition of additional metal to the reservoir.

Source: Adapted from Ref. [6]

permanent mold in this method of casting. The solidified ingot is continually chopped into pieces and removed from the process as soon as it is cast, as opposed to the process ending after a set amount of time. In comparison to more conventional ingot casting techniques, this process has significant financial benefits.[11]

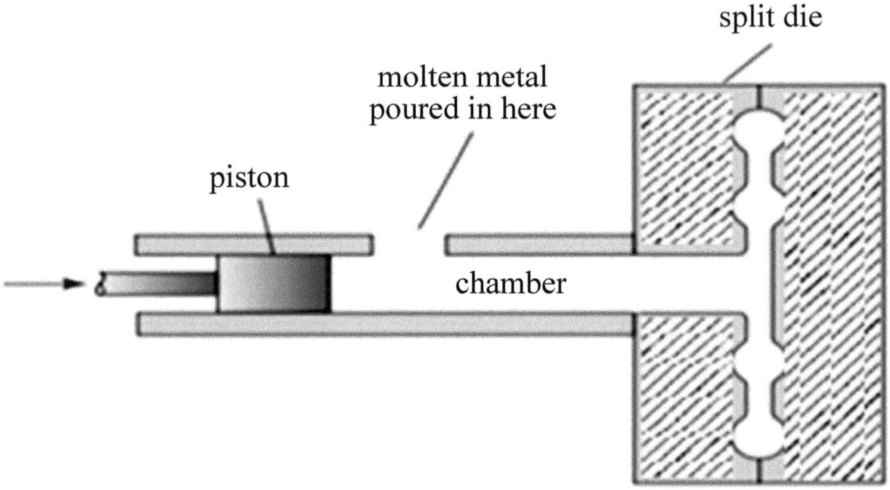

FIGURE 7.4 The process of cold chamber die casting is shown schematically.

Source: Adapted from Ref. [3]

The method by which the melt is fed into the mold, the molding process, or the molding material are commonly used to categorize mold casting. Sand casting, permanent mold casting, pressure casting, and centrifugal casting are the four primary casting types used in forming. When casting into sand molds, the mold is made up of sand grains, water, clay, and other elements that have been compacted along the sample to a density of 20 to 80% of the bulk of its combined substance. The two components of the internal cavities, into which the metal is cast, are created by fastening together the mold's tongue and thrust. Because cast iron, steel, or bronze molds may be reused, the process is referred to as permanent mold casting. In this instance, machining is used to create the mold cavity. The advantages of this method over sand casting include a smoother surface, tighter tolerances, and higher production. It is used for cast iron and nonferrous alloys. Injection molding is a refinement of the permanent molding technique. Under pressure ranging from 0.7 to 700 MPa, the molten metal is driven through a plate having a hole shaped in the desired way, known as a die (100–100,000 psi). There are two primary forms of injection molding. There are cold chamber machines and hot chamber machines. A part of the molten metal is pushed into the cavity of a hot chamber machine at a pressure of up to 14 MPa (2000 psi). Lead, zinc, and tin alloys, which have low melting points, are cast using this method.[12]

Molten metal is injected into the injection barrel and driven into the cavity at a pressure that is around ten times that of the hot chamber process in the cold chamber process (see Figure 7.4). High melting point alloys, including those made of aluminum, magnesium, and copper, are used in this process. The benefits of injection molding include high production, high quality and strength, good surface smoothness, and near tolerances. Last but not the least, during centrifugal casting, inertial forces of rotation distribute molten metal into the mold cavities. In centrifugal casting, the rotating speed is chosen to provide 40 to 60 gravitational constants of acceleration. Die materials include cast iron or forged steel. Evaporative casting and zero gravity casting are two additional varieties of casting methods.[12]

7.3 SOLIDIFICATION

Three basic types of microstructures may develop during the solidification of an ingot or casting (see Figure 7.5). We will only discuss the final structures here. Chapter 2 covered the liquid–solid phase transition's thermodynamics. The cooling zone, which is the term for the ingot's top layer, is made up of a thin layer of equiaxed crystals that are randomly oriented. This is brought on by the melt's intense supercooling next to the mold wall. Remember that a high level of supercooling generates a driving force for a high rate of nucleation. A large number of nuclei grow on the mold's walls, resulting in a fine grain size.[13]

7.3.1 RAPID SOLIDIFICATION

We examined the connection between grain size, magnetic domain size, and magnetic characteristics in ferrous materials. We observed the impact of the cooling rate on metal grain size in the preceding section. The magnetic characteristics of ferrous alloys are closely related to the solidification processes, as one might anticipate, and this is indeed the case for high rates of cooling. Such amorphous materials are created via a group of techniques collectively referred to as fast cure processing, in which the quench rate, the technical term for solidification rate, is at its highest. The second zone, which

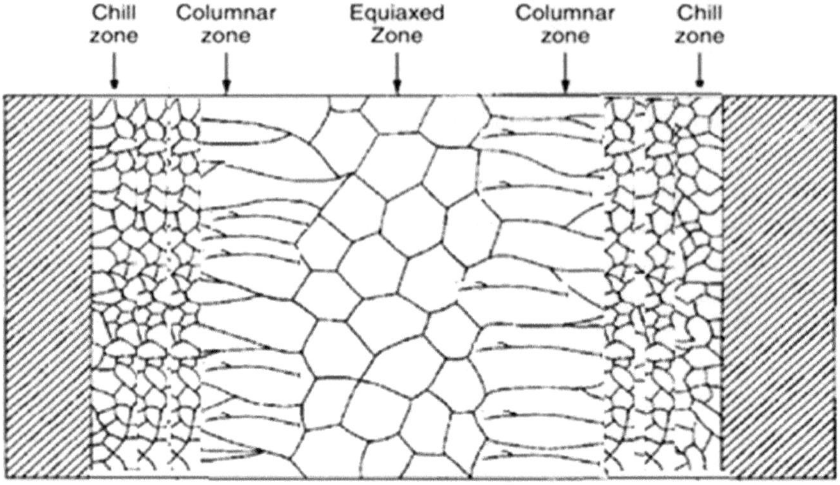

FIGURE 7.5 Crystallization zones in a solidified metal ingot are shown schematically.
Source: Adapted from Ref. [2]

makes up the majority of the ingot, is composed of elongated grains that are frequently dendritic in shape, long, and parallel to the ingot's direction of heat flow. The columnar zone is where this is located. The core region of the ingot also solidifies last, which results in a substantially slower rate of heat transfer and a larger equiaxed grain size. The alloy's composition and the rate of solidification (cooling) can both affect the relative amount of grains in each zone. The size of the ingot, the temperature of the melt, and the thermal characteristics of the mold material all affect how quickly the ingot cools. The redissolution of the cooled grains can cause the cooling zone to totally vanish if the melt temperature is too high. Columnar structures can advance toward the center of the ingot where grains advancing from the opposing wall meet in pure metals and alloys with very little supercooling. There is no equiaxed zone as a result. A fine-grained, equiaxed microstructure can be formed in the ingot when cooling rates are extremely high. This typically happens for smaller cold molds and ingots. Constitutional hypothermia, a process, is what creates the center-equiaxed zone. This promotes nucleation in the melt at the casting's center prior to the advance of the solid boundary. The addition of various nucleating agents can also encourage the expansion of the equiaxed zone. In general, equiaxed crystallization predominates in solid solution alloys, whereas columnar development predominates in pure metals. The zone of an equiaxed crystal expands as the number of dissolved atoms and the hard alloy's freezing interval increase, and in rare situations, columnar growth can be totally abolished. Amorphous metals and alloys are produced when nucleation and crystallization are totally abolished by rapid cooling.[14] That is the subject of the following section. Segregation and shrinkage are two typical curing-related processes. Segregation may result from changes in concentration that might happen during the solidification process in particular regions of the alloys. Macrosegregation and microsegregation are the two basic types of segregation that can happen. In macrosegregation, compositional changes result from gravitational, normal, or reverse segregation at relatively large sizes. Due to variations in the melt component densities, gravitational segregation occurs. Due to differing diffusion speeds in the solid and the liquid, the solute is normally eliminated at the approaching solid–liquid interface. The impurity-rich interdendritic liquid moves outward as a result of inverse segregation, increasing the average impurity level at the interface of the ingot's crest. Compositional differences over relatively small distances are referred to as microsegregation. During solidification, insoluble foreign particles may become trapped, causing segregation at grain borders. Metals vary in volume when cooled, just like polymers do (see Figure 7.6). The volume difference between liquid at casting temperature and solid metal at room temperature is equivalent to the shrinkage that results. The combined action of compression of a liquid on cooling from the casting temperature to freezing, compression on cooling from freezing to a solid state, and compression of a solid metal on cooling from a solid state to room temperature results in the overall volume change. The ingot's characteristics may be negatively impacted by cavities brought on by shrinkage between 105 and 106°C/s. Since the minimum cooling rates at the liquid–solid transition must be at least 1014°C/s, it is currently impossible to obtain amorphous pure metals; therefore, metallic glasses always contain at least two or more different types of atoms. For some alloy compositions, rapid solidification may not result in a completely amorphous (noncrystalline) material but the crystallite sizes of rapidly solidified crystalline. Typically, alloys have a thickness of 500 nm or less, as opposed to 50 m for traditionally machined alloys. In alloys like iron–silicon, the lack of grain boundaries enhances their magnetic characteristics, making them particularly desirable for soft magnet applications like transformer cores. Another typical alloying element that encourages the production of noncrystalline structures in ferrous alloys is boric acid. In Table 7.1, the compositions of a few typical quick-solidification amorphous ferroalloys are mentioned. In addition to the aforementioned iron–boron, platinum–phosphorus, and niobium–rhodium systems, other known glass-forming metal systems include transition metal or noble metal alloys having between 10 and 30 percent semimetals, as well as alloys comprising Group II metals like magnesium, zinc, rare earth, and alloys made of transitional metals like gadolinium and cobalt with yttrium iron.[15–17]

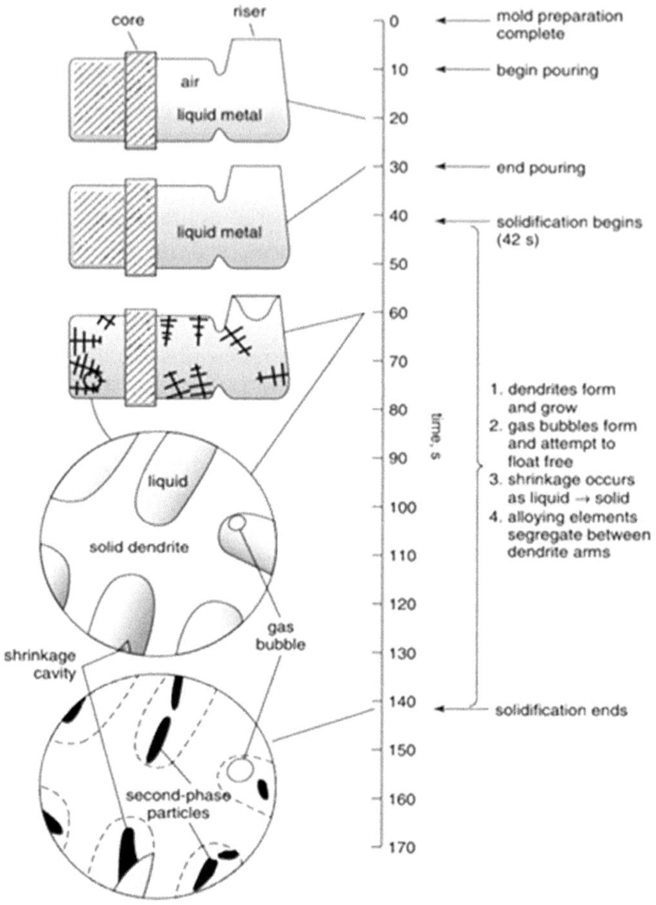

FIGURE 7.6 The shaped component's history of solidification.

Source: Adapted from Ref. [3]

TABLE 7.1 Some Amorphous Ferrous Alloys' Composition.[3]

Composition (wt % balance Fe)					
B	**Si**	**P**	**Mn**	**Ni**	**Cr**
20	–	–	–	–	–
10	10	–	–	–	–
28	–	–	6	–	6
6	–	14	–	40	–

The solute's ability to form glass and maintain stability can both be significantly increased. For example, tri-alloy glass with a cooling rate of 102°C/s or less, such as platinum, nickel, phosphorus, palladium, copper, and silicon, is shaped into cylindrical bars with a diameter of 2.5 mm. An amorphous alloy's X-ray is extremely similar to that of the same alloy when it is liquid. The coordination number in metallic glasses is near 12, as in a closely packed crystal structure, as shown by the X-ray diffraction pattern.[18]

7.4 EXTRUSION

A small piece of metal, referred to as a billet or billets, is forced through a die during extrusion. The first way the billet differs from an ingot is in its smaller size. In extrusion processing, the die is sometimes referred to as an orifice or nozzle. A slider presses against the workpiece, creating compressive and shear forces, which are used to push the metal through the hole. Failure is prevented because there are no tensile forces present. Direct extrusion is the process of forcing a workpiece through a hole with a punch (see Figure 7.7). The identical change in cross-sectional shape can alternatively be achieved by mechanically compressing the die around the workpiece in a process known as indirect extrusion. In any event, the ratio of the work necessary to deform to the actual work done is used to determine the extrusion process' efficiency. Extrusion is often used for nonferrous metals with lower melting temperatures (see Table 7.2); nevertheless, steel extrusion is now possible because of lubricants and advancements in presses.[19] The majority of carbon and stainless steels is extruded using glass as a lubricant at temperatures of about 1200°C.

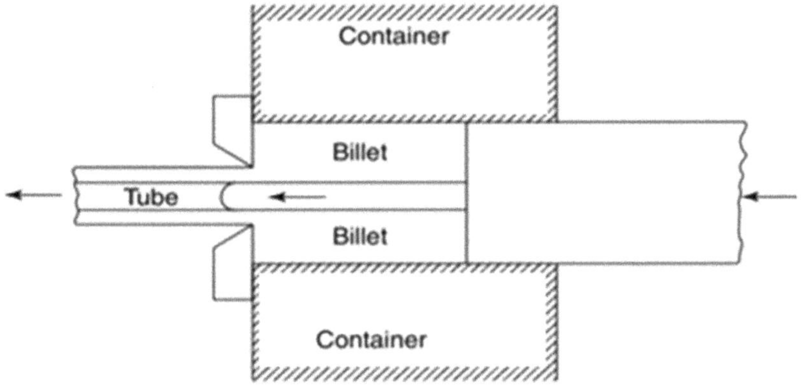

FIGURE 7.7 Diagrammatic representation of a metallic billet being directly extruded.

Source: Adapted from Ref. [4]

TABLE 7.2 A Few Common Nonferrous Metal Extrusion Processing Variables.[3]

Materials	Minimum section thickness (in)	Extrusion pressure (ksi)	Extrusion temperature (°F)	Extrusion ratio ln $(A_0/A_1)^a$	Extrusion exit speed (fpm)
Aluminum alloys	0.04–0.300	40–130	550–1050	4–7	4–300
Copper alloys	0.05–0.300	30–130	1200–1650	5–6.5	80–1000
Magnesium alloys	0.040	100–130	570–800	4–5.3	4–100
Zinc alloys	0.060	90–110	400–660	4–5	75–100

The cross-sectional areas of the billet and the extruded shape are represented by the letters A0 and A1, respectively. Metal Forming: Fundamentals and Applications by T. Altan, S.-I. Oh, and H. L. Gegel is the original source (By the American Society for Metals, 1983). It promotes the dissolution of oxide deposits on the metal surface, is easy to use, and is chemically stable. Extrusion is often used to create pipes and other continuous structures with complex cross-sections (Figure 7.7). In the context of polymer processing, we will discuss the extrusion procedure.[20]

7.5 PROCESSING OF CERAMICS AND GLASSES

Perhaps more than any other class of materials, ceramics and glass can be molded using a larger variety of processes. The product's intended application affects processing variations the most. While some techniques are employed for semiconductors, superconductors, and magnetic materials, others are used for structural ceramics. We will generally focus on three different forms of processing: chemical processing, melt processing, and powder molding. Glass and ceramic melt processing is sufficiently comparable to polymer melt processing to postpone the discussion. This chapter views the latter category, which includes the vapor phase and sol-gel processes, as optional. Here, roasting and sintering, two subjects relevant to powder processing, will be covered in a powdery substance. These processes resemble the powder metallurgy procedures that were discussed in the previous section in major part. For instance, pressing is a common technique for processing ceramics, but you can press ceramic powders either dry or wet. They can also be cast in a variety of ways while wet and extruded like metals. The nominal forming pressures and shear rates are connected to several of these processing techniques.[21] When each of the numerous processing techniques has been thoroughly explained, you can come back to this table.

7.5.1 PRESSING

Many ceramic powders can be sintered and pressed to create a densified product, just like metal powders can. Remember that pressing is the process of compacting and forming a powder inside a flexible mold or die. We will examine each phase of the pressing process, including powder generation (known as granulation for ceramics), die filling, and compaction, similar to how we do with metal processing. The final stage is firing because it applies to both pressing operations and casting operations.[13]

7.5.2 METAL PARTICLES

Many of the same processes that are used to create metal particles can also be used to create glass and ceramic powders. This is especially true for milling processes, for which both metals and ceramics are covered. For ceramic materials, however, wet grinding is far more typical, particularly when smaller particle sizes are desired. Slurry is the name given to the mixture of dry particles and dispersion, such as water. Slurries have a much higher solid concentration than solutions or slurries, often 50 weight percent or more solids. The amount of solids in the mixture can have an impact on how effectively it grinds depending on the type of grinding machinery utilized.[22]

To ascertain the tum-bling ball mills' breakage distribution and breakage rate functions, a total of 14 small size intervals, ranging from 0.106–0.075 mm to 4.75–4.00 mm, were covered by the simulated particle sizes. Figure 7.8 displays typical simulation results. A comparison shows that while power increased as ball diameter increased, the frequency of collisions between balls and between balls and geometric objects dropped. Figure 7.9 illustrates the good agreement between the measurements and the total power from DEM simulations. Deviations from the measured power draw were less than 12%, demonstrating the contact parameters' capacity to account for the absence of slurry in the simulations.[5,23]

7.5.3 CASTING

Similar to the casting of metals, the casting of ceramics involves pouring a liquid substance into a mold, followed by a transformation that results in the development of a solidified product and raises the yield strength. Ceramic casting typically employs a slurry as the fluid, as opposed to metal casting,

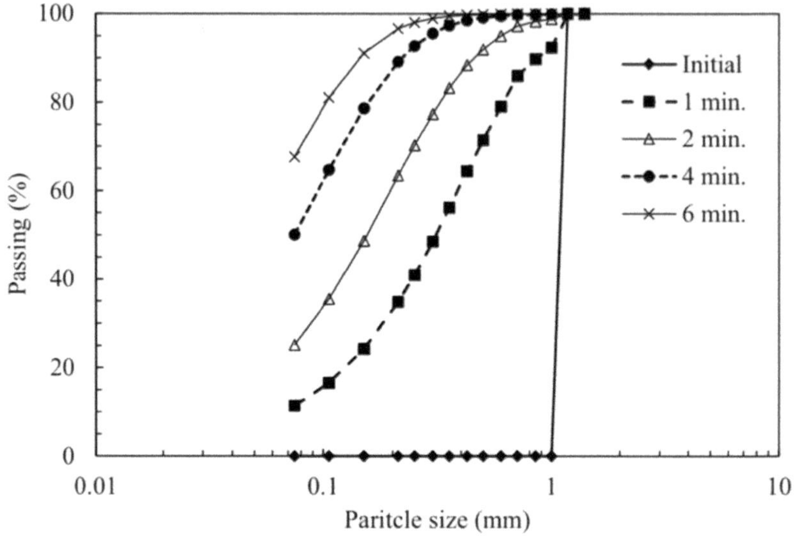

FIGURE 7.8 Simulated batch grinding tests for copper ore with a feed of 1.18 to 1.00 mm, 30 m/s stirrer speed, 50 weight percent solid slurry, and 20 mm balls.

Source: Reprinted with permission from Ref. [5]. Copyright 2021. Elsevier

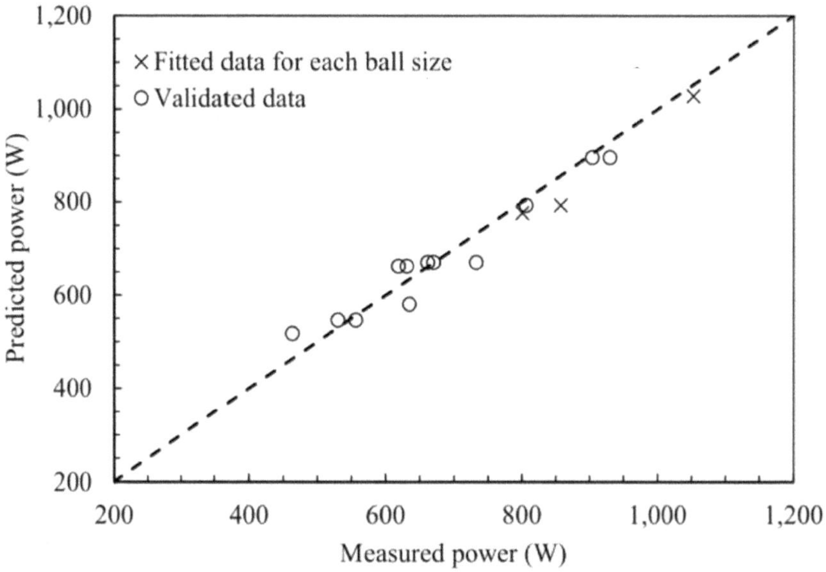

FIGURE 7.9 For all experimental data from Hasan et al. [43], including different mill speeds, different percentages of solids, and different ball sizes for the mill illustrated, a comparison between measured and anticipated power is made.

Source: Reprinted with permission from Ref. [5]. Copyright 2021. Elsevier

which uses molten metal. Furthermore, in contrast to metal casting, where the transformation required to obtain the desired yield strength in the casting is generally a physical transformation (solidification), the transformation in ceramic casting might be a physical, chemical, thermal, or a combination of all three. Let us go on to examine some of the variations between the most popular types of ceramic casting now that the fundamental casting techniques have been previously discussed.[24]

7.5.3.1 SLIP CASTING

Slip is the name for an aqueous solution of fine-grained clay. When slip casting occurs, a porous plaster mold is typically used. Sludge particles coagulate at the mold surface as a result of the capillary absorption of water from the sludge into the mold, which produces the casting. Utilizing pressure (injection casting), vacuum (vacuum-aided casting), or centrifugal force (centrifugal casting) are all variations of the slip casting process that speed up casting. There are two primary varieties of slip casting: drain casting and solid casting, as depicted in Figure 7.10. The molds are initially filled with mortar before being filled with drainage. The excess suspension is drained from the molds after some initial hardening. This process is appropriate for creating hollow objects such as pipes or crucibles as the core. In continuous casting, a dense finished mold is created by utilizing the full volume of suspension in the mold. Preparing the slurry, filling the mold, draining it, and partially drying it inside the mold are all crucial processes in the slip casting procedure that will be covered in detail below. The separation of the slurry from the mold after drying and the postmold finishing and surface finishing procedures before the final firing or sintering step are additional steps that are not covered in this article.[25]

The ingredients used to make suspensions are a liquid (often water), powders or granules that have been pulverized, and additives. Two typical ceramic slurries' compositions are displayed.[25] Typically, casting slurries are made to be shear-thinned, with an apparent viscosity of less than 2000 cP (2000 mPa.s).

7.5.3.2 TAPE CASTING

A strip casting solution, similar to slip casting, is made up of deflocculated powders or granules, binders and/or plasticizers, and a solvent like water.

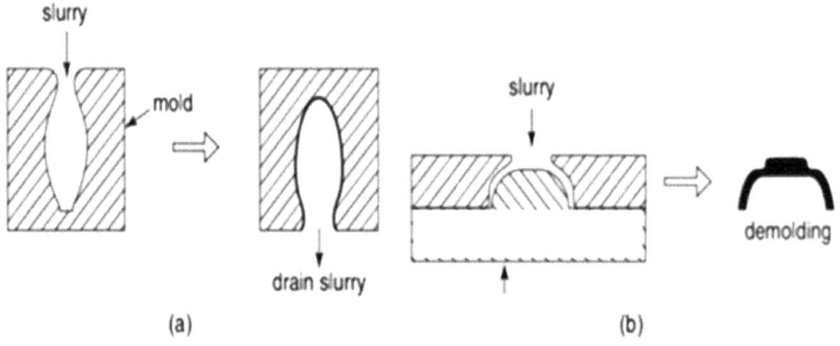

FIGURE 7.10 Schematic depicting the drainage and continuous slip casting processes.
Source: Adapted from Ref. [6]

Slurry, in contrast to slip casting, can be used to create "green sheets," which are thin, flexible sheets that can be cut and dried. The term "green" has nothing to do with color; rather, it describes the fact that the product is prefired or presintered. Due to the high amount of binder in the dried tape, it is elastic and has a very smooth surface. The main applications of tape-cast ceramics are as substrates for electrical sensors, solar cells, resistors, capacitors, and solid battery electrolytes. On the tape's surface, electronic materials can be printed, and after being fired, they form a permanent part of the device, a piece of the ceramic element. We will quickly go over the slurry formulation, casting, and drying concerns related to this crucial processing technique, much like with slip casting. The maximum particle size and specific surface area of ceramic powders used for tape casting are typically 1 to 5 m and 2 to 5 m^2/g, respectively. Belt surfaces are smoother thanks to finer particles. Chemical homogeneity is essential for the production of electronic components, which frequently employ tapes. During the drying and fire operations, particle size and dispersion also have an impact on dimensional stability. Since the additives must be dissolved, the liquid–solvent system must also ensure their adsorption on the ceramic particles. As a result, several solvents are frequently used. Low viscosity, low boiling point, low heat of vaporization, and high vapor pressure of organic solvents all help to shorten drying periods. Table 7.3 lists a few popular nonaqueous suspension formulations. Longer drying durations and higher drying temperatures are needed when employing an aqueous system. In processing techniques including compression and extrusion the common molecules with greater flexibility are acrylic and vinyl (Table 7.3). The orientation of these macromolecules during casting gives the casting its toughness and strength.[28]

TABLE 7.3 Composition of Some Nonaqueous Tape Casting Solutions.[7]

Function	Composition	Vol %	Composition	Vol %
Powder	Titanate	2.80	Alumina	27
Solvent	Methylethyl ketone	3.30	Trichloroethylene	42
	Ethyl alcohol	16	Ethyl alcohol	16
Deflocculant	Menhaden oil	1.70	Menhaden oil	1.8
Binder	Acrylic emulsion	6.7	Polyvinyl butyral	4.4
Plasticizer	Polyethylene glycol	6.7	Polyethylene glycol	4.8
	Butylbenzylphthalate	6.7	Octyl phthalate	4.0
Wetting agent	Cyslohexanone	1.20		

In order to provide the necessary mechanical qualities while still enabling a high concentration of ceramic particles in the slurry, the binder and plasticizer combination must be carefully managed.[29]

TABLE 7.4 Commonly Used Binders and Plasticizers in Tape Casting Solutions.[7]

	Plasticizer	Binder
Nonaqueous	Dioctyl phthalate	Polyvinyl butyral
	Dibutyl phthalate	Polymethylmethacrylate
	Benzyl butyl phthalate	Polyvinyl alcohol
	polyethylene glycol	Polyethylene
Aqueous	Glycerine	Acrylics
	Polyethylene glycol	Methyl cellulose
	Dibutyl phthalate	Polyvinyl alcohol

An excessive amount of binder might decrease the PF of the ribbon's particles while increasing particle separation. A wetting agent to aid in spreading the slurry across the substrate, a homogenizer to enhance surface finish, and defoamers are further additives. The tape casting slurry for slip casting needs to be shear-thinned and have a viscosity of at least 4000 cps at the tape shear rate (Table 7.4).

As demonstrated in Figure 7.12, a high solution viscosity and a carrier velocity greater than or equal to 0.5 cm/s are preferred for thickness uniformity as the solvent evaporates in the drying tube. The solvent concentration on the surface, as well as the temperature and solvent content of the drying medium, which is typically air, all directly affect the drying rate. The solvent is transported to the drying surface by capillary forces. Due to the presence of moisture in the incoming air, the initial evaporation causes distortion. Hard ferrite magnets with a strongly oriented microstructure are produced via a type of injection molding. A matrix with permeable pistons is drilled in this procedure to allow fluid passage. An evenly distributed slurry of ferrite particles is used to fill the mold set. The pistons are forced into the mold after a magnetic field is used to align the particles, creating a dense casting of aligned particles.[28]

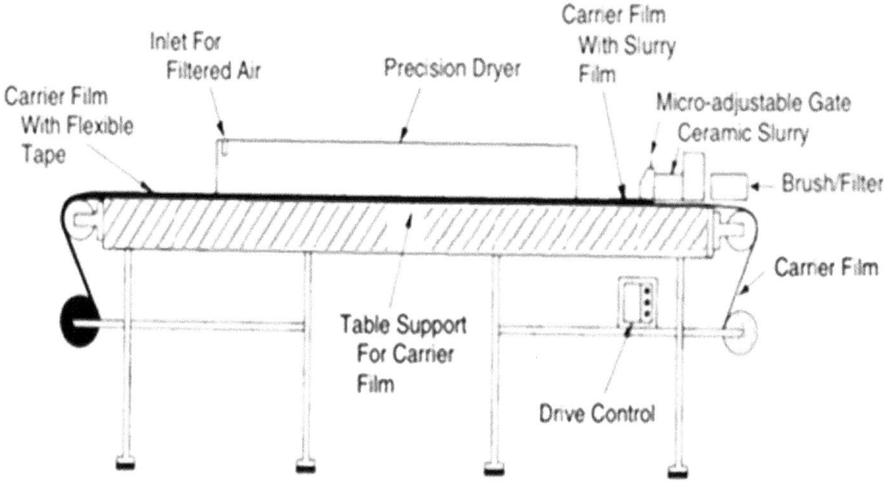

FIGURE 7.11 An example of the typical tape-casting procedure.
Source: Adapted from Ref. [6]

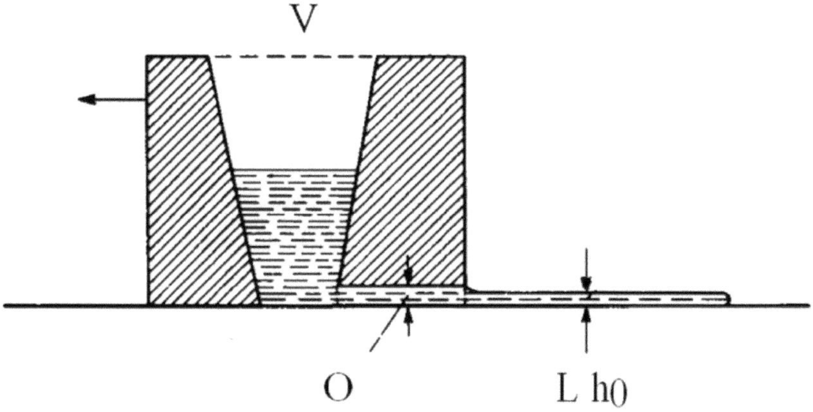

FIGURE 7.12 The tape casting flow model (7.22).
Source: Adapted from Ref. [7]

Highly porous refractory insulating materials with complicated forms are frequently created via vacuum casting. Chopped refractory fiber and partially deflocculated ceramic powder are frequently seen in slurry. The fiber raises the casting's porosity, increases viscosity, and necessitates more fluid. A vacuum line is attached to the disposable permeable preform and then submerged in the remedy. When the wall has been cast to the desired thickness, the workpiece is removed and taken somewhere else to dry. Vacuum casting is another term for the technique of introducing a vacuum to the outside of a plaster mold during slip casting in order to speed up casting. Modern ceramics with complex shapes can be molded by centrifugal casting when a very high casting density is needed. With centrifugation, the liquid is ejected to the top of the centrifuge chamber, and the driving power for particle settling is increased. The Stokes law describes the settling rate for a reasonably diluted dispersion, and it states that larger or denser particles settle more quickly and produce a structural gradient. It was cast. Speed rises as both the liquid flow rate from the deposition surface and the centrifugal force velocity increase. This method makes it possible to create castings with intricate shapes, homogeneous densities, and deposition rates of 0.6 mm/min. Beyond the scope of this text are further casting process possibilities and extra postprocessing stages like shape, trimming, laminating, and printing. For more detail, the curious reader can consult Reed's textbook on ceramic processing. Beyond the scope of this text are further casting process possibilities and extra postprocessing stages like shape, trimming, laminating, and printing. For more detail, the curious reader can consult Reed's textbook on ceramic processing.[29]

7.5.3.3 FIRING

The green ceramic component is typically fired after it has been cast using one of the techniques mentioned in the previous section, dried, and finished. To get the necessary microstructure and characteristics, raw ceramic must be heated in a high-temperature furnace called a kiln during the firing process. Depending on the ceramic being burned, kiln designs vary widely, but they are typically built to run either continuously or occasionally. The heating components such as silicon carbide (up to 1150 °C), molybdenum disilicide (up to 1700 °C), refractory metals like molybdenum or tungsten (over 1700 °C), graphite (above 1700 °C in inert gases), plasma heating sources, microwave heating, or induction can all be used to produce heat. It is combustion heating. Also, frequently used lysed, a process in which methane, typically in the form of natural gas, is heated to more than 1700°C by air combustion. A number of processes take place during firing, including sintering, binder burnout, and additional drying, beyond that accomplished during casting. We shall discuss each of these subjects in brief.[30]

7.5.3.3.1 Presintering Processes

Before sintering, material changes during firing may include drying, the breakdown of organic additives, and gradual evaporation. Calcination, resulting from the decomposition of carbonates and sulfates, is introduced as additives or raw materials. In addition to the pyrolysis of particulate organic matter and the decomposition of water from the surface of the particles, the oxidation state of some transition and rare metals changes. Thrombolysis, a crucial process before densification during sintering, involves the breakdown of organic additives like binders. Uncontrolled thermolysis and incomplete binder removal might lower yield or worsen product performance. The green ceramic should ideally tolerate thermolysis without expanding pores, warping, or cracking. The composition of the binder, as well as the composition, affects how long the binder lasts and furnace atmosphere flow rate. It also heavily depends on the dynamic changes in the microstructure that occur as the binder is removed, as well as the microstructure of the organic, powder, and porosity phases. Table 7.5 displays a few of the chemical and physical modifications that take place during thermolysis. The majority of cast ceramic components feature pore channels that are large enough to ensure the transportation of vapors and gases from the binder's burnout zone to the product's surface. Instead of the typical phase size of the binder, the diffusion length for vapor phase transfer determines the thermolysis time. Depending on the rate of outgassing, the degradation and evaporation of organic materials will result in a compact internal gas pressure, gas permeability, and size. The loading of the binder determines the permeability, and the volume fraction and pore radius determine the pore structure. Therefore, in dense compacts of extremely small particles, the rate of gas penetration is significantly reduced. Take the burnout of polyvinyl alcohol caused by the compaction of barium titanate in the air as an illustration of thermolysis.[31]

Heating vinyl binders with other typical pendant groups and cellulosic binders gives similar results, although the range of burnout may be smaller. Wax and polyethylene glycol additives, for example, melt at relatively low temperatures and evaporate over a more limited temperature range. The temperature rises faster as the binder oxidizes. The initial oxidation of the binder may include absorbed oxygen. The graph of thermogravimetric analysis (TGA) of methylcellulose in Figure 7.13, which traces the weight loss of a ceramic sample as a function of temperature, shows that evaporation predominates, while thermolysis is generally endothermic and shifts toward higher temperatures. The starting temperature in nitrogen for PEG binders is approximately 300°C. Without oxygen, some residual carbon can be expected. According to Table 7.6, the burnout residue of synthetic polymerized glycols and acrylic binders is very modest. Carbon is again a frequent residue after firing nitride and carbide ceramics in an inert atmosphere; this residue can help with sintering. To make sure the mill was running in steady-state circumstances, DEM simulations were run for 2 seconds. The information needed to apply the mechanistic mill model was generated at this stage by tracking collision events for 10 seconds. At one of the lowest rotational frequencies utilized in the simulations, this is comparable to at least 32 stirrer revolutions. Additionally, collision events that registered total energy losses below the 10–8 J threshold were disregarded because they are considered to be just numerical leftovers from the model, which further reduced the volume of data that needed to be stored.[5,32]

TABLE 7.5 Many Chemical and Physical Changes That Take Place during Thermolysis.[7]

Chemical reactions	Decomposition
	Depolymerization
	Carburization
	Oxidation
Physical changes	Melting/softening (solid–liquid)
	Sublimation (solid–gas)
	Evaporation (liquid–gas)
Reaction zone	External
	Internal Uniform displacement Nonuniform displacement
Transport	Gas diffusion
	Liquid flow

Carbon is again a frequent residue after firing nitride and carbide ceramics in an inert atmosphere; this residue can help with sintering. The rate of heating must be watched since gas pressure can lead to volume expansion, which is entirely in control of bodies with a lot of clay. The dehydration of aluminum hydrates in the region of 320–560°C is another dehydration event associated with gas production. The polymer must be formable, just like metal, in order to be shaped. The physical characteristics of polymers, including viscosity, crystallinity, and hardness, are included in the definition of formability, despite the fact that this is just as difficult to define for polymers as it is for metals.[5,31]

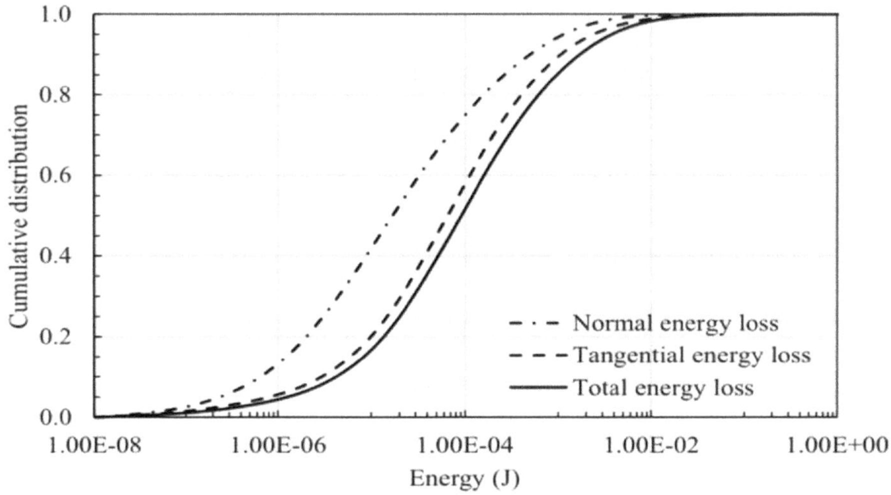

FIGURE 7.13 Energy spectra of collisions in a vertically stirred medium rotating at 2.5 m/s. A 20 mm grinding medium charge of 86.8 kg is placed within the machinery. The collision energies of normal, tangential, and total are distinguished.

Source: Reprinted with permission from Ref. [5]. Copyright 2021. Elsevier.

Most pure polymers are difficult to process due to their high molecular weight or cohesive strength. Thermal conductivity as well as temperature and pressure are influencing factors during processing. Due in part to their high molecular weight or cohesive strength, the majority of pure polymers are challenging to process. As a result, low-molecular-weight additives are frequently added to polymers to enhance their formability (or for other purposes, as we shall see) in order to create plastics.[33]

We have not really distinguished between polymers and plastics up until now, and we will not change that. However, be aware of the connection between the two and understand that it is possible that the term "plastic" is more relevant when the adjective "polymer" is used to describe a particular type of processing. However, whether a process is pure or not, the fundamental ideas are the same. For our purposes, it is not necessary to distinguish between polymer and plastic formation (Figure 7.14).

TABLE 7.6 Metal Alkoxides Utilized in Ceramic Sol-Gel Processes.[6]

		Aloxide
Single-cation alkoxides		
I A (1) group	Li, Na	$LiOCH_3$ (s), $NaOCH_3$ (s)
I B (11) grp.	Cu	$Cu(OCH_3)_2$ (s)
II A (2) grp.	Ca, Sr, Ba	$Ca(OCH_3)_2$ (s), $Sr(OC_2H_5)_2$, $Ba(OC_2H_5)_2$ (s)
II B (12) grp.	Zn	$Zn(OC_2H_5)_2$ (s)
III A (3) grp.	B, Al, Ga	$B(OCH_3)_3$(l) $Al(i-OC_3H_7)_3$ (s) $Ga(OC_2H_5)_3$ (s)
III B (13) grp.	Y	$Y(OC_4H_9)_3$
IV A (4) grp.	Si, Ge	$Si(OC_2H_5)_4$ (l) $Ge(OC_2H_5)_4$ (l)
IV B (14) grp.	Pb	$Pb(OC_4H_9)_4$ (s)
V A (5) grp.	P, Sb	$P(OCH_3)_3$ (l) $Sb(OC_2H_5)_3$ (l)
V B (15) grp.	V, Ta	$VO(OC_2H_5)_3$ (l) $Ta(OC_3H_7)_5$ (l)
VI B (16) grp.	W	$W(OC_2H_5)_6$ (s)
Lanthanide	La, Nd	$La(OC_3H_7)_3$ (s) $Nd(OC_2H_5)_3$ (s)
Alkoxides with various alkoxyl groups		
	Si	Si $(OCH_3)_4$ (l) $Si(OC_2H_5)_4$ (l) $Si(i-OC_3H_7)_4$ (l) $Si(t-OC_4H_9)_4$
	Ti	Ti $(OCH_3)_4$ (s) $Ti(OC_2H_5)_4$ (l) $Ti(i-OC_3H_7)_4$ (l) $Ti(OC_4H_9)_4$ (l)
	Zr	$Zr(OCH_3)_4$ (s) $Zr(OC_2H_5)_4$ (s) $Zr(OC_3H_7)_4$ (s) $Zr(OC_4H_9)_4$ (s)
	Al	$Al(OCH_3)_3$ (s) $Al(OC_2H_5)_3$ (s) $Al(i-OC_3H_7)_3$ (s) $Al(OC_4H_9)_3$ (s)
Double-cation alkoxides		
	La–Al	$La[Al(i-OC_3H_7)_4]_3$
	Mg–Al	$Mg[Al(i-OC_3H_7)4]_2$, $Mg[Al(s-OC_4H_9)_4]_2$
	Ni–Al	$Ni[Al(i-OC_3H_7)_4]_2$
	Zr–Al	$(C_3H_7O)_2Zr[Al(OC_3H_7)_4]_2$
	Ba–Zr	$Ba[Zr_2(OC_2H_5)_9]_2$

Premolding, formation, and postshaping are the three essential processes in any polymer molding procedure. Melting, mixing, and transferring the material to the next stage are all parts of the process. The characteristics at this point may be impacted by the thermomechanical history. The effects of time and temperature are significant. The deformation of the mass after premolding gives the product with the desired shape during the molding procedure. Anisotropic features can be obtained at this point. Finally, while curing is the primary function of postforming, ancillary activities such as printing, painting, gluing, and electroplating can also be conducted here. Heat transfer is a significant contribution to solidification, and dimensional tolerances are established at this point. We will concentrate on the shaping process because it is where the majority of the engineering principles are applied, but we will also discuss the features of the other two steps as necessary. Let us begin with some of the most important shaping techniques. Continuous processing and batch processing are the two basic polymer processing processes.[31]

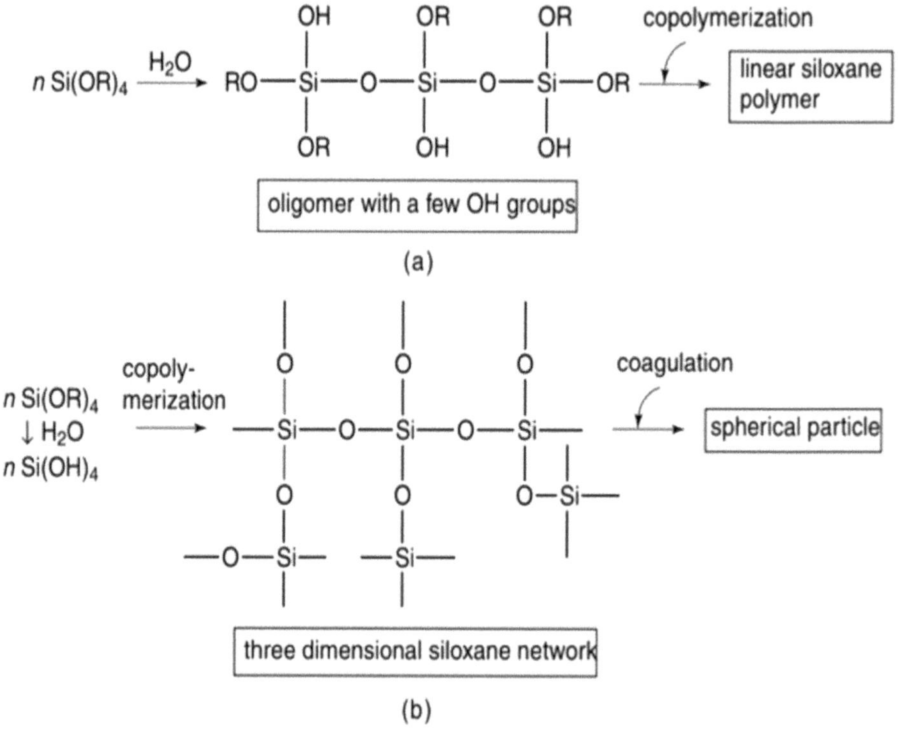

FIGURE 7.14 Both the acidic (a) and basic (b) techniques were used to hydrolyze and polymerize the whole alkoxide $Si(OR)_4$.

Source: Adapted from Ref. [6]

Rubber stamping and molding and mold casting are cyclical operations, but calendering is a continuous process. We will go over the procedures in each of these categories, as well as secondary molding techniques like thermoforming. Finally, some of the more specialized methods of polymer processing will be discussed.

7.5.4 CONTINUOUS PROCESSING

Stamping (extrusion) and calendering are the two basic forms of continuous polymer processing processes. These methods have the same advantages as any continuous process: high productivity and product homogeneity. The product geometry that can be generated utilizing these technologies, however, is limited. As with metals, the goods that can be produced constantly are largely sheets and rods. In addition to extrusion and calendering, we will discuss fiber spinning, a sort of extrusion technique used to create synthetic fibers for a variety of key applications. We shall detail the engineering principles that can affect the structure and performance of the polymer in these continuous activities, where suitable.[32]

7.6 CONCLUSIONS

- Various types of mineral processing techniques have been identified and described.
- The best grinding procedure has been determined to create optimally sized metal and ceramic powders.
- Familiarity with a suitable grinding technique for the preparation of metal and ceramics.
- Powders in the required size.
- Determination of ceramic processing methods.
- Multiple and continuous polymer processing methods have been described for their microstructure and mechanical properties.

KEYWORDS

- **metals**
- **ceramics**
- **polymers**
- **processing methods**
- **engineered materials**

REFERENCES

1. Howard, H.; Gall, T. *Metals Handbook American Society for Metals*; Metals Park, OH, 1985.
2. Schanda, J. (Ed.). *Colorimetry: Understanding the CIE System*; John Wiley & Sons, 2007.
3. McGraw-Hill. *Encyclopedia of Science and Technology*, Vol. 11, 8th edn.; McGraw-Hill, 1997; p 35.
4. Jastrzebski, Z. D. Nature and Properties of Engineering Materials. 1976.
5. de Oliveira, A. L.; de Carvalho, R. M.; Tavares, L. M. Predicting the Effect of Operating and Design Variables in Grinding in a Vertical Stirred Mill Using a Mechanistic Mill Model. *Powder Technol.* **2021,** *387*, 560–574.
6. Yanagida, H.; Koumoto, K.; Miyayama, M. *The Chemistry of Ceramics*; John Wiley & Son Limited, 1996.
7. Reed, J. S. *Principles of Ceramics Processing*, 2nd edn.; Wiley-Interscience: New York, 1995.
8. Wolf, S.; Tauber, R. N. *Silicon Processing*, Vol. 1, Lattice Press: Sunset Beach, 1986.
9. Tadmor, Z.; Gogos, C. G. *Principles of Polymer Processing*; John Wiley & Sons: New York, 1979.
10. Middleman, S. *Fundamentals of Polymer Processing*; McGraw-Hill: New York, 1977.
11. Powell, P. C.; Ingen Housz, A. J. *Engineering with Polymers*; S. Thornes, Cheltenham, 1998.
12. Crawford, R. J. *Plastics Engineering, Butterworth-Heinemann*; Boston, 1997; *Modern Plastics Encyclopedia*; McGraw-Hill: New York, 1977.
13. Mallick, P. K.; Newman, S. (Ed.). Composite Materials Technology: Processes and Properties; Hanser: Munich, 1990.
14. Lubin, G. *Handbook of Fiberglass and Advanced Plastics Composites*; Van Nostrand Reinhold: New York, 1969.
15. Powell, P. C.; Ingen Housz, A. J. *Engineering with Polymers*; Stanley Thornes: Cheltenham, 1998.
16. Suchanek, W.; Yoshimura, M. Processing and Properties of Hydroxyapatite-Based Bio-Materials for Use as Hard Tissue Replacement Implants. *J. Mater. Res.* **1998,** *13* (1), 94.
17. Knott, J.; Withey, P. *Fracture Mechanics: Worked Examples*; Institute of Materials: London, 1993.
18. *Metals Handbook, Desk Edition*; American Society for Metals, 1985.
19. Wachtman, J. B. *Mechanical Properties of Ceramics*; John & Sons, Wiley: New York, 1996.
20. McCrum, N. G.; Buckley, C. P.; Bucknall, C. B. *Principles of Polymer Engineering*; Oxford Science: Oxford, 1997.
21. Kelly, A. *Composite Materials*; American Elsevier: New York, 1966.
22. Dyson, B. F.; Lohr, R. D.; Morrell, R. (Eds.). *Mechanical Testing of Engineering Ceramics at High Temperatures*; Elsevier Applied Science: London, 1989.
23. Lee, W. E.; Rainforth, W. M. *Ceramic Microstructures: Property Control by Processing*; Chapman & Hall: London, 1994.
24. Vigo, T. L.; Kinzig, B. J. (Eds.). *Composite Applications: The Role of Matrix, Fiber and Interface*; VCH Publishing: New York, 1992.

25. *The Mechanical Properties of Biological Materials, Symposium, Society of Experimental Biologists*; Cambridge University Press: New York, 1980.
26. Biomaterials Properties Database, University of Michigan, Dr. William J. O'Brien, Director. http://www.lib.umich.edu/dentlib/Dental tables/toc.htm
27. Evans, F. E. *Mechanical Properties of Bone*; C. Charles: Springfield, 1973.
28. Muccio, E. A. *Plastics Processing Technology*; ASM International, Materials Park: OH, 1994.
29. Powell, P.C.; Housz, A. J. *Engineering with Polymers*, 2nd ed.; Nelson Thornes: Cheltenham, 1998.
30. Rosen, S. L. *Fundamental Principles of Polymeric Materials*, 2nd edn.; Wiley: New York, 1993.
31. Rudin, A. *The Elements of Polymer Science and Engineering*, 2nd edn.; Academic Press: San Diego, 1998.

PART III

Physical Properties of the Materials

CHAPTER 8

Morphological Analysis and Optical Studies of ZnO Nanostructures

Y. AL-DOURI[1,2,3], RIYADH A. AL-SAMARAI[4], and VIPUL SRIVASTAVA[5]

[1]*College of Health and Medical Techniques, Al-Bayan University, Baghdad, Iraq*

[2]*Department of Mechanical Engineering, Faculty of Engineering, Piri Reis University, EflatunSk, No: 8, 34940 Tuzla, Istanbul, Turkey*

[3]*Nanotechnology and Catalysis Research Centre, University of Malaya, 50603 Kuala Lumpur, Malaysia*

[4]*Electromechanical Engineering Department, Engineering College, University of Samarra, Samarra, Iraq*

[5]*School of Chemical Engineering & Physical Sciences, Lovely Professional University Phagwara, Punjab, India*

ABSTRACT

ZnO nanostructure is grown on a thin aluminum (Al) layer and deposited on a silicon (Si) substrate using the sol–gel technique. The morphology of zinc oxide nanostructures at different precursor concentrations is investigated utilizing scanning electron microscopy (SEM). Raman spectroscopy suggested that nanorods have resumed to grow up including nanoflakes at a specific concentration of 50 mM, while the nanorod's density increases as the concentration increases to 75 mM. Raman spectra are intensified and red-shifted as concentration increases. The optical studies of the optical dielectric constant and refractive index are researched. However, the structural properties at lower levels of the precursor are attributed to the

Advanced Materials for Next-Generation Technologies: Challenges and New Prospects. Shrikaant Kulkarni, Vipul Srivastava, Rabah Khenata, & Yarub Al-Douri (Eds.)

hypoxic environment, while the Raman spectra's red-shift is attributed to the structural characteristics of ZnO nanocrystals.

8.1 INTRODUCTION

The fundamental properties of II–VI semiconductors have been understood through experimental and theoretical efforts. The II–VI-based materials are embodied in sensors, laser diodes, and light-emitting diodes.[1–3] ZnO nanostructures have wide applications due to their unique physical properties in light-emitting diodes (LEDs),[4] sensors, and storage.[6–9] One-dimensional (1D) morphology is exhibited as nanoneedles, nanobelts, nanotubes, and nanorods.[10–13] While nanoporous nanowalls, nanoplates, and nanosheets are available in two-dimensional (2D),[14–17] they are applicable in gas sensors, nanofilters, and catalysts.[18]

ZnO nanostructures have been fabricated using different processes such as thermal evaporation based on vapor–liquid–solid (VLS) using metal catalysts or ZnO thin film catalysts,[19] the metal organic chemical vapor deposition (MOCVD) technique without catalysts,[20] the plasma-assisted molecular beam epitaxy (P-MBE) technique,[21] and the hydrothermal method.[22]

In the current work, the spin-coating technique has been used to fabricate ZnO nanoflakes in terms of different precursor concentrations (12.5, 75 mM). The morphology of ZnO nanoflakes is characterized via scanning electron microscopy (SEM). The optical properties are investigated by Raman spectroscopy, while the structural properties are analyzed by X-ray diffraction (XRD). On the other hand, higher precursor concentrations affect the formation quality of nanorods and the crystallization of ZnO. The intrinsic defects are inflicted by the fabricated nanostructures, possibly by forming a hypoxic atmosphere at a higher level of precursor.

Also, the ability of energy conversion for ZnO is still limited, so this is a hybrid application, which means there is an exciting scope for further exploration. From the previous works, nano- to miniaturized-scale devices/gadgets have shown advantages in their designs due to the need of one active area; therefore, their turn is simple for the fabrication process. The ZnO nanostructures prepared by the electrospinning technique are appended to transparent glass that has conductivity by screen-printing.[23]

The nanoparticles have attracted more interest because of their attractive features applicable to multifield applications in catalysts, biomedicine, field photoelectron devices, cosmetics, gas sensors, and solar energy conversion. These properties of nanomaterials are wide in thermal, structural, and electronic devices. Zinc oxide has a big energy gap of 3.37 eV (observed binding energy) and 60 meV (high mobility of carriers at 25°C) within the hexagonal structure of the lattice constants a = 3.249 Å and c = 5.205 Å. There are many techniques to synthesize zinc oxide nanostructures, such as spray pyrolysis, zinc oxidation, reactive sputtering, electrodeposition, pulsed laser deposition, radio frequency (RF) magnetron sputtering, chemical vapor deposition (CVD), chemical bath deposition (CBD), metal organic CVD (MOCVD), and plasma-enhanced CVD (PECVD).[24]

Various efforts have been put to growing ZnO thin films utilizing different techniques like thermal evaporation, chemical vapor deposition, and pulsed laser deposition. However, these are very expensive techniques, which add to the high cost of the products made from such materials. The search for low-cost-effective techniques has led to chemical-related techniques, namely, electrochemical deposition (ECD). The advantages of ECD in comparison with other methods are morphology and thickness, which could be controlled by growth parameters; the rate of deposition is high; experimental setup is cost-effective; the process temperature is low; and the ease of impurity doping. The ZnO films have been reported by several researchers. Yang et al. have studied the concentration effect of zinc nitrate on the optical and structural properties of zinc oxide. The surface condition of the substrate effect on ZnO formation is investigated by Yiwen Tang et al. They used three-electrode electrochemical cells.[25]

8.2 ZnO

ZnO is categorized as an inorganic material; it has a white color, could be dissolved in water, and has wide applications in producing first-aid tapes, fire retardants, ferrites, batteries, foods, pigments, sealants, adhesives, ointments, paints, lubricants, cement, glass, ceramics, plastics, and rubbers. The synthesis of ZnO is industrial, but it happens naturally like the mineral zincite.[11]

ZnO is attributable to the wide energy gap of II–VI semiconductors. The n-type refers to native doping because of zinc interstitials or vacancies of oxygen.[13] This is attractive for applications due to its strong room-temperature luminescence, wide band gap, high electron mobility, and good transparency. The emerging applications are light-emitting diodes, electronics such as thin-film transistors, heat-protecting windows, or energy-saving and transparent electrodes in liquid crystal displays. ZnO has a big energy gap of 3.37 eV and an observed binding energy of 60 meV, in addition to excellent thermal and chemical stability, which leads to excellent applications in gas sensors, piezoelectrics, catalysis, and ultraviolet devices. Different techniques are utilized for fabricating ZnO nanostructures like chemical bath deposition (CBD), r.f. magnetron sputtering, thermal oxidation, and electrochemical deposition techniques. While pulsed laser deposition (PLD) has offered advantages such as the controlled composition of the deposited structure and lessened contaminations,[26]zinc oxide has attracted interest due to its electronic properties, that is, its semiconducting band gap of about 3.4 eV, and the possibilities of application in optoelectronics, either directly or as a substrate for the growth of other semiconductors such as GaN and SiC. Moreover, because of its superior electronic, optical, and piezoelectronic properties, it is an attractive candidate for applications in ultraviolet light emitters, transparent field-effect transistors, ultraviolet nanolasers, photodetectors, surface acoustic wave devices, sensors, and piezoelctronic devices. Low-resistive zinc oxides have been achieved by doping with different group-III elements like Al, B, In, and Ga or with group-VII elements like fluorine. The essential phase of crystallization for ZnO is the hexagonal wurtzite phase. Theoretical and experimental studies have been focused on ZnO, with an interest in studying the properties of ZnO in its cubic phase. In the last few years, there have been tremendous efforts to understand the optical features of zinc oxide in specific fabrications such as UV lasers.[27]

Many techniques are available for preparing ZnO such as synthesis using microwaves, the sonochemical method, the pyrolysis spray method, and growing from a gas phase. Grasza et al.[28] have synthesized ZnO in pure crystal form. The ZnO crystals are prepared from the gas phase (arsenic, gaseous zinc, atmospheric oxygen, nitrogen, and air). Different external effects of temperature and time are utilized with an emphasis on characterizing surfaces within their interactions with gaseous zinc, oxygen, and air. The morphology's homogeneity and crystal surface purity are characterized via XRD and AFM. It is confirmed that heating has led to the same modifications in the surface of the crystal in spite of observed differences in the change's rate. They[28] have displayed that heating leads to a roughness of the surface for < 1 nm, but it causes degradation in the surface of the crystal. The study has shown that the porosity of the surface of a crystal increases as temperature and time of heating increase. It is shown that there is a milky surface of crystal as a result of imperfections that arise within surface preparation. Wei et al.[29] have obtained ZnO in a thin layer in an O_2 atmosphere under 1.3 Pa pressure via PLD. They have determined the temperature effect on optical and structural properties utilizing PL, FTIR, SEM, and XRD. They have indicated that better optical and structural properties are obtained for ZnO at 700°C utilizing ZnO powder and at 400°C utilizing ceramic ZnO. The PL has indicated that the emission of UV increases as temperature increases. It is proven that the UV emitted by ZnO made from powder is smaller than that of ceramic ZnO.

Zhao et al.[30] have obtained ultrapure ZnO via the aerosol pyrolysis technique. They have utilized $Zn(CH_3COO)_2 \cdot 2H_2O$ as a zinc precursor for its low temperature and high solubility decomposition. It has been made up of the kinetics and mechanism of zinc acetate dihydrate thermal decomposition and a correlation of the mechanism with the results of the aerosol pyrolysis process. The analysis of TGA and DTA shows that a crystallization of thermal decomposition is lost under 200°C and then anhydrous zinc acetate is formed. At 210–250°C, the anhydrous zinc acetate is decomposed into organic and ZnO compounds via reactions of exothermic and endothermic nature. The decomposition process of $Zn(CH_3COO)_2 \cdot 2H_2O$ is achieved at 400°C. The synthesis of ZnO via the aerosol pyrolysis technique consists of nanoparticles in the range of 20–30 nm. While Hu et al.[31] have grown ZnO rods utilizing microwave heating and sonochemical processes (exposure to ultrasound in ambient conditions), in the sonochemical technique, the $Zn(NO_3)_2 \cdot 6H_2O$ is added to $(CH_2)6N_4$ (hexamethylenetetramine, HMT). The resulting solution is displayed to ultrasound for 30 min, where the temperature of the reaction reached 80°C. For the second technique, the solution is heated via microwaves at 90°C for 2 min. The obtained precipitates are dried at 60°C for 2 h, washed by ethanol and water, and centrifuged. The PL, EDS, TEM, XRD, SEM, and SAED (selected area electron diffraction) have been employed to analyze ZnO. These are enabling high yields for ZnO in excess of 90%. Briefly, they have stated that the microwave heating and sonochemical methods do not need surfactants applicable to mass scale and are low-cost, energy-efficient, and simple. ZnO rods are successfully utilized in optoelectronics and electronics. The sonochemical technique could be utilized in the future for synthesizing the one-dimensional structure of metal oxides.

8.3 SOL–GEL PROCESS

The sol–gel process uses small molecules to produce solid materials. The method is used for the fabrication of metal oxides. The procedure includes the conversion of monomers to a colloidal solution (sol), which acts as a precursor for an incorporated network (or gel) of polymers or discrete particles. The metal alkoxides are typical precursors. In a chemical procedure, the "sol" (a colloidal solution) is created, which gradually evolves across the formation of a gel-like diphasic system including both solid and liquid phases where morphology ranges from discrete particles to continuous polymer networks. In the case of colloid, the particle friction volume (particle density) is so low that a specific number of fluids need to be removed initially for gel-like properties to be recognized. It can be achieved in any way. The simplest technique is to allow sedimentation time to occur and then pour off the remaining liquid. Centrifugation can be utilized for accelerating phase separation process. The remaining liquid (solvent) phase removal requires a drying process that is accompanied by a specific amount of densification and shrinkage. The removal rate of solvent is determined by the porosity distribution in the gel. The final component's microstructure will be influenced by changes imposed by the structural template during the processing phase.[14]

Then, a firing process or thermal treatment is required for further grain growth, densification, structural stability via final sintering, enhancing mechanical properties, and polycondensation. The densification is a distinct advantage of utilizing this method as opposed to the traditional processing method, which is achieved at low temperatures. The sol can be deposited on a substrate to form a thin film via spin coating or dip coating, casting into an appropriate container with a desired shape to get aerogels, membranes, fibers, glasses and monolithic ceramics, or synthesizing powders such as nanospheres or microspheres. The sol–gel method is a low-temperature technique that is cheap and allows controlled chemical composition. For a few dopants, like rare-earth elements and organic dyes, it can be present in sol and uniformly dispersed in a final product. It can be utilized in ceramics manufacturing and processing as a casting material or in producing metal oxide thin films. The sol–gel has multiapplications in separation (e.g., chromatography), technology, reactive material, medicine (e.g., controlled drug release), space, (bio)sensors, energy, electronics, and optics.[15]

The interest in processing of sol–gel can trace back to the mid-1800s with an observation of tetraethyl orthosilicate (TEOS) hydrolysis under acidic conditions, which has led to the formation of SiO_2 in monoliths and fibers. The sol–gel process became important in the 1990s, when around 35,000 papers were published worldwide on the subject.[17]

The p-type silicon wafer (100) is used as a substrate for the growth of the n-type nanostructure ZnO. To remove the native oxide layer, silicon substrates (1 cm × 1.5 cm) are washed in acetone, ethanol, and DI water with an ultrasonic cleaner before being cleaned with BOE. Using a thermal evaporator, a very thin coating of aluminum (Al) (about 25 nm) is applied to a silicon substrate. The hydrothermal approach is used to manufacture zinc oxide nanoflakes on Si substrate coated with Al.[22] Under vigorous stirring for 30 min, 2-methoxyethanolamine is dissolved in 0.35 M of zinc acetate dehydrate solution to create the seed solution. The monoethanolamine was then added to the solution, which had been continuously stirred for 2 h at 60°C to make it transparent and uniform. The final product is spin-coated on a silicon substrate for 20 s at 3000 rpm. Three coats are applied and then the drying process is reiterated.

Zinc nitrate hexahydrate and hexamethyl tetra-amine (HMT) solution are used as precursors in varied quantities (12.5–75 mM) in 150 mL deionized water to develop ZnO nanoflakes on seeded Al–Si substrate. The substrate is put in a Teflon sample holder that faces downward, and it is incubated at 93°C for 2 h in the prepared oven. ZnO nanoflakes are then dried with a N_2 gun and rinsed with deionized (DI) water. A scanning electron microscope (JEOL 6460) is used to characterize the morphology of zinc oxide nanoflakes. Raman spectroscopy (Jobin Yvon Horiba HR800UV system) and Philips PW 1710 X-ray diffractometer are used, respectively, to study the optical characteristics and structural attributes.[32]

The synthesized zinc oxide nanoparticles via spin-coating technique have attracted more interest due to their relatively mild conditions of synthesis, repeatability, reliability, low-cost, and simplicity, which enable the modification of surfaces with specific organics that improves their properties and widen their applications. The preferred optical properties have become a hot topic of research in many published papers and books.[33]

Benhebal et al.[34] have synthesized ZnO powder via the spin-coating technique using ethanol as a solvent, oxalic acid, and zinc acetate dihydrate. It has been analyzed and characterized by UV-Vis, SEM, XRD, and nitrogen adsorption isotherms. The synthesized ZnO has a hexagonal structure of spherically shaped particles. While Risti et al.[35] have employed the BET method to obtain a surface area for calcined zinc oxide equal to 10 m^2/g, crystallized materials, and low porosity, the tetramethylammonium hydroxide (TMAH) solution has been added to the zinc 2-ethylhexanoate (ZEH) solution in propan-2-ol. The resulting colloidal suspension is left for 30 min (alternatively for 24 h) and then washed with water and ethanol. TMAH has a strong base of organics that is comparable to an inorganic base like NaOH that is analyzed via pH ~14. A high pH means metal oxides are not contaminated with cations from bases that have an observed effect on the ohmic conductance of oxides. A determination is made of the maturing time of the colloidal solution and the quantity effect of ZEH. The images of TEM have shown ZnO nanoparticles of sizes 20–50 nm. The ZEH quantity has a negligible effect on the size of the nanoparticle. And Yue et al.[36] have prepared zinc oxide via spin-coating technique to prepare ordered ZnO nanotubes, which are uniform and high-filling into the membrane of ultrathin AAO. Incorporating the spin-coating technique with ultrathin AAO membranes could help to manufacture high-quality one-dimensional nanomaterials that are applicable in different fields.

8.4 CHEMICAL STUDIES OF ZnO

ZnO is complex and has a variety of one-dimensional (1D) structures: nanorods, nanotubes, and nanowires; two-dimensional (2D) structures: nanodisks and nanosheets; and three-dimensional (3D) structures: nanodendritics. The nanostructure of ZnO has unique properties like electron mobility capability and scattering of electrons that influence the future application performance.[29] The ZnO structure is hexagonal because of its stable thermodynamic phase as compared to zinc blende (ZnS), which has a face-centered facet and is metastable.[38,39] Meanwhile, another rock salt structure can only be obtained after being applied with high pressure. ZnO possesses ionic bonds due to its total charge density, which maintains high ionicity and has no covalent degree.[30] A large amount of ionicity of zinc oxide happens due to having high electronegativity between oxygen and zinc atoms; thus, this causes a repulsion between nearby atom clouds with electric charge, making their structure more stable in the wurtzite structure.[40]

The ZnO surface shows the existence of a dipole moment (in the case of zincite) that leads to two different polar surfaces with (0001) being zinc-terminated and (000) being terminated by an atom of oxygen. The mentioned surfaces of polar are unstable, achieving phase stability, atoms of Zn and O at the surface of ZnO performing metallization, geometrical rearrangement, and charge transfer.[41,42] The process in the formation of ZnO nanostructures influences the surface tension, particle size, and surface energy. So, it is observed that analysis affects ZnO properties. The surface energy differs between two different planes ({100} and {0001}) of ZnO, indicating the kinetic control of anisotropic crystal growth along the c-axis. Thus, ZnO nanoparticles with rod-like structures are often observed.[43]

At the same time, the ripening, crystal growth, and nucleation theories of zinc oxide nanostructures illustrate big supersaturation and guide an increasing rate of nucleation to reduce crystal growth and produce small zinc oxide nanoparticles. On the other hand, the increase of interfacial tension or surface tension increases the dielectric constant (referring to the liquid phase synthesis procedure) to increase solution solubility to proceed with low supersaturation and decrease the rate of nucleation. So, the increase in surface tension produces a larger particle size.[44] This intrinsic polarity of structure, which arises from the tension of the surface, while the energy of the surface encourages crystal growth leads to piezoelectricity, plasticity, and spontaneous polarization properties.[45] The morphology and particle size of zinc oxide nanostructures affect the electronic and electrical properties; these characteristics control electron scattering and electron mobility, thus differentiating the electronic performance from one structure to another.

Meanwhile, the adsorbed interstitial zinc and oxygen at the surface of ZnO influence the photoconductivity and properties of zinc oxide nanostructures.[46] It is proven that the charge density of the surface is influenced by zinc oxide with sizes smaller than 10 nm. The nanostructure produces a high surface charge density.[47,48] In addition to nanostructure size, the area of the surface and number of edges, defect site density, and kinks increase, resulting in dissolution "hot spots" that are important for long-term stability in health and environmental applications.[47]

For PL, some defects like oxygen antisites, interstitial oxygen, oxygen vacancies, interstitial zinc, and zinc vacancies are associated. The exciton process of recombination into 350–450 nm is known to be attributed to free exciton recombination, which indicates good crystalline zinc oxide.[49,50] ZnO nanorods have the highest crystalline arrangement; their minimum emissions are between the 50–450 nm region (UV area), but small emissions are found in the 600–700 nm region (orange–red region). Small intrinsic defects are caused by excess oxygen presence like that of interstitial oxygen.[51] But synthesizing ZnO from a higher voltage electrodeposition method has presented more than one defect combination into the atomic structure. These defect combinations can be vacancies of oxygen and interstitials of zinc.[52] The green emission of PL is typically because of oxygen-based defects.[52–55] Also, ZnO has broad peaks covering the orange, yellow, and green regions produced by both defects' mixtures, oxygen vacancies, and zinc interstitial defects.[52,56] Even though others[51,57] have produced zinc oxide, it includes a large quantity of $Zn(OH)_2$ to produce a wide orange, yellow, and green area[51,57] that is attributed to insufficient annealing.

For zinc oxide with few defects, PL in the 500–800 nm region illustrates excessive performance with a high rate of recombination in the 350–450 nm region.[37] ZnO has a dense structure. Also, because of the existence of high vacancies of oxygen in the zinc oxide bulk, the electronic achievement for amorphous ZnO is weak, indicating an increased number of recombination centers.[58] The recombination centers, which are causing enhanced states of traps, are leading to reduced photoinduced charge carrier mobility back into contact with zinc itself. Furthermore, the zinc oxide amorphous phase comprises a high level of concentration of defects like dangling bonds, impurities, and microvoids, which also functioned as a center for the recombination of charge carriers, which led to low electronic achievement.[59] Dai et al.[60] and Alim et al.[61] have determined that the Raman peak shift is for impurities or defects in the tensile strain effect, laser irradiation heating, optical phonon confinement, and nanocrystals.

The main science interest is drawn by dynamics in metal oxide structure. It has an enormous nanoscaled architecture for a limited absorption of visible light and quick recombination of charge carriers into electronic applications. The utilization of visible light of solar energy is to reduce the charge carriers' recombination and increase the effectiveness of electronics. The metal oxide modification is required to ameliorate specific properties. Many possible modification techniques have been researched and reported to enhance metal oxide research studies like semiconductor-modified metal oxide, polymer-modified metal oxide, nonmetal-modified metal oxide, and metal-modified metal oxide.[61]

In the context of defects, Ding and Wang[62] have investigated defects of planarity in zinc oxide nanowire and nanobelts created from doping by impurities, where they have displayed (01 1), (0 12), (01 3), and (2 112) 1/6 (202 3) prismatic plane stacking faults and twins, I1, and I2 basal plane stacking faults to provide piezoelectric properties, unique semiconducting properties, and biosafety. In addition, Deng et al.[63] have studied Sn metal in zinc oxide to improve near-band edge (NBE) emission at 3.24 eV so that the peak becomes broader and slightly shifted to lower energy and, therefore, results in minimized structural strain. Therefore, due to surface energy order rearrangement of {0001}, {0110}, and {2110} surfaces, it causes nanobelts to grow up along direction <0110>. Information on thelimitations of planar defects in metal oxide is available.

At the same time, Abd Samad et al.[58] have illustrated defects like oxygen antisites, oxygen interstitials, oxygen vacancies, zinc interstitials, and zinc vacancies in the photoluminescence studies for the hybrid TiO_2–ZnO. The PL of ZnO has shown UV-Vis, which is centralized in the 380 nm width (determined intrinsic defects) of the emission band for large visibility.[49] With the increasing TiO_2–ZnO dipping cycles, PL continued to decrease. This is attributed to the rise of the TiO_2 layer as a result of the separation impact of the charge carrier for TiO_2 and ZnO (heterojunction) band alignment type-II.[50] In order to perform the separation state of excitonic charge, the charge carriers are photogenerated in a stunned-band offset, separated, and connected into ZnO and TiO_2. The charge carrier separation efficiency decreases the recombination rate of hole/electron, but PL decreases for the TiO_2–ZnO hybrid films.[64–66] Therefore, the combination of semiconductor-modified metal oxide may contribute to the improved transportation of charge carriers and reduced recombination rates.

8.5 METAL OXIDES

The metal's availability on earth is not stable, but it has metal oxide forms with higher stability and unique functionalities than pure ones. Oxygen ions and metals play an important role to present different properties for industries. The metal oxides are solid in powder form with an anion as oxygen and a cation as metal. The oxides are abundant with different properties like chemical properties with high strength, electrical, optical, physical, composition, phase transitions, various crystal structures, and high stability that are appropriate for memories, gas sensors, energy storage devices, solid-state microbatteries, catalysts, transistors, electronic components, transducers, optoelectronic devices, membrane reactors, and bioapplications. In nature, metal oxides have a brittle property at 25°C.[67] Anyway, metal oxides' electrical properties have a wide range; insulators toward superconductors within conductors and semiconductors make them competitive economically. So, their use has been confined to insulators. The reason is attributed to their difficulty in maintaining the desired oxygen levels in the compound, crystalline transformation, and brittleness. The phase changes and crystalline transformations are regarded as disadvantages but meanwhile, they have advantages due to unexpected properties drawn from phase transformations that make them useful in thermochromic devices, electrochromic photochromic devices, and microbatteries. Also, at certain temperatures or pressures, the transition of the metal insulators is to draw attention for the widespread theme before the high-temperature superconductivity sprang onto the center stage. Due to localized electron availability, the oxides are utilized as ferroelectric and ferromagnetic materials to exhibit photoelectric and thermoelectric effects.

Based on their physical properties, metal oxides can be categorized into nontransition metal oxides and transition metal oxides.[68,69] The band gap separates the valence and conduction bands in nontransition metal oxides. Hence, in the ordinary conditions, they will act as diamagnetic insulators. The magnesium oxide (MgO) and silicon dioxide (SiO_2) are nontransition metal oxides. At high temperatures, intrinsic activation energy is higher than the energy required for the creation and migration of defects, so ionic conduction is predominant over electronic conduction. For transition metal oxides, they illustrate unusual electronic structures based on transition metals.[70,71] They are interesting and enormous because they exhibit a variety of magnetic, electrical, optical, and structural properties. In turn, it will be divided into the following:

i) A metal ion which has an electronic configuration (d0). For example: Tantalum oxide (Ta_2O_5), Niobium oxide (Nb_2O_5), Zirconium dioxide

(ZrO_2), Chromium trioxide (CrO_3), Tungsten trioxide (WO_3), Molybdenum trioxide (MoO_3), and Vanadium pentoxide (V_2O_5).

ii) Shell (d) is filled partially.

For example: Rhenium trioxide (ReO_3), Rhodium oxide (RhO_2), Ruthenium oxide (RuO_2), Rhenium dioxide (ReO_2), Chromium dioxide (CrO_2), Niobium monoxide (NbO), Titanium monoxide (TiO), Osmium dioxide (OsO_2), Tungsten dioxide (WO_2), and Molybdenum dioxide (MoO_2).

The metal oxide of electronic configuration (d0) at octahedral shape exhibits antiferroelectric distortions and spontaneous ferroelectricity. Many are losing oxygen at high temperatures because they are nonstoichiometric. Oxygen loss, or the insertion of metal atoms of electropositives into oxide, places electrons in the conduction band. The electronic conduction nature depends on electron–phonon coupling strength and conduction bandwidth derived from metal states (d). The transition metal oxide contains partially filled states (d) to show unusual properties for behaving as a semiconductor or metal.[72] Metal oxides and their composites have become attractive subjects due to their new applications and characteristics for exploring novel challenges in experimental and theoretical investigations.[73–76] Also, one can protect the oxides of the environment, that is, the protection of environmental gases, chemical reactions, and heat.[77,78]

The thin layer of material deposited on the metal oxide surface is to protect it from any exposure. On the other hand, the coating protects devices and materials from degradation via a barrier created between the environment and materials. The ceramics, nonmetals, and metals can be utilized as protective coatings. Also, these coatings are utilized for changing the properties of the surface for metal oxides like structure and morphology. Different methods have been used to grow up layers for avoiding corrosion. But now, the synthesis, analysis, and characterization of metal oxides have such importance that they are increasingly used in different scientific and industrial fields for devices and technologies. Furthermore, the nano-oxides can be utilized in the mentioned applications, which increase the ratio of surface area to volume. Metal oxides, or nanoforms of metal oxides, have been chosen to be employed in multiapplications due to significant improvements in electrical, physical, and chemical properties compared with bulk ones.

The metal oxide that can be synthesized by different physical and chemical technologies is necessary to improve their properties on nanoscale with the widespread development and fabrication of miniaturized devices as a part of nanoscience and nanotechnology in material science. The ability to develop different versatile materials to exhibit interesting properties has resulted in microoptoelectronic devices, drug delivery, photonics, energy storage devices, catalysis, and sensors. The metal oxide is sensitive to the environment, especially under temperature. Because of coatings of thermal barrier, one could protect components from destruction at high temperatures by acting as insulators and layers of oxidation resistance as well. So, the device and system efficiencies can be improved. Also, thermal barriers need resistance against corrosion when exposed to high temperatures. The coating of AlTiN is utilized for increasing the oxidation resistance and surface hardness. The thermal absorption capacity is available for a few materials that are required for application in nuclear fusion and thermal conduction that transfers heat to surfaces. It is supposed to select the material and deposition technique for the surface's coating of metal oxide. The coating of thermal barriers for protection applications is needed to have porosity that can be changed with thermal expansion within cooling or heating processes. So, the stability of the phase is necessary, or any crack could damage the devices.

For all of these, one can list effective coatings of thermal protection for sustainable service over a greater number of cycles as follows:

1. The material should be thermally stable and chemically inert.
2. There is no phase change between room temperature and operating temperature.
3. The materials and metallic surface have the same thermal expansion.
4. Protective material is supposed to be firmly adherent to substrates.
5. Should have oxidation and corrosion-resistant properties
6. Should hold hardness in the optimum range.

The metallic layer of oxidation resistance is to serve as a bond coating that is deposited on a metallic substrate with a thickness of 100 μm. Thermally, the grown-up oxide is coated on the top of the ceramic barrier. This layer has produced a large thermal gradient and keeps the bottom layers at a low temperature. The degradation occurs in the coating as attributed to porosity and voids of failure in bond coats, open interfaces, and grown oxides.

The most common materials to protect coatings are metal-glass composites (a mixture of metal and glass), glass ceramics, yittria-stabilized zirconia, zirconates of rare-earth metals (BaO.ZrO2, SrO.ZrO2, etc.), alumina, and mullite (compounds of alumina and silica). The coatings of thermal protection are useful in different applications, like diesel engines and gas turbines, for enhancing the strength of bonds. This part of the book tries to give a novel idea about metal oxide's importance and coatings for thermal protection. It aims to help readers and specialists evaluate the multifunctional metal oxide's role in technologies for protecting oxide powder by utilizing different coatings using varied techniques. Hence, it is necessary to know the surface layer property of metal oxide for controlling properties including further discussion on protection of coatings for environment and temperature. The high thermal stability coatings protect metal oxide in bulk or nanoscale form at different temperatures to help and maintain properties for thermal barrier coatings or thermal protection coatings. The coatings are grown up on metallic surfaces and oxide with a suitable thickness, 200 m to 1 mm for insulating components from heat. The stable and sustainable materials are considered for thermal coatings. The protection coatings allow high operating temperatures without altering the thermal stability of the component, fatigue strength, oxidation states, structure, and composition. As the industry demands for engines of high efficiency at high temperatures, there is a necessity to improve new coatings of thermal protection with less thickness, cyclability, durability, response, and a better lifetime. All depend on electrical, chemical, and physical properties in addition to deposition methods for developing coatings for surfaces for protecting layers of oxide.

8.6 METAL OXIDE IN DIELECTRICS

ZnO has widespread applications in industrial products, electrical products, and household products. In spite of occurrence of mineral zincite, the production originates from anthropogenic sources. Nano-ZnO can be prepared by grinding ZnO that is in a wurtzite structure for use in different applications. Nano-ZnO can be grown up via different techniques as follows:

- Polyol medium hydrolysis technique
- Gas condensation technique
- Polymeric precursor technique
- Pyrolysis (aerosol spray) technique
- Hydrothermal technique
- Sol–gel method
- Solid-state microwave decomposition method

All the techniques mentioned are traditional ones that require high energy consumption, making them uneconomical. The most economical and sustainable technique of ZnO synthesis is the precipitation method.[79] Nano-ZnO serves as an n-type semiconductor due to its wide bandwidth (~3.37 eV).[80] Based on its superconductive properties, piezoelectricity, chemical stability, thermal and electrical properties, nano-ZnO has been applied in piezoelectric cells,[81] cholesterol and glucose biosensing,[82] solar cells,[83] and photovoltaic transducers.[84] In powder form, nano-ZnO can be mixed into pigments, paints, foods and adhesives,[85] medical disinfection, and environmental remediation.[86] Due to its UV-vis properties, nano-ZnO is utilized for cosmetics and sunscreens[87] with an estimated 1000 tons per year during 2005–2010.[88]

It is interesting to object ZnO nanoparticles into MIM structures (metal–insulator–metal). It is focused on altering oxide layers that lack homogeneity, porosity, and an amorphous nature. Verbakel[89] has investigated the resistance effect of diode switching containing nanometric ZnO covered with an active layer of polystyrene matrix. These diodes consist of two PEDOT: PSS electrodes. Using an impedance spectroscope, it was found that the effect of electronic memory in nanostructured metal oxides can be affected by surface modification of nanoparticles utilizing coordinating ligands of thiols and amines depending on the temperature of the voltage measurements. This process is to provide a new prospect for ecological modification of the ZnO surface utilizing inorganic hybrids.

Others have tried to correlate the size of the constituent particles with the biological activity of inorganic antibacterial agents.[90,91] Inorganic nanometal oxides are interesting due to their high surface areas and greater suitability for biological applications.[92] ZnO has been studied as an antimicrobial agent due to its photocatalytic activity under UV-Vis.[93,94] These antimicrobial substances based on inorganic chemicals have been researched to be effective for therapy.[95] Padmavathy et al.[96] have illustrated that nano-ZnO is more abrasive than ZnO (particle size is 0.11 μm), and this contributes to greater mechanical damage to a cell membrane to enhance the bactericidal effect produced by nano-ZnO.

8.7 CHARACTERIZATION AND ANALYSIS

The substrate of Al has been used to create the porous structure of ZnO nanoplates.[15,22] While Al is used as an assistant to facilitate porous formation by preventing nanorod growth,[22,81] there are a few potential problems to grow ZnO porous structures utilizing Al substrate: (1) Al is rapidly oxidized by atmospheric oxygen; (2) Al-oxide removal could not be readily made without destroying the structural integrity of substrate; and (3) Al is a good conductor, and its thick layer of Al under the nanostructured ZnO might be problematic when applied to electrical devices. On the other hand, Si substrate is easily available at cheaper prices. The oxide layer of Si substrate can be easily removed by diluted hydrofluoric (HF) acid without destroying its structural integrity. In the current study, ZnO nanoflakes are grown on a thin layer (25 nm) of Al deposited on Si substrate. The thin Al layer is dissolved into Al $(OH)_4$ in the alkaline environment of HMT, and 2D ZnO originates from its anisotropic structure. The additive Al $(OH)_4$ ions play a regulatory role that promotes the formation of porous structures and prevents nanorod formation.[22] As a result, the drawbacks of using an Al substrate were completely avoided when using an Al thin layer.

The consequence of precursor concentration on nanostructured ZnO morphologies is shown in Figure 8.1. Uniform nanoflake distribution on Si substrate is observed at growth that is made with 12.5 mM precursor (Figure 8.1a). The SEM images have revealed a wall thickness of 80–100 nm and a diameter of 300–500 nm of grown nanoflakes. When the precursor's concentration is raised to 50 mM, the growth of ZnO nanorods is noticed on the surface and that of nanoflakes on the walls (Figure 8.1b). When the concentration increases to 75 mM, the nanorods' density increases and their length decreases (Figure 8.1c).[82] It is interesting to note that the little hexagonal nanorods on the surface of the nanoflakes appear to be granules due to the length decrease (Figure 8.1c). This illustrates the significant influence that precursors have on the morphological characteristics of nanoflakes and nanorods.

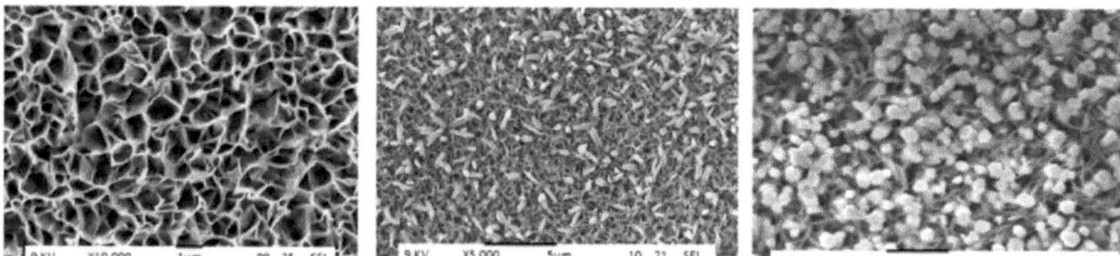

(a) (b) (c)

FIGURE 8.1 SEM images of ZnO nanoflakes showing surface morphology at: (a) 12.5 mM, (b) 50 mM, and (c) 75 mM of precursor.

The accurate mechanisms of structural transition from nanoflakes to nanorods were not clearly depicted in the previous studies. It has been seen that the relaxation of lattice-mismatch-induced strain has a great role in the initial phase of nanostructure growth.[97] Anyway, this cannot be the causative parameter in the present study as Al favors the growth of sheet-like structures and prevents the formation of nanorods or nanowires by capping the growing ZnO surface.[22] The preferred development of nanorods may be attributed to the higher density of dangling bonds at crossing points or the ends of nanowalls than at other locations. In order to maintain a low energy level across the entire system, the concentration of these bonds leads to the accumulation of adatoms or molecules, after which nanorods are created in the succeeding growth.[97] Whereas phase refining is derived from concentration-dependent surface tension or atomic deposition on the surface, the outcomes of the competition have had two effects: the phases occasionally choose stable, nanometric, or larger sizes.[98,99] Figure 8.2 displays the XRD patterns of nanostructured ZnO at various precursor concentrations. Each peak is consistent with ZnO's hexagonal structure. With an increase in precursor concentration, significant changes have been seen, including an increase in peak intensity that corresponds to the (002) crystal plane. But at 12.5 mM, a lower precursor concentration, there are two separate peaks at angles of around 39 and 45 degrees, which are most likely caused by a thin Al layer. It has been shown that at lower concentrations than at higher concentrations, the thin Al layer is not completely dissolved by nanostructured ZnO. Although earlier experiments[15,22] used Al substrate and produced peaks at lower precursor concentrations, these results are consistent for angles of 39 and 45 degrees.

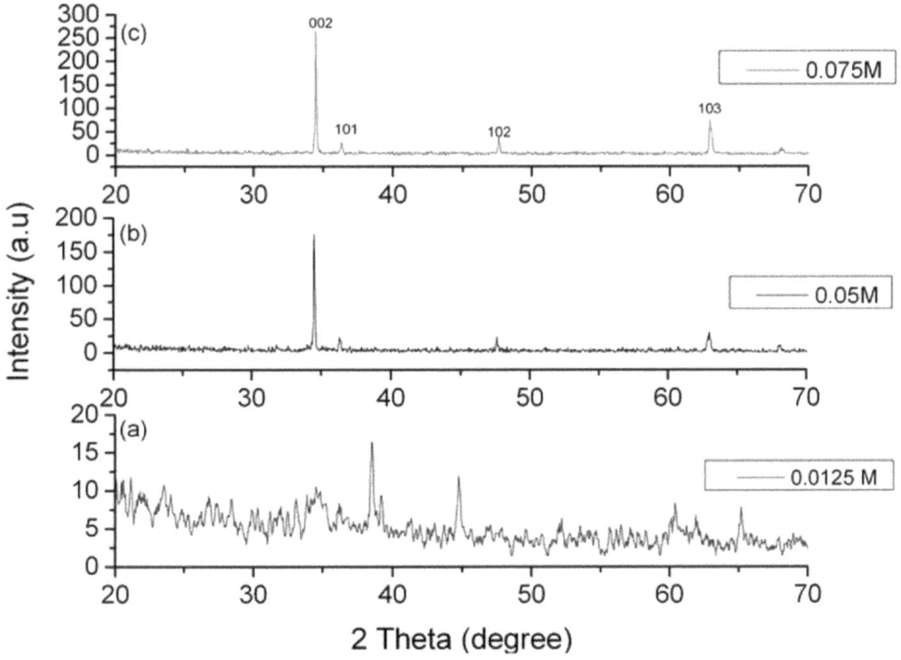

FIGURE 8.2 XRD patterns of ZnO nanoflakes at: (a) 12.5 mM, (b) 50 mM, and (c) 75 mM of precursor.

Useful information on crystal quality and ZnO nanoflakes and nanorods could be provided by Raman spectroscopy. The fundamental theoretical background of the Raman data explanation can be found elsewhere.[100] Figure 8.3 illustrates the Raman scattering spectra of ZnO nanoflakes prepared by the sol–gel method. The spectra are collected by exciting a laser line at 488 nm with an excitation power of 20 mW. One sharp peak at 520 cm^{-1} and two weak peaks at 300 and 615 cm^{-1} (Figure 8.3a) can be attributed to the TO phonon mode originating from the Si substrate.[84] When the light penetration depth exceeds the product thickness, such peaks might appear.[101] Therefore, they are not affected by precursor concentration and could not provide beneficial information about ZnO qualities. The peak appeared at 433 cm^{-1} with significantly high intensity, which is a typical characteristic of ZnO hexagonal wurtzite. At 12.5 mM precursor, its intensity is incredibly low, but it dramatically increased as the precursor level increased (green and blue dotted lines). This peak is allocated to the Raman active optical phonon in E2H mode.[102–106]

There are two weak peaks appearing at 330 and 380 cm^{-1} that are assigned to the multiphonon scattering process (E2H–E2L) and A1T mode, respectively.[107] They were totally absent at 12.5 mM precursor and became clearly visible at 75 mM precursor, clarifying the precursor concentration influence in ZnO nanoflake synthesis. Figure 8.3b shows a magnified view of the Raman spectrum in the (400–480 cm^{-1}) window. It is observed that when the precursor concentration is increased to 50 and 75 mM, the peak location of the E2H mode moves 4 cm^{-1} from 429 to 433 cm^{-1}. The photoluminescence spectroscopy data are used to compute the energy gaps (E_g) of nanostructured ZnO (results not given here).

$$E_g = hc/\lambda \tag{8.1}$$

where E_g stands for energy gap, h for the Planck constant, c for light speed, and λ for wavelength. The predicted values for precursor concentrations of 12.5, 50, and 75 mM are 3.2191, 3.2375, and 3.37 eV, respectively.

There have been several efforts to establish straightforward correlations between the refractive index and the energy gap E_g.[108–113] These n-relations, however, are not affected by the temperature or the energy of the incoming photons. This section will explore the many relationships between n and E_g. An equation for n as a function of E_g has been submitted to Ravindra et al.[109]

$$n = \alpha + \beta E_g \tag{8.2}$$

where $\alpha = 4.048$ and $\beta = -0.62$ eV^{-1}. Light refraction and dispersion will be inspired. Herve and Vandamme[110] proposed an empirical relation as follows:

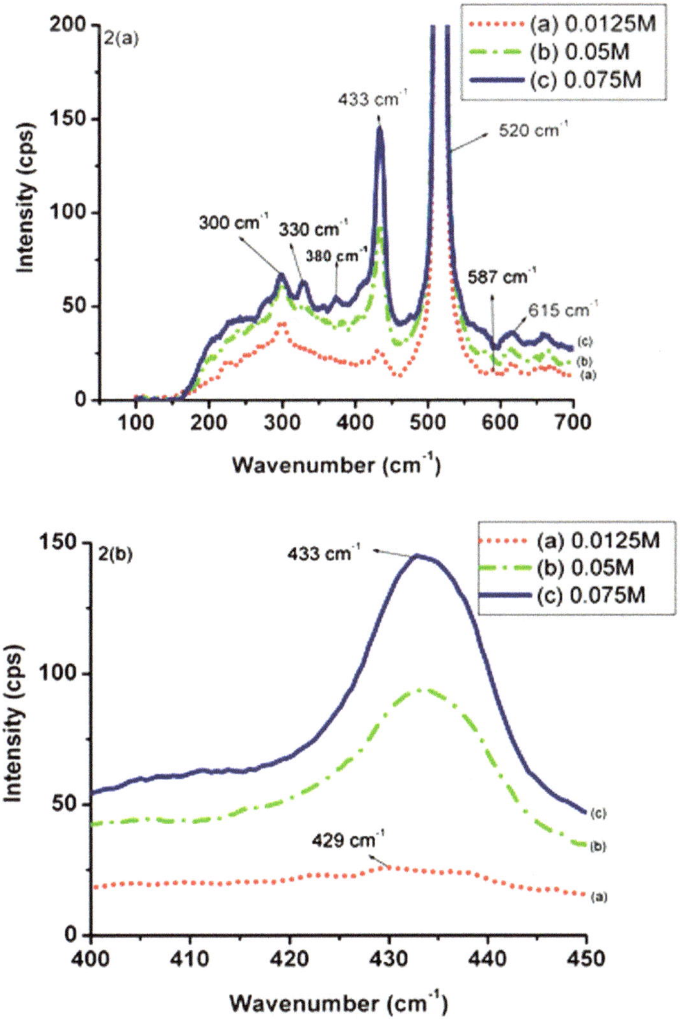

FIGURE 8.3 ZnO nanoflakes under Raman measurement. The spectrum is (a) between 100 and 700 cm^{-1} and (b) the magnified view showing the Raman shift is near the 430 nm region.

$$n = \sqrt{1 + \left(\frac{A}{E_g + B} \right)^2} \tag{8.3}$$

in which A is 13.6 eV and B is 3.4 eV. An empirical formula for group-IV semiconductors has been reported by Ghosh et al.[111] on the basis of band structure and Van Vechten[113] and Penn[112] quantum dielectric considerations:

$$n^2 - 1 = \frac{A}{(E_g + B)^2} \tag{8.4}$$

where A=8.2Eg+134, B=0.225Eg+2.25, and (Eg+B) denotes the proper average energy gap. The end-point compounds' computed refractive indices and energy gaps are shown in Table 8.1, and they show good agreement with experimental[102] and theoretical[103–107] values.

TABLE 8.1 Energy Gaps and Calculated Refractive Indices of Nanostructure ZnO Using Ravindra et al.,[109] Herve and Vandamme,[110] and Ghosh et al.[111] Models Corresponding to Optical Dielectric Constants.

Concentration (mM)	Energy (eV)	n	ε_∞
12.5	3.2191 3.44[a] 1.57[b] 0.73[c] 0.88[d] 2.26[e]	2.052[f] 2.285[g] 2.276[h] 2.008[i]	4.210[f] 5.221[g] 5.180[h]
50	3.2375	2.040[f] 2.279[g] 2.270[h]	4.161[f] 5.193[g] 5.152[h]
75	3.37	1.954[f] 2.244[g] 2.230[h]	3.833[f] 5.035[g] 4.973[h]

[a]Ref. [114] Exp, [b]Ref. [115] Theor, [c]Ref. [116] Theor, [d]Ref. [117] Theor, [e]Ref. [118] Theor, [f]Ref. [109], [g]Ref. [110], [h]Ref. [111], [i]Ref. [119] Exp.

Calculating the optical dielectric constant, which depends on the refractive index, serves to confirm this. The optical dielectric constant is found to equal n2.[120] It is obvious that nanostructured ZnO requires researched refractive indices utilizing the Ravindra et al.[109] models to improve detection and sensing. High absorption may, thus, contribute to improved sensor performance.

When nanoflakes turn into nanorods, the morphology of the nanostructured ZnO surface undergoes a significant shift. The walls of the tubular nanoflakes (Figure 8.1a) functioned as resonant cavities and underwent structural changes that caused the Raman spectra to shift to red.[32] We obtained ZnO nanoflakes with morphologies (Figure 8.1a) that resembled the tubular forms of ZnO nanotubes. Due to structural modification, the walls of nanoflakes most likely act as resonant cavities. It is identified that the appearance of a 587 cm^{-1} LO phonon peak is indicative of an oxygen crisis at the lower precursor (25 mM) (Figure 8.3a), which is in good accordance with the theoretical model of Fonoberov et al.[119] At a high precursor concentration of 75 mM, such a peak cannot be noticed, indicating a very low level of intrinsic defects because of the oxygen crisis. As identical excitation energy from laser power is utilized in measurements, the disappearance of the LO phonon peak cannot be attributed to the thermal destruction of the ZnO nanocrystal. Higher intensities of the E2H phonon than the LO phonon mode are suggestive of little defects in ZnO nanocrystals. Figure 8.3a clearly illustrates that at high concentrations, 75 mM of precursor E2H phonon at 433 cm^{-1} is intensified, and LO phonon at 587 cm^{-1} completely disappears. This clearly reflects the better quality of ZnO nanocrystals in developed nanoflakes. So, we are successful in demonstrating how different morphological characteristics of zinc oxide nanoflakes can be identified with a change in concentration and how these features have a big impact on optical qualities.

8.8 CONCLUSIONS

Researchers looked at how the precursor concentration affected the morphological, optical, and crystalline characteristics of nanostructured ZnO. It was discovered that the concentration of the precursor was crucial in the conversion of nanoflakes into nanorods. Nanoflake configuration predominated at low precursor concentrations. Anyway, when concentration was increased, nanorods protruded from the surfaces of nanoflakes. At increasing levels of precursor, Raman scattering experiments revealed enhanced crystal quality and fewer structural flaws. Raman spectra were red-shifted, and the structural transition from ZnO nanoflakes to nanorods was attributed. The optical characteristics were examined, and it was demonstrated that the model developed by Ravindra et al. with increasing precursor concentrations was better suitable for nanostructured ZnO that may be used as sensors.

KEYWORDS

- **ZnO**
- **SEM**
- **XRD**
- **refractive index**
- **sol–gel**

REFERENCES

1. Gunshor, R. L.; Nurmikko, A. V.; Otsuka, N. The Molecular Beam Epitaxial Growth of Wide Gap II–VI Injection Lasers and Light-Emitting Diodes. *Thin Solid Films* **1993,** *231*, 190.
2. Nurmikko, A. V.; Gunshor, R. L. Optical Physics and Laser Devices in II–VI Quantum Confined Heterostructures. *Physica B* **1993,** *185*, 16.
3. Al-Douri, Y. Electronic and Optical Properties of ZnxCd$_{1-x}$Se. *Mater. Chem. Phys.* **2003,** *82*, 49.
4. Guo, H.; Lin, Z.; Feng, Z.; Lin, L.; Zhou, J. White-Light-Emitting Diode Based on ZnO Nanotubes. *J. Phys. Chem. C.* **2009,** *113*, 12546.
5. Umar, A.; Rahman, M. M.; Kim, S. H.; Hahn, Y.- B. Zinc Oxide Nanonail Based Chemical Sensor for Hydrazine Detection. *Chem.Commun.* **2008,** *2*, 166.
6. Wan, Q.; Lin, C. L.; Yu, X. B.; Wang, T. H. Room-Temperature Hydrogen Storage Characteristics of ZnO Nanowires. *Appl. Phys. Lett.* **2004,** *84*, 124.
7. Mridha, S.; Basak, D.; Investigation of a p-CuO/n-ZnO Thin Film Heterojunction for H_2 Gas-Sensor Applications. *Semicond. Sci. Tech.* **2006,** *21*, 928.
8. Zeng, Y.; Zhang, T.; Yuan, M.; Kang, M.; Lu, G.; Wang, R.; Fan, H.; He, Y.; Yang, H.; Growth and Selective Acetone Detection Based on ZnO Nanorod Arrays. *Sensor. Actuat. B-Chem.* **2009,** *143*, 93.
9. Yang, Z.; Li, L.-M.; Wan, Q.; Liu, Q.-H.; Wang, T.-H. High-Performance Ethanol Sensing Based on an Aligned Assembly of ZnO Nanorods. *Sens. Actuators B: Chem.* **2008,** *135*, 57.
10. Manekkathodi, A.; Lu, M.-Y.; Wang, C. W.; Chen, L.-J. Direct Growth of Aligned Zinc Oxide Nanorods on Paper Substrates for Low-Cost Flexible Electronics. *Adv. Mater.* **2010,** *22*, 4059.
11. Xi, Y.; Song, J.; Xu, S.; Yang, R.; Gao, Z.; Hu, C.; Wang, Z. L. Growth of ZnO Nanotube Arrays and Nanotube Based Piezoelectric Nanogenerators. *J. Mater. Chem.* **2009,** *19*, 9260.
12. Cao, B. Q.; Liu, Z. M.; Xu, H. Y.; Gong, H. B.; Nakamura, D.; Sakai, K.; Higashihata, M.; Okada, T. Catalyst/Dopant-Free Growth of ZnO Nanobelts with Different Optical Properties from Nanowires Grown via a Catalyst-Assisted Method. *Cryst. Eng. Commun.* **2011,** *13*, 4282.
13. Cho, S.; Lee, K.-H. Synthesis of ZnO Nanostructures Composed of Nanosheets with Controllable Morphologies. *Cryst. Growth Des.* **2009,** *10*, 1289.
14. Wang, N.; Jiang, L.; Peng, H.; Li, G. Synthesis of ZnO Nanostructures Composed of Nanosheets with Controllable Morphologies. *Cryst. Res. Technol. 44*, 341.
15. Cheng, J. P.; Liao, Z. M.; Shi, D.; Liu, F.; Zhang, X. B. Oriented ZnO Nanoplates on Al Substrate by Solution Growth Technique. *J. Alloy Compd.* **2009,** *480*, 741.
16. Mäder, M.; Gerlach, J. W.; Höche, T.; Czekalla, C.; Lorenz, M.; Grundmann, M.; Rauschenbach, B. ZnO Nanowall Networks Grown on DiMPLA Pre-Patterned Thin Gold Films. *Phys. Status Solidi RRL* **2008,** *2*, 200.
17. Kashif, M.; Ali, S. M. U.; Foo, K. L.; Hashim, U.; Willander, M. ZnO Nanoporous Structure Growth, Optical and Structural Characterization by Aqueous Solution Route. In: *Enabling Science and Nanotechnology International Conference (EsciNano)*; Kuala Lumpur, Malaysia, (AIP Conference Proceedings), Vol. 92, 2010; p 1341.
18. Chang, J. F.; Kuo, H. H.; Leu, I. C.; Hon, M. H. The Effects of Thickness and Operation Temperature on ZnO: Al Thin Film CO Gas Sensor. *Sens. Actuat. B-Chem.* **2002,** *84*, 258.

19. Kim, S.-W.; Park, H.-K.; Yi, M.-S.; Park, N.-M.; Park, J.-H.; Kim, S.-H.; Maeng, S.-L.; Choi, C.-J.; Moon, S.-E. Epitaxial Growth of ZnO Nanowall Networks on GaN/Sapphire Substrates. *Appl. Phys. Lett.* **2007,** *90*, 033107.
20. Lee, W.; Jeong, M.-C.; Myoung, J.-M. Catalyst-Free Growth of ZnO Nanowires by Metal Organic Chemical Vapour Deposition (MOCVD) and Thermal Evaporation. *Acta. Mater.* **2004,** *52*, 3949.
21. Su, S. C.; Lu, Y. M.; Zhang, Z. Z.; Li, B. H.; Shen, D. Z.; Yao, B.; Zhang, J. Y.; Zhao, D. X.; Fan, X. W. Structural, Optical, and Hydrogenation Properties of ZnO Nanowall Networks Grown on a Si (111) Substrate by Plasma-Assisted Molecular Beam Epitaxy. *Physica B* **2008,** *403*, 2590.
22. Cheng, J. P.; Zhang, X. B.; Luo, Z. Q. Oriented Growth of ZnO Nanostructures on Si and Al Substrates. *Surf. Coat. Technol.* **2008,** *202*, 4681.
23. Al-Douri, Y. ZnO Nanostructures: Analysis and Characterization by Electrospinning Technique. *Emerg. Mater.* **2020,** *3*, 947–953.
24. Gherab, K.; Y. Al-Douri, Voon, C. H.; Hashim, U.; Ameri, M.; Bouhemadou, A. Aluminium Nanoparticles Size Effect on the Optical and Structural Properties of ZnO Nanostructures Synthesized by Spin-Coating Technique. *Results Phys.* **2017,** *7*, 1190–1197.
25. Hassan, N. K.; Hashim, M. R.; Al-Douri, Y. Morphology and Optical Investigations of ZnO Pyramids and Nanoflakes for Optoelectronic Applications. *Optik* **2014,** *125*, 2560–2564.
26. Al-Douri, Y.; Haider, A. J.; Reshak, A. H.; Bouhemadou, A.; M. Ameri "Structural investigations through cobalt effect on ZnO nanostructures" Optik 127, 10102–10107.
27. Y. Al-Douri, Reshak, A. H.; Ahmed, W. K.; Ghazai, A. J. Structural and Optical Investigations of In Doped ZnO Binary Compound. *Mater. Express* **2014,** *4*, 159–164.
28. Grasza, K.; Łusakowska, E.; Skupiński, P.; Kopałko, K.; Bąk-Misiuk, J.; Mycelski, A. Effect of Annealing Atmosphere on the Quality of ZnO Crystal Surface. *Phys. Status Solidi* **2007,** *244*, 1468–1472.
29. Wei, X. Q.; Zhang, Z.; Yu, Y. X.; Man, B. Y. Comparative Study on Structural and Optical Properties of ZnO Thin Films Prepared by PLD Using ZnO Powder Target and Ceramic Target. *Opt. Laser Technol.* **2009,** *41*, 530–534.
30. Zhao, X.; Zheng, B.; Li, C.; Gu, H. Acetate-derived ZnO Ultrafine Particles Synthesized by Spray Pyrolysis. *Powder Technol.* **1998,** *100*, 20–23.
31. Hu, X. L.; Zhu, Y. J.; Wang, S. W. Sonochemical and Microwave-Assisted Synthesis of Linked Single-Crystalline ZnO Rods. *Mater. Chem. Phys.* **2004,** *88*, 421–426.
32. Kashif, M.; Al-Douri, Y.; Hashim, U.; Ali, M. E.; Ali, S. M. U.; Willander, M. Characterisation, Analysis and Optical Properties of Nanostructure ZnO Using the Sol–Gel Method. *Micro Nano Lett.* **2012,** *7*, 163–167.
33. Mahato, T. H.; Prasad, G. K.; Acharya, B. S. J.; Srivastava, A. R.; Vijayaraghavan, R. Nanocrystalline Zinc Oxide for the Decontamination of Sarin. *J. Hazard. Mater.* **2009,** *165*, 928–932.
34. Benhebal, H.; Chaib, M.; Salomon, T.; Geens, J.; Leonard, A.; Lambert, S. D.; Crine, M.; Heinrichs, B. Photocatalytic Degradation of Phenol and Benzoic Acid Using Zinc Oxide Powders Prepared by Sol-Gel Process. *Alex. Eng. J.* **2013,** *52*, 517–523.
35. Ristić, M.; Musić, S.; Ivanda, M.; Popović, S. Sol–Gel Synthesis and Characterization of Nanocrystalline ZnO Powders. *J. Alloy. Compd.* **2005,** *39*, L1–L4.
36. Yue, S.; Yan, Z.; Shi, Y.; Ran, G. Synthesis of zinc oxide nanotubes within Ultrathin Anodic Aluminum Oxide Membrane by Sol-Gel Method. *Mater. Lett.* **2013,** *98*, 246–249.

37. Abd Samad, N. A., Lai, C. W.; Abd Hamid, S. B. Influence Applied Potential on the Formation of Self-Organized ZnO Nanorod Film and Its Photoelectrochemical Response. *Int. J. Photoenergy* **2016**.
38. Alghamdi, G. S.; Alzahrani, A. Z. Bonding Formation and Orbitals Nature of ZnO Structure. *Middle-East J. Sci. Res.* 2013. *13* (9), 1144–1149.
39. Morkoç, H.; Özgür, Ü. Zinc Oxide: Fundamentals, Materials and Device Technology. John Wiley & Sons, 2008.
40. Aguilar, G. V. et al. Photoluminescence Studies on ZnO Thin Films Obtained by Sol-Gel Method. *Recent App. Sol-Gel Synth.* 2017, 195.
41. Ludi, B.; Niederberger, M.; Zinc Oxide Nanoparticles: Chemical Mechanisms and Classical and Non-Classical Crystallization. *Dalton Trans.* **2013,** *42* (35), 12554–12568.
42. Wöll, C. The Chemistry and Physics of Zinc Oxide Surfaces. *Progress Surface Sci.* **2007,** *82* (2–3), 55–120.
43. Laudise, R.; Ballman, A. Hydrothermal Synthesis of Zinc Oxide and Zinc Sulfide1. *J. Phys. Chem.* **1960,** *64* (5), 688–691.
44. Lizandara-Pueyo, C. et al. Nucleation and Growth of ZnO in Organic Solvents-an In Situ Study. *J. Am. Chem. Soc.* **2008,** *130* (49), 16601–16610.
45. Özgür, Ü. et al. A Comprehensive Review of ZnO Materials and Devices. *J. Appl. Phys.* **2005,** *98* (4), 11.
46. Morrison, S. Miller, P. Jr. Adsorption of Oxygen on Zinc Oxide. *J. Chem. Phys.* **1956,** *25* (5), 1064–1065.
47. Mudunkotuwa, I. A. et al. Dissolution of ZnO Nanoparticles at Circum neutral pH: A Study of Size Effects in the Presence and Absence of Citric Acid. *Langmuir* **2011,** *28* (1), 396–403.
48. Abbas, Z. et al. Size-Dependent Surface Charging of Nanoparticles. *J. Phys. Chem. C* **2008,** *112* (15), 5715–5723.
49. Chandrinou, C. et al. PL Study of Oxygen Defect Formation in ZnO Nanorods. *Microelectron. J.* **2009,** *40* (2), 296–298.
50. Shen, X.-Y., Zhai, Y.-C.; Zhang, Y.-H. Preparation and Characterization of Ultrafine Zinc Oxide Powder by Hydrothermal Method. *Trans. Nonferrous Metals Soc. China* **2010,** *20,* s236–s239.
51. Djurišić, A. et al. Defect emissions in ZnO nanostructures. *Nanotechnology* **2007,** *18* (9), 095702.
52. Alvi, N. H. *Luminescence Properties of ZnO Nanostructures and Their Implementation as White Light Emitting Diodes (LEDs)*; Linköping University, 2011.
53. Kröger, F.; Vink, H. The Origin of the Fluorescence in Self-Activated ZnS, CdS, and ZnO. *J. Chem. Phys.* **1954,** *22* (2), 250–252.
54. Studenikin, S., Golego, N.; Cocivera, M. Fabrication of green and Orange Photoluminescent, Undoped ZnO Films Using Spray Pyrolysis. *J. Appl. Phys.* **1998,** *84* (4), 2287–2294.
55. Yamauchi, S., Goto, Y.; Hariu, T. Photoluminescence Studies of Undoped and Nitrogen-Doped ZnO Layers Grown by Plasma-Assisted Epitaxy. *J. Crystal Growth* **2004,** *260* (1), 1–6.
56. Janotti, A.; Van de Walle, C. G. Native Point Defects in ZnO. *Phys. Rev. B* **2007,** *76* (16), 165202.
57. Zhou, H. et al. Behind the Weak Excitonic Emission of ZnO Quantum Dots: ZnO/Zn $(OH)_2$ Core-Shell Structure. *Appl. Phys. Lett.* **2002,** *80* (2), 210–212.
58. Abd Samad, N. A., Lai, C. W.; Abd Hamid, S. B. Easy Formation of Nanodisk-Dendritic ZnO Film via Controlled Electrodeposition Process. *J. Nanomater.* **2015**.

59. Lai, C. W. and Sreekantan, S. Photoelectrochemical properties of TiO_2 Nanotube Arrays: Effect of Electrolyte pH and Annealing Temperature. *J. Exp. Nanosci.* **2014,** *9* (3), 230–239.
60. Dai, K., et al., Photocatalytic Degradation and Mineralization of Commercial Methamidophos in Aqueous Titania Suspension. *Environ. Sci. Technol.* **2008,** *42* (5), 1505–1510.
61. Alim, K. A., Fonoberov, V. A.; Balandin, A. A. Origin of the Optical Phonon Frequency Shifts in ZnO Quantum Dots. *Appl. Phys. Lett.* **2005,** *86* (5), 53103–53103.
62. Ding, Y.; Wang, Z. L. Structures of Planar Defects in ZnO Nanobelts and Nanowires. *Micron* **2009,** *40* (3), 335–342.
63. Deng, R. et al. Planar Defects in Sn-Doped Single-Crystal ZnO Nanobelts. *J. Phys. Chem. C* **2007,** *111* (35), 13013–13015.
64. Shao, D. et al. High Quality ZnO–TiO_2 Core–Shell Nanowires for Efficient Ultraviolet Sensing. *Appl. Surface Sci.* **2014,** *314*, 872–876.
65. Greene, L. E. et al. ZnO-TiO_2 Core-Shell Nanorod/P3HT Solar Cells. *J. Phys. Chem. C* **2007,** *111* (50) 18451–18456.
66. Panigrahi, S.; Basak, D. Core–Shell TiO_2@ ZnO Nanorods for Efficient Ultraviolet Photodetection. *Nanoscale* **2011,** *3* (5), 2336–2341.
67. Tsuda, N.; Nasu, K.; Yanase, A.; Siratori, K. Electronic Conduction in Oxides. *Springer Ser. Solid—State Sci.* **1991,** *94*.
68. Rao, C. N. R.; Gopalakrishnan, J. *New Directions in Solid State Chemistry*. Cambridge University Press, 1997.
69. Adler, D. in: *Treatise on Solid State Chemistry*; Hanny, N. B., Ed., Vol. 2; Plenum Press: New York, 1975.
70. Rao, C. N. R.; Subba Rao, G. V. *Transitional Metal Oxides, Crystal Chemistry, Phase Transition and Related Aspects*. NSRDS—NBS Monograph, National Bureau of Standards, Washington, 1974; p 49.
71. Goodenough, J. B. *Prog. Solid State Chem.* **1971,** *5*, 149.
72. Rao, C. N. R. *Solid State Chemistry*; Marcel Dekker: New York, 1974.
73. Bunshah, R. F. *Deposition Technologies for Films and Coatings*; Noyes Publications: New Jersey, 1982.
74. George, J. *Preparation of Thin Films*; Marcel Dekker: New York.
75. Maissel, L. I.; Glang, R. *Handbook of Thin Film Technology*; Mc Graw-Hill: New York, 1970.
76. Chopra, K. L. *Thin Film Phenomena*; Mc Graw-Hill: New York, 1969.
77. David Clarke, R.; Phillpot, S. R. Thermal Barrier Coating Materials. *Mater. Today* **2005,** 22–29.
78. Wright, P. K.; Evans, A. G. Current Opinion in Solid State. *Mater. Sci.* **1999,** *4* (3), 255–265.
79. Dang, Z. M.; Fan, L. Z.; Zhao, S. J.; Nan, C. W. Preparation of Nanosized ZnO and Dielectric Properties of Composites Filled with Nanosized ZnO. In: *Materials Science and Engineering B: Solid-State Materials for Advanced Technology*; 2003.
80. Khan, S. B.; Faisal, M.; Rahman, M. M.; Jamal, A. Low-Temperature Growth of ZnO Nanoparticles: Photocatalyst and Acetone Sensor. *Talanta* **2011,** *85* (2), 943–949.
81. Minne, S. C.; Manalis, S. R.; Quate, C. F. Parallel Atomic Force Microscopy Using Cantilevers with Integrated Piezoresistive Sensors and Integrated Piezoelectric Actuators. *Appl. Phys. Lett.* **1995,** *67* (26), 3918–**3920**.
82. Umar, A.; Rahman, M. M.; Vaseem, M.; Hahn, Y.-B. Ultra-Sensitive Cholesterol Biosensor Based on Low-Temperature Grown ZnO Nanoparticles. *Electrochem. Commun.* **2009,** *11* (1), 118–121.

83. Keis, K.; Magnusson, E.; Lindström, H.; Lindquist, S.-E.; Hagfeldt, A.; A 5% Efficient Photoelectrochemical Solar Cell Based on Nanostructured ZnO Electrodes. *Sol. Energy Mater. Sol. Cells* **2002,** *73* (1), 51–58.
84. Martin, P. M.; Good, M. S.; Johnston, J. W.; Posakony, G. J.; Bond, L. J.; Crawford, S. L. Piezoelectric Films for 100-MHz Ultrasonic Transducers. *Thin Solid Films* **2000,** *379* (1–2), 253–258.
85. Morris, M. A.; R. S.; Walsh, I. V. Zinc Oxide Photoactive Material. U. S. Patent No. 6,063,849, 2000.
86. Thurian, U.; Kaczmarczyk, P.; Siegle, G.; Heitz, H.; Hoffmann, R.; Broser, A.; et al. Local Vibrational Modes of 3d Elements in WurtzrrE Type ZnO and GaN Crystals. 1995.
87. Franklin, N. M.; Rogers, N. J.; Apte, S. C.; Batley, G. E.; Gadd, G. E.; Casey, P. S. Comparative Toxicity of Nanoparticulate ZnO, Bulk ZnO, and ZnCl 2 to a Freshwater Microalga (*Pseudokirchneriella subcapitata*): The Importance of Particle Solubility. *Environ. Sci. Technol.* **2007,** *41* (24), 8484–**8490**.
88. Pitkethly, M. J. Nanomaterials—The Driving Force. *Mater. Today* **2004,** *7* (12), 20–29.
89. Verbakel, F.; Meskers, S. C. J.; Janssen, R. A. J. Surface Modification of Zinc Oxide Nanoparticles Influences the Electronic Memory Effects in ZnO-Polystyrene Diodes. *J. Phys. Chem. C* **2007,** *111*, 10150–10153.
90. Brayner, R.; Ferrari-Iliou, R.; Brivois, N.; Djediat, S.; Benedetti, M. F.; Fievet, F. Toxicological Impact Studies Based on *Escherichia coli* Bacteria in Ultrafine ZnO Nanoparticles Colloidal Medium. *Nano Lett.* **2006,** *6*, 866–870.
91. Stoimenov, P. K.; Klinger, R. L.; Marchin, G. L.; Klabunde, K. J. Metal Oxide Nanoparticles as Bactericidal Agents. *Langmuir* **2002,** *18*, 6679–6686.
92. Ahmad, M.; Zhu, J. ZnO Based Advanced Functional Nanostructures: Synthesis, Properties and Applications. *J. Mater. Chem.* **2011,** *21*, 599–614.
93. Fortuny, A.; Bengoa, C.; Font, J.; Fabregat, A. Bimetallic Catalysts for Continuous Catalytic Wet Air Oxidation of Phenol. *J. Hazard. Mater.* **1999,** *64*, 181–193.
94. Rana, S.; Rawat, J.; Sorensson, M. M.; Misra, R. D. K. Antimicrobial Function of Nd3+ -Doped Anatase Titania-Coated Nickel Ferrite Composite Nanoparticles: A Biomaterial System. *Acta Biomater*. **2006,** *2*, 421–432.
95. Shanmugam, S.; Viswanathan, B.; Varadarajan, T. K. A Novel Single Step Chemical Route for Noble Metal Nanoparticles Embedded Organic–Inorganic Composite Films. *Mater. Chem. Phys.* **2006,** *95*, 51–55.
96. Padmavathy, N.; Vijayaraghavan, R. Enhanced Bioactivity of ZnO nanoparticles—An Antimicrobial Study. *Sci. Technol. Adv. Mater.* **2008,** *9*, DOI: 10.1088/1468-6996/9/3/035004.
97. Zhang, B. P.; Wakatsuki, K.; Binh, N. T.; Segawa, Y.; Usami, N. Low-Temperature Growth of ZnO Nanostructure Networks. *J. Appl. Phys.* **2004,** *96*, 340.
98. Li, F.; Li, Z.; Jin, F. J. Structural and Luminescent Properties of ZnO Nanorods Prepared from Aqueous Solution. *Mater. Lett.* **2007,** *61*, 1876.
99. Walgraef, D.; Nanostructure Initiation During the Early Stages of Thin Film Growth. *Physica E* **2002,** *15*, 33.
100. Willander, M.; Yang, L. L.; Wadeasa, A.; Ali, S. U.; Asif, M. H.; Zhao, Q. X.; Nur, O. Zinc Oxide Nanowires: Controlled Low Temperature Growth and Some Electrochemical and Optical Nano Devices. *J. Mater. Chem.* **2009,** 19, 1006.

101. Zhaochun, Z.; Baibiao, H.; Yongqin, Y.; Deliang, C. Electrical Properties and Raman spectra of Undoped and Al-Doped ZnO Thin Films by Metalorganic Vapor Phase Epitaxy. *Mater. Sci. Eng. B-Adv*. **2001,** *86*, 109.
102. Xing, Y. J.; Xi, Z. H.; Xue, Z. Q.; Zhang, X. D.; Song, J. H.; Wang, R. M.; Xu, J.; Song, Y.; Zhang, S. L.; Yu, D. P. Optical Properties of the ZnO Nanotubes Synthesized via Vapor Phase Growth. *Appl. Phys. Lett.* **2003,** *83*, 1689.
103. Fan, H. J.; Scholz, R.; Kolb, F. M.; Zacharias, M.; Gösele, U.; Heyroth, F.; Eisenschmidt, C.; Hempel, T.; Christen, J. On the Growth Mechanism and Optical Properties of ZnO Multi-Layer Nanosheets. *Appl. Phys. A—Mater*. **2004,** *79*, 1895.
104. Moss, T. S. A Relationship Between the Refractive Index and the Infra-Red Threshold of Sensitivity for Photoconductors. *Proc. Phys. Soc. B* 63, 167.
105. Gupta, V. P.; Ravindra, N. M. Comments on the Moss Formula. *Phys. Stat. Sol. B* 100, 715.
106. Al-Douri, Y.; Feng, Y. P.; Huan, A. C. H. Optical Investigations Using Ultra-Soft Pseudopotential Calculations of $Si_{0.5}Ge_{0.5}$ Alloy. *Solid State Commun.* **2008,** *148*, 521.
107. Al-Douri, Y.; Reshak, A. H.; Baaziz, H.; Charifi, Z.; Khenata, R.; Ahmad, S.; Hashim, U. An Ab Initio Study of the Electronic Structure and Optical Properties of $CdS_{1-x}Tex$ Alloys. *Solar Energy* **2010,** 84, 1979.
108. Herve, P.; Vandamme, L. K. J. General Relation Between Refractive Index and Energy Gap in Semiconductors. *Infrared Phys. Technol.* **1993,** *35*, 609.
109. Ravindra, N. M.; Auluck, S.; Srivastava, V. K. On the Penn Gap in Semiconductors. *Phys. Stat. Sol.* (B) **1979,** *93*, K155.
110. Herve, P. J. L.; Vandamme, L. K. J. Empirical Temperature Dependence of the Refractive Index of Semiconductors. *J. Appl. Phys.* **1995,** *77*, 5476.
111. Ghosh, D. K.; Samanta, L. K.; Bhar, G. C. A Simple Model for Evaluation of Refractive Indices of Some Binary and Ternary Mixed Crystals. *Infrared Phys*. **1984,** *24*, 34.
112. Penn, D. R. Wave-Number-Dependent Dielectric Function of Semiconductors. *Phys. Rev.* **1962,** *128*, 2093.
113. Van Vechten, J. A. Theory of Electronegativity in Covalent Systems. I. Electronic Dielectric Constant. *Phys. Rev.* **1969,** *182*, 891.
114. Schroer, P.; Kruger, P.; Pollmann, J. First-Principles Calculation of the Electronic Structure of the Wurtzite Semiconductors ZnO and ZnS. *Phys. Rev. B* **1993,** *47*, 6971.
115. *Handbook of Chemistry & Physics*; 53th edn.; CRC Press, 1973.
116. Samara, G. A. Température and Pressure Dependences of the Dielectric Constants of Semiconductors. *Phys. Rev. B* **1983,** *27*, 3494.
117. Israr, M. Q.; Sadaf, J. R.; Yang, L. L.; Nur, O.; Willander, M.; Palisaitis, J.; Persson, P. O. A. Trimming of Aqueous Chemically Grown ZnO nanorods into ZnO Nanotubes and Their Comparative Optical Properties. *Appl. Phys. Lett.* **2009,** *95*, 073114.
118. Yang, R. D.; Tripathy, S.; Li, Y.; H.-Sue, J. Photoluminescence and Micro-Raman Scattering in ZnO Nanoparticles: The Influence of Acetate Adsorption. *Chem. Phys. Lett.* **2005,** *411*, 150.
119. Fonoberov, V. A.; Balandin, A. A. Interface and Confined Optical Phonons in Wurtzite Nanocrystals. *Phys. Rev. B* **2004,** *70*, 233205.
120. Abdulgafour, H. I.; Hassan, Z.; N. Al-Hardan, Yam, F. K. Growth of Zinc Oxide Nano Flowers by Thermal Evaporation Method. *Physica B* **2010,** *405*, 2570.

CHAPTER 9

First-Principles LDA+U Prediction of Electronic, Magnetic, and Structural Characteristics of the Ternary Iron Arsenide $CaFe_2As_2$

L. BELDI[1], F. SEMARI[2], VIPUL SRIVASTAVA[3], B. BOUHAFS[1], and R. KHENATA[2]

[1]*Laboratoire de Modélisation et Simulation en Sciences des Matériaux, Université DjillaliLiabès de Sidi Bel-Abbès, Sidi Bel-Abbès, Algeria*

[2]*Laboratoire de Physique Quantique et de Modélisation Mathématique, Université de Mascara, Algeria*

[3]*Department of Physics, School of Chemical Engineering & Physical Sciences, Lovely Professional University, Punjab, India*

ABSTRACT

Electronic structure estimations using density functional theory of $CaFe_2As_2$ ternary compound have been predicted through the full-potential augmented plane wave as well as the local orbitals method. The LSDA and LSDA+U schemes have been implemented to parameterize the effects of exchange and correlation. The equilibrium volumes for the nonmagnetic, ferromagnetic, and antiferromagnetic orderings are determined via the total energy approach. Further, the electronic band profile has been analyzed as well. Results mainly show that the $CaFe_2As_2$ compound has the AFM2 spin configuration. The band structures and the density of states for $CaFe_2As_2$ exhibit a metallic behavior due to the 3*d* orbital of iron atom.

Advanced Materials for Next-Generation Technologies: Challenges and New Prospects. Shrikaant Kulkarni, Vipul Srivastava, Rabah Khenata, & Yarub Al-Douri (Eds.)

9.1 INTRODUCTION

It is well known that high-temperature superconductivity has been receiving attention of researchers and scientific community. As a result, it has produced a whole new family of superconducting materials for technological and fundamental issues, as well as new trends for comprehending the physics of atypical superconductivity in other materials, such as the heavy-fermion inter-metallics, copper-oxides, etc. Kamihara et al.[1,2] discovered the first iron arsenide based superconductor (i.e., LaFeAs $(O_{1-x}F_x)$), with a transition temperature T_c=26–43K, followed by SmFeAs$(O_{1-x}F_x)$, with a T_c=55 K.[3] Basically, the ZrCuSiAs-type structure constructs both the compounds, and the layers $(LnO)^+$ and $(FeAs)^-$ are the main composition, where Ln=La and Sm. On the other hand, similar layers (i.e., FeAs) are figured out in other new family, $AEFe_2As_2$, where *AE*=alkaline earth metal (Ba, Ca, Eu, and Sr), with the $ThCr_2Si_2$-type structure.[4–16] It is acknowledged that the magnetic properties of the Fe_2As_2 layer have a significant impact on the mechanism of the superconductivity in these compounds. In general, under specific conditions, the $AEFe_2As_2$ materials with the $ThCr_2Si_2$-type structure (122 phase) have been determined to be superconductors. Superconductivity in this family can be induced by hole doping up to 38 K, for example, partial substitution of K or Na for Ca, Ba, or Sr atoms, and can be also induced by applied hydrostatic pressure.[17–22]

Among the most noteworthy superconductors of this family, it is the $CaFe_2As_2$ system, because Sr and Ba materials need 2.8 and 2.5 GPa, respectively, to induce superconductivity; it can be done at only 0.4 GPa (non-hydrostatic).[23,24] It is observed that a transition from tetragonal to orthorhombic symmetry is carried out for all these compounds, and accompanied by anti-ferromagnetic spin ordering.[5–16] Moreover, a strong coupling of the magnetic and structural phase transitions is perceived.[7,25] In this context, at ambient pressure and below approximately 170 K, a transition from a NM-tetragonal phase to AFM-orthorhombic phase has been noticed for $CaFe_2As_2$.[5,7–12] For the orthorhombic phase, Fe moments order in the known AFM2 structure with moments directed through the a-axis of the orthorhombic structure (Fe atoms is 0.80 μ_B).[7,26] The results showed that when p is greater than 0.35 GPa (at T=50 K)[9] diffraction measurements of $CaFe_2As_2$ under hydrostatic pressure, the AFM-orthorhombic phase transforms to a new NM ordered tetragonal structure with a significant decrease for both c/a ratio (11%) and the unit cell volume (5%). Park et al.[27] explain that the transition to the collapsed tetragonal phase takes place near to the pressure at which superconductivity is first seen.

A large number of experimental studies have been carried out on $CaFe_2As_2$.[13,15,16,28–35] The powder X-ray diffraction spectra on synthesized single crystals of $CaFe_2As_2$ have been presented by Ronning et al.[15] and Ni et al.[28] Wu et al.[16] have studied the anisotropy of resistivity and susceptibility of $CaFe_2As_2$. Liu et al.[29] have used the angle-resolved photoemission spectroscopy (ARPES) to study the electronic properties of $CaFe_2As_2$. Recently, Poirier et al.[30] have performed a pulsed ultrasonic experiment on thin pnictide single crystals $CaFe_2As_2$. Choi et al.[31] have reported inelastic light-scattering experiments on $CaFe_2As_2$. Neutron scattering experiments on $CaFe_2As_2$ were performed by Kreyssig et al.[13] Adhikary et al.[32] have employed the high-resolution photoemission spectroscopy to investigate the temperature evolution of the electronic structure of $CaFe_2As_2$. Dioguardi et al.[33] have reported the NMR spectra in the orthorhombic phase of $CaFe_2As_2$. Recently, Ortenzi et al.[34] have presented the temperature varying magnetic moment (MM) in $CaFe_2(As_{1-x}P_x)_2$. Mishra et al.[35] have conducted additional research to study the magnetic phase transition in $CaFe_2As_2$ using the time differential perturbed angular distribution (TDPAD) spectroscopy. Many theoretical studies using the state-of-the-art computational methods have been carried out to examine the magnetic properties and the electronic structure of $CaFe_2As_2$.[12,14,36–42]

Based on a first-principles calculations work, it has been proposed that the Fe moment is not quenched in the collapsed tetragonal phase and orders in an alternative Nèel state (the supposed AFM1 structure).[12] A lot of investigations into the electronic and magnetic properties on $CaFe_2As_2$ have been made both experimentally and theoretically,[7,9,12,25–27,43–45] Nakamura et al. suggested that a negative on-site energy U (Hubbard) can be adopted in such a system to solve the problem of the overestimated MM of Fe-atom in iron-based superconductors.[46] Theoretical studies on the electronic structure of $CaFe_2As_2$ have been performed by several authors using the projected augmented waves (PAW) pseudopotentials method within GGA.[14,36–38] Ji et al.[36] show that strong Hund's rule coupling on Fe 3d-orbitals is the primary cause of the creation of a local moment on the Fe ion. The high-pressure phases study and the uniaxial pressure structural relaxations for $CaFe_2As_2$ have been reported by Colonna et al.[37] and Tomić et al.,[14] respectively. With theoretical studies on the structural aspects of $CaFe_2As_2$ using the same approach, Widom and Quader[38] have found that both magnetic-orthorhombic and NM-tetragonal phases remain mechanically stable and that the Fe moment is larger ($\mu_{Fe} \approx 1.78\ \mu_B$ at $P = 0$) than the experimental results (0.8 μ_B). Some theoretical studies have focused directly on the band structures of tetragonal and collapsed tetragonal phases of $CaFe_2As_2$ using a new generation of the LMTO method, the so-called Nth-order muffin-tin orbital (NMTO) method.[39] There have also been a number of theoretical studies discussing the phase transition of $CaFe_2As_2$ using the full potential linearized augmented plane wave method.[40–42] The electronic structure of $CaFe_2As_2$ has been investigated by Goldman et al.[40] for the tetragonal, and antiferromagnetic orthorhombic phases, by Diehl et al.[41] for the tetragonal and collapsed tetragonal phases and by Zhang et al.[42] under pressure.

In the present investigation, a density functional calculation has been performed to study the electronic structures and magnetic orders of parent-undoped $CaFe_2As_2$ in order to understand this class of materials using the WIEN2k code. The local density approximations LSDA and LSDA+U were used to treat the exchange and correlation functional, since the studied material is strongly correlated. A brief description of the theoretical method used in our study is given in the next section. For the low- and high-temperature phases of $CaFe_2As_2$, the results of the magnetic properties and the electronic structure are presented in Section 3. Finally, in Section 4 a brief summary is given.

9.2 THEORETICAL METHOD

Basically, at room temperature, $CaFe_2As_2$ is stable in the $ThCr_2Si_2$-type structure (SG *I4/mmm*). There are two formula units in the unit cell, as shown in Figure 9.1(a). According to a review of the literature, the experimental lattice parameters are 3.879 and 11.740 Å.[7] The crystallization of the compound (a=5.506 Å, b=5.450 Å, and c=11.664 Å) is found in orthorhombic (SG *Fmmm*) structure at low temperature. The unit cell has four formula units in it (Fig. 9.1(b)). The optimizations started with the crystallographic data from Goldman et al.[7] as their starting point.

The FP-LAPW method was used to perform the band structure calculations as executed in the WIEN2K code,[47] which is an efficient method for the precise calculations and modeling of the ground-state properties of the materials.[48,49] This is regarded as a density functional theory[51,52] implementation for the hybrid FP-LAPW and local orbitals (LAPW+lo) approach.[50] In essence, core states are treated using a multi-configuration relativistic Dirac–Fock technique, whereas valence states are treated using a scalar relativistic approach. The exchange-correlation energy was estimated using the LDA,[53] but the LSDA with additional Hubbard correlation components (i.e., using the LSDA +U technique) described e-e repulsion linked to the 3d bands of Fe.[54] A highly meticulous step analysis is used to ensure convergence of the total energy in terms of the variational cutoff-energy parameter. An appropriate set of k points to compute the total energy have been used in the same time. In general, for our calculations, Ca:[Ar]4s^2, Fe:[Ar]3$d^6$4s^2, and As:[Ar]3d^{10}4$s^2$4p^3 were used as electronic configuration. The equilibrium lattice constants as well as bulk modulus have been computed through fitting the total energy versus volume to the Murnaghan[55] equation. Present calculations use 405 and 729 k-points for tetragonal and orthorhombic phases, respectively, and for the cutoff-energy, the value of 9 Ry is used for all configurations (NM, FM, and AFM phases) considered in the investigation. Additionally, the self-consistent calculations are only regarded to be converged when the calculated total energy of the crystal converged to less than 1 mRy. The values of 2.5, 2.17, and 1.92 Bohr have been adopted for 'Ca', 'Fe,' and 'As' as MT radii. Finally, a modified tetrahedron method of Blöchl et al.[56] was used to predict the total density of states (DOS).

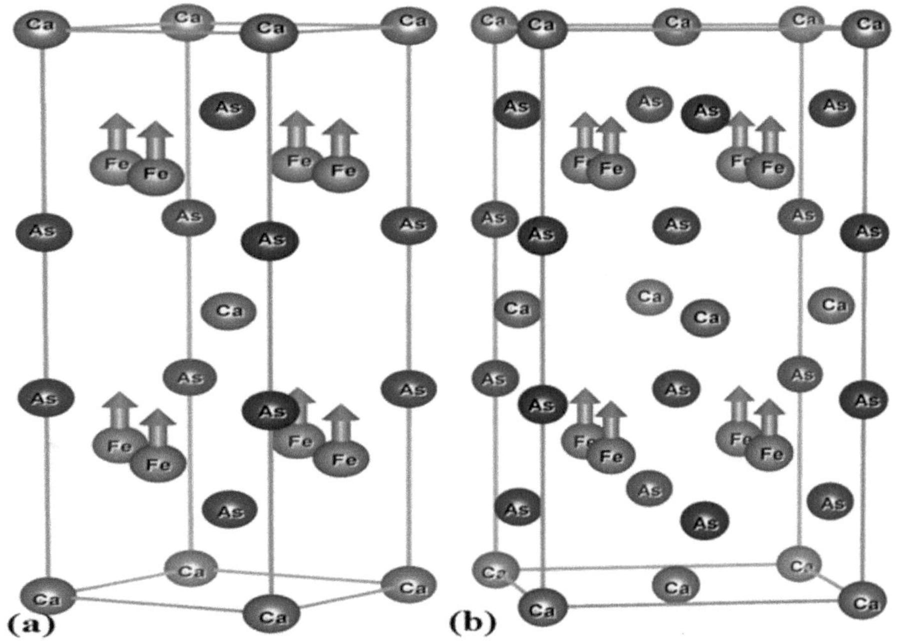

FIGURE 9.1 Crystal structures of $CaFe_2As_2$ compound in ferromagnetic spin configuration for (a) tetragonal, (b) orthorhombic structures.

9.3 RESULTS AND DISCUSSION

a. Structural properties

In general, an optimization for the nonmagnetic (NM), ferromagnetic (FM), and antiferromagnetic (AFM) states is done for the structure of $CaFe_2As_2$ in orthorhombic *Fmmm* symmetry. As illustrated in Figure 9.2, a prediction of the AFM spin ordering is done.

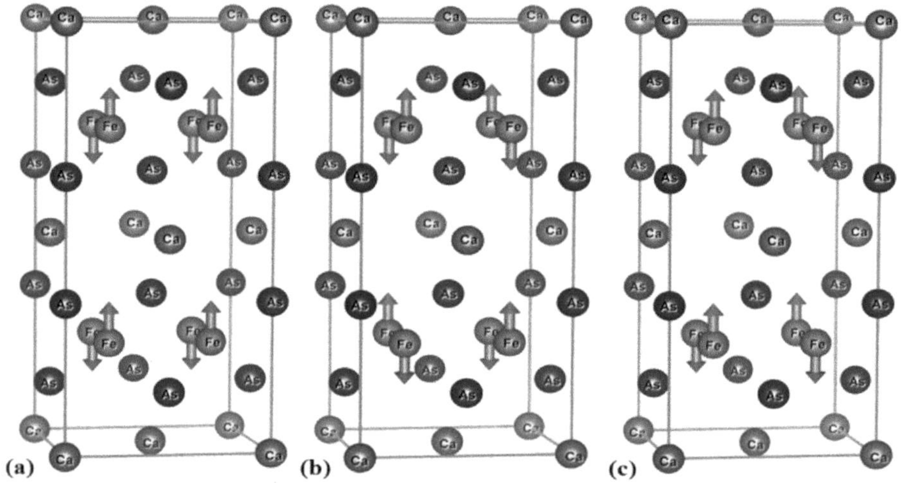

FIGURE 9.2 Crystal structures of orthorhombic $CaFe_2As_2$ compound in antiferromagnetic spin configuration: (a) AFM1, (b) AFM2, and (c) AFM3.

The total energy with cell volume has been calculated for the NM, FM, and AFM orderings by LDA and LSDA, as depicted in Figure 9.3.

Besides, clear evidence can be observed that the FM state is the most stable state for *Fmmm* orthorhombic phase that has the lowest energy. In this context, energies higher than the ground state are observed for the other excited configurations (i.e., NM, AFM_1, AFM_2, and AFM_3). It can be interpreted based on our calculations that there is continuous convergence between the FM state in $CaFe_2As_2$ and the NM state. This indicates that there is no existence of the AFM ordering in the orthorhombic phase, which is in inconsistency with the experimental results.[7] To resolve this issue, LSDA+U calculation has been conducted. For $CaFe_2As_2$, the standard DFT functional cannot produce precise charge densities. It will not be possible to forecast structural characteristics, magnetic moments, or orbital occupancy with

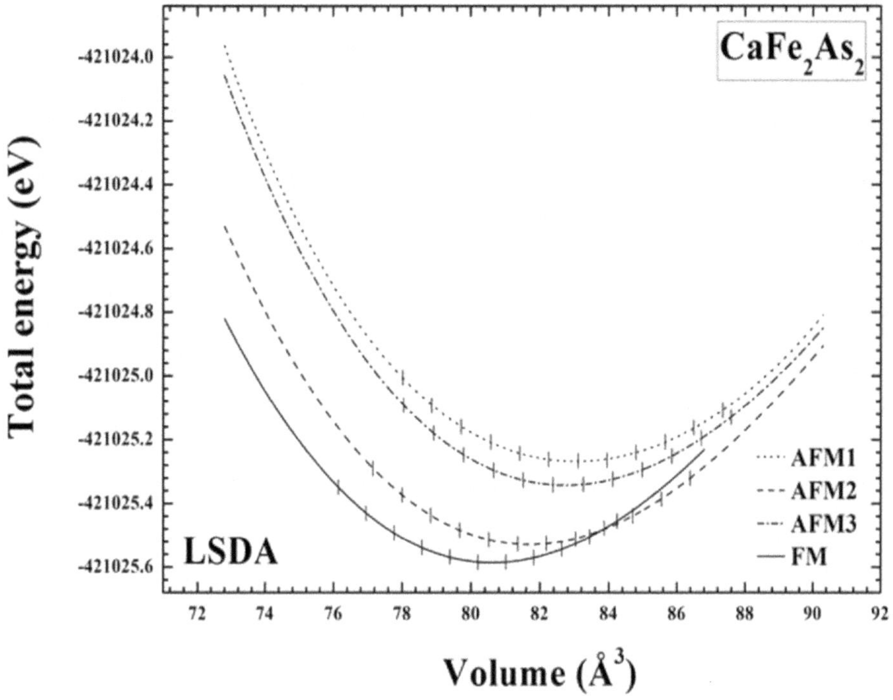

FIGURE 9.3 Variation of total energy of orthorhombic $CaFe_2As_2$ compound (LSDA) with unit cell volume for different magnetic and no-magnetic phases.

sufficient accuracy. In order to enforce the correct orbital occupancy (and hence increase the charge density) for this system, it is essential to include the Hubbard U correction. This will allow for a more accurate forecast of the system's electronic, magnetic, and structural properties. A Hubbard-like component is added to the effective potential to tackle the high Coulomb repulsion between localized Fe 3d states, which improves the description of correlation consequences in $CaFe_2As_2$. In actuality, there is not a widely recognized and acknowledged approach for how to include a Hubbard term in the DFT framework. The precise value of U, which corresponds to the Ueff parameter (Ueff = U-J), is still up for debate because it significantly affects the expected physical characteristics. This research demonstrates how the Hubbard U correction enhances the description of the electronic structure and has a major impact on the correct occupancy of the Fe 3d-orbitals, which results in a fairly accurate charge density of the complex $CaFe_2As_2$. The change of the magnetic moment of $CaFe_2As_2$ per Fe atom as a function of U_{eff} is shown in Figure 9.4. We begin with the lattice parameters as optimized in the first step (FM state) and vary U_{eff} from positive values (1 eV) to negative values (− 1 eV) to find out the best U_{eff} value to fit the experimental magnetic moment value of Fe atoms (0.80 μ_B)[7] (see Figure 9.4). We find the best fitting value at $U_{eff} = -0.18$ eV. Goldman et al.[7] clarified that the negative U correction is crucial for matching the estimated Fe magnetic moment with the seen one. Besides, an agreement with the experimental results in a negative U_{eff} region has been observed as the moment decreases with the decreasing of U_{eff}.

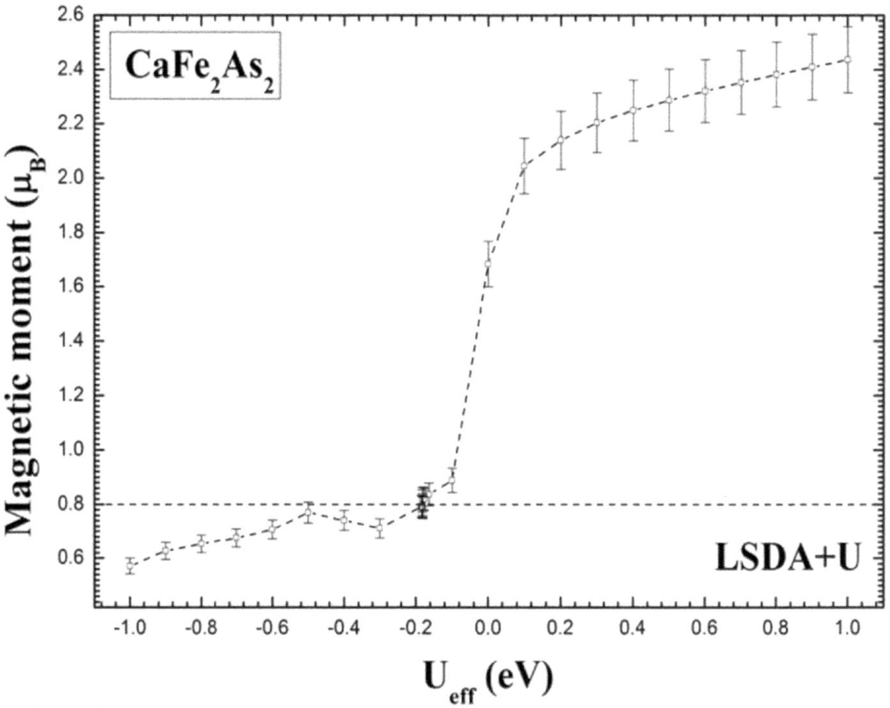

FIGURE 9.4 Variation of magnetic moment per Fe atom in the FM spin configuration of orthorhombic $CaFe_2As_2$. The Fe atoms' experimental magnetic moment value is shown by the horizontal dashed line.

On the other hand, for the positive U_{eff} region, U_{eff} stresses the localized character of Fe 3*d* orbital electrons, and overemphasizes the discrepancy with the experiments. In fact, there have been several reports about the negative Hubbard *U*. Using ab-initio calculations, DFT predicts a negative value for U_{eff} in iron-based superconductors $LaFeAsO_{1-x}F_x$,[46] carbon vacancy in 4H-SiC,[57] and iron-based double perovskite Sr_2FeMoO_6.[58]

Finally, we fix the U_{eff} and optimize the compound in orthorhombic phase for the FM and AFM states using LSDA+U. The energies versus volume are plotted in Figure 9.5. It is clearly seen that the AFM_2 ordering represented the ground state; there is a good agreement with the theoretical calculations[8] and experimental observations.[7]

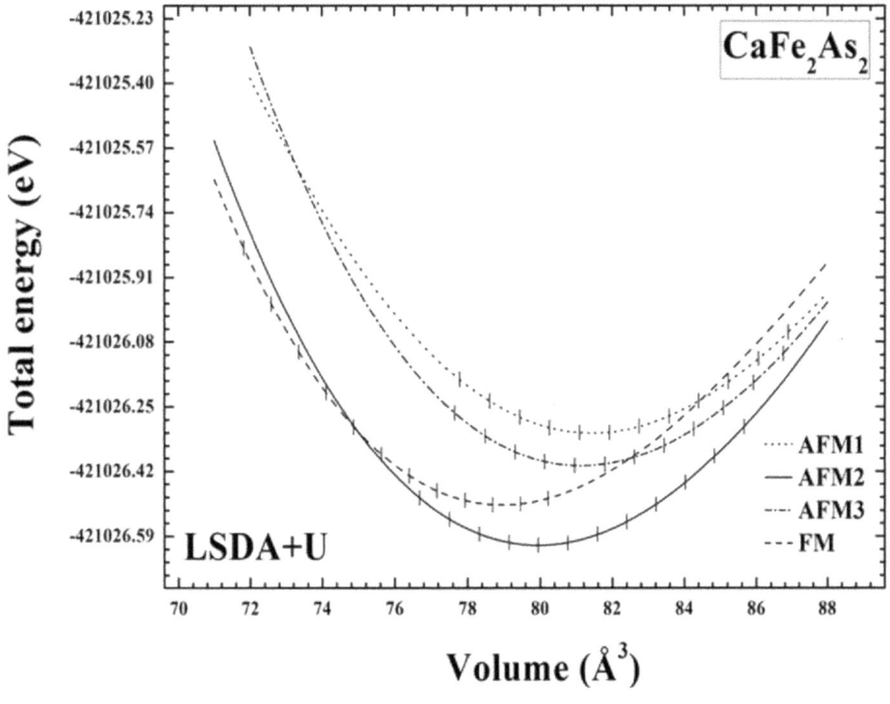

FIGURE 9.5 Variation of total energy of orthorhombic $CaFe_2As_2$ compound (LSDA+U) with unit cell volume for FM and AFM spin ordering.

Table 9.1 lists the calculated lattice constants (a, b, and c in Å), internal coordinate (z_{As}), the bulk modulus, and its derivative of the $CaFe_2As_2$ in *Fmmm* symmetry for different magnetic states. A comprehensive comparison has been done with the other researchers in terms of theoretical analysis as well as experimental results.

For completeness purpose, we have also estimated the structure of $CaFe_2As_2$ with tetragonal I4/mm symmetry. Tetragonal structure is described by the two lattice parameters a and c, which are optimized for the NM and FM states using LDA and LSDA. For the configurations NM and FM, the total energy as a function of volume is depicted in Figure 9.6.

TABLE 9.1 The Calculated Lattice Constants (*a*, *b*, and *c*), Internal Coordinate (z_{As}), Bulk Modulus (B), and its Pressure Derivative (B') of Orthorhombic $CaFe_2As_2$ Compound for Different Magnetic States, Obtained in our LDA, LSDA, and LSDA+U Calculations.

			a (Å)	*b* (Å)	*c* (Å)	z_{As}	*B* (GPa)	*B'*
LDA	This work	NM	5.395	5.265	11.429	0.358	137.620	4.329
LSDA		FM	5.408	5.262	11.330	0.360	138.269	4.110
		AFM1	5.286	12.050	5.213	0.392	124.903	3.923
		AFM2	5.330	5.343	11.470	0.357	136.547	4.691
		AFM3	5.232	5.379	11.758	0.356	134.706	4.773
LSDA+*U*		FM	5.389	5.279	11.102	0.364	140.779	4.379
		AFM1	5.205	11.628	5.379	0.389	128.746	4.246
		AFM2	5.367	5.454	10.922	0.366	138.887	4.643
		AFM3	5.272	5.398	11.401	0.376	137.597	4.653
Exp.	Ref. [7]	Exp.	5.506	5.450	11.664	0.366		
	Ref. [13]		5.5312	5.4576	11.683	0.3689		
GGA	Ref. [14]		5.60	5.48	11.40		70	
	Ref. [8]	FM	5.580	5.476	11.505	0.365	60.734	3.664

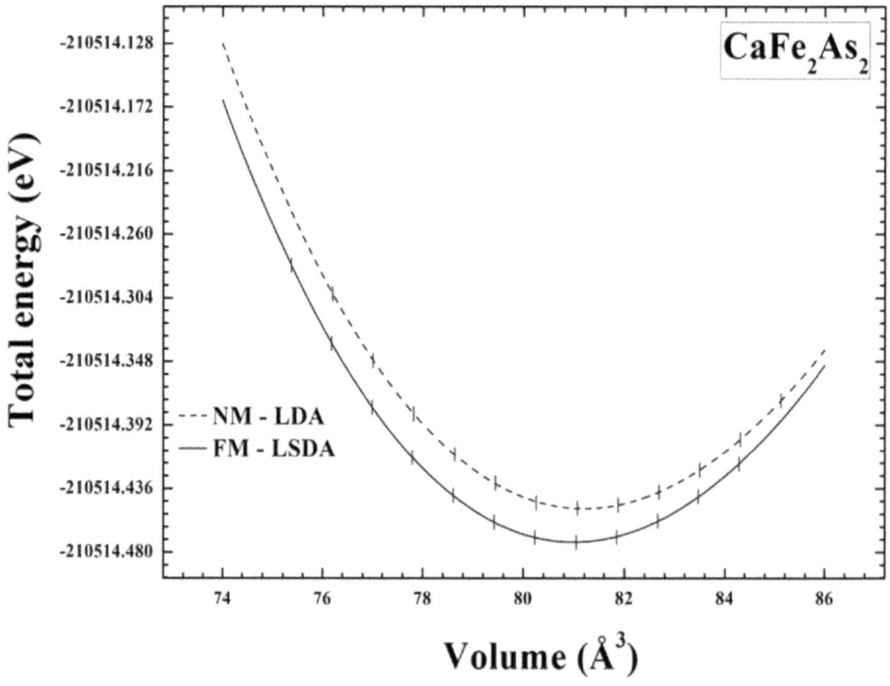

FIGURE 9.6 Variation of total energy of tetragonal $CaFe_2As_2$ compound with unit cell volume for the NM-LDA and AFM-LSDA configurations.

The energy difference between the NM and FM states is quite modest (approximately 0.2 meV), as can be seen in the graph. Consequently, from the structural perspective, the ferromagnetic case is considered analogous to the nonmagnetic case. According to the experimental side, $CaFe_2As_2$ in the tetragonal structure is not magnetic, as demonstrated by Ni et al.,[5] Goldman et al.,[7] and Kreyssig et al.[9] Table 9.2 lists the estimated lattice constants for the two magnetic states (a and c), internal coordinate (zAs), bulk modulus, and its derivative for the $CaFe_2As_2$ in I4/mm symmetry. They are compared with the existing experimental data as well as the theoretical works.

TABLE 9.2 The Calculated Lattice Parameters (a and c), Internal Coordinate (z_{As}), Bulk Modulus (B), and its Pressure Derivative (B') of Tetragonal $CaFe_2As_2$ Compound for Different Magnetic States, Obtained in our LDA, LSDA, and LSDA+U Calculations.

		***a* (Å)**	***c* (Å)**	z_{As}	***B* (GPa)**	***B'***
LDA	This work	3.766	11.448	0.358	137.905	4.208
LSDA		3.779	11.341	0.358	138.856	4.031
Exp.	Ref. [5]	3.912	11.667			
	Ref. [7]	3.879	11.740			
	Ref. [9]	3.8915	11.690	0.372		
	Ref. [13]	3.9797	10.628	0.365		
	Ref. [15]	3.887	11.758			
	Ref. [16]	3.872	11.730	0.3665		
LDA	Ref. [10]	3.910	10.264			
	Ref. [11]	3.914	11.108			
GGA	Ref. [12]	3.982	10.39			
	Ref. [8]	3.985	10.426	0.362	79.492	7.697

b. Band structure and density of states

Using the optimized lattice parameters, the electronic band structures have been calculated, the density of states (DOS) and charge densities of the compound in both phases tetragonal and orthorhombic in the nonmagnetic, ferromagnetic, and antiferromagnetic states. Additionally, the direction of the iron spin for the majority spin and the minority spins is taken into consideration when determining the spin direction for up and down.

Figure 9.7 displays the DOS of the AFM2 spin configuration using LSDA+U_{eff} in the orthorhombic phase. It is noticeable that the compound exhibits a metallic behavior. It has been observed that at the bottom of the valence band (at ~ − 13.3 and − 11.2 eV), the As4*s* states are concentrated. Besides, the As4*p* band demonstrates two peaks (at ~ − 5.9 and − 2.6 eV) and the upper band is strongly mixed with Fe3*d* states. The Fe3*d* states are hybridized with As4*p* states positioned around 3–4 eV lower than the Fermi level.

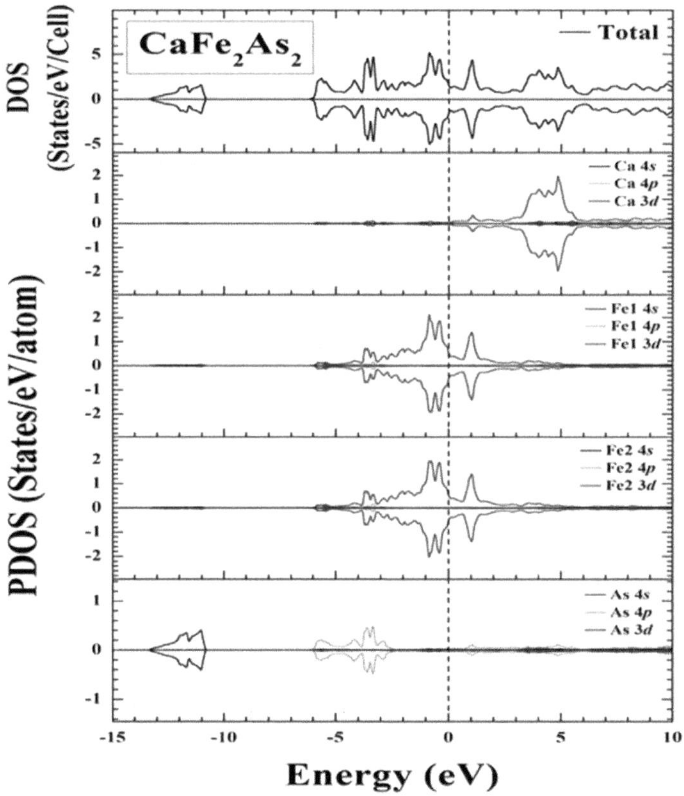

FIGURE 9.7 Spin-polarized total and partial DOS of the orthorhombic $CaFe_2As_2$ compound (LSDA+U) in the AFM2 spin configuration. On X-axis scale value "0" represents Fermi level.

Figure 9.8 shows that the bands at the Fermi level (at − 4 to 2.7 eV) for the tetragonal structure are primarily produced by Fe3d states. The As 4p states that hybridize with the Fe3d states produce the bands that range in energy from − 5.4 to − 2.6 eV. The Ca atom makes a small contribution in the valence bands; however, it is placed at the bottom of the conduction band. We therefore observe the analogy to the nonmagnetic states.

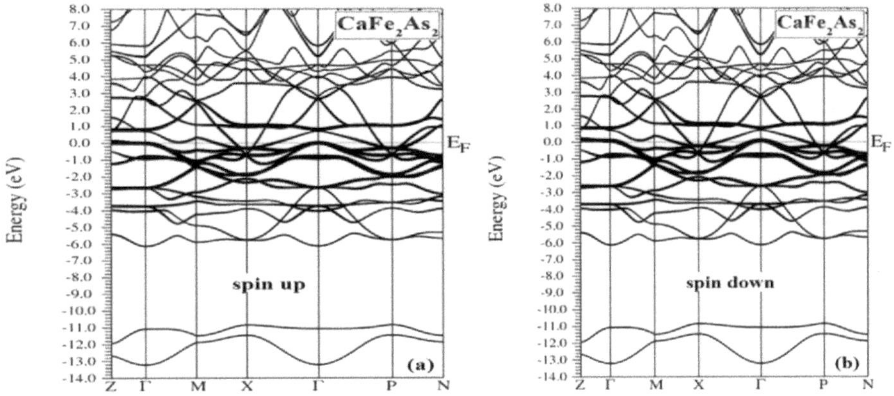

FIGURE 9.8 Spin-polarized band structures of the tetragonal $CaFe_2As_2$ compound (LSDA) at the equilibrium lattice constant (*a*) spin up, and (*b*) spin down states.

Figures 9.9 and 9.10 display the predicted non-spin-polarized and spin-polarized DOS using LDA and LSDA, respectively. Figure 9.10 indicates that the DOS is dominated (at −13.2 to −10.8eV) by As 4*s* states. The presence of three zones can be seen in the valence band when it is stretched from −6 eV to the Fermi level. The first one (between −6 and −4 eV) is mostly caused by As4p states, with Fe3d states having a little influence. As 4p and Fe 3d states make up the majority of the second area (at −4 to −2.5 eV), which is dominated by their contributions. The iron and arsenic atoms' covalent bonds are the result of this sp-d hybridization. The third region which ranges from −2.5 eV up to the Fermi level is mainly due to Fe 3*d* states. Above the Fermi level up to 6 eV, the contribution of Ca 3*d* states is dominated. Both configurations (NM) and (FM) exhibit a metallic behavior in the tetragonal structure.

In resume, the Fe3*d* states dominate the density of states near the Fermi level, which could be related to the superconductivity of the material. Our calculations approve the other first-principle studies[59–61] for the electronic structure of $CaFe_2As_2$.

c. *Charge density and bonding*

Mainly, it has been proved that our adopted method is capable of fulfilling a precise charge density from the highly converged wave functions. Therefore the bonding nature of the solid can be studied through charge density. Accordingly, the AFM configuration has been used for the orthorhombic phase of our compound. The complete valence charge density is demonstrated in theAFM$_2$ configuration using LSDA+U, as shown in Figure 9.11. It is clear that the atoms of Ca and As do not have any directional (covalent) bonds. In actuality, a metallic band exists between the two Ca atoms. The Fe-3d orbitals are responsible for the charge density that is significant surrounding the Fe ions. Covalent bonds bind Fe and As. According to the density of states, each Fe atom creates Fe-As bonds in tetrahedral directions by hybridizing its 3d and 4p orbitals with those of As. Ca, Fe, and As have electronegativity values of 1, 1.85, and 2.2, respectively.[62] Both from Fe to As atoms and from Ca to Fe-As layers, electrons are transported.

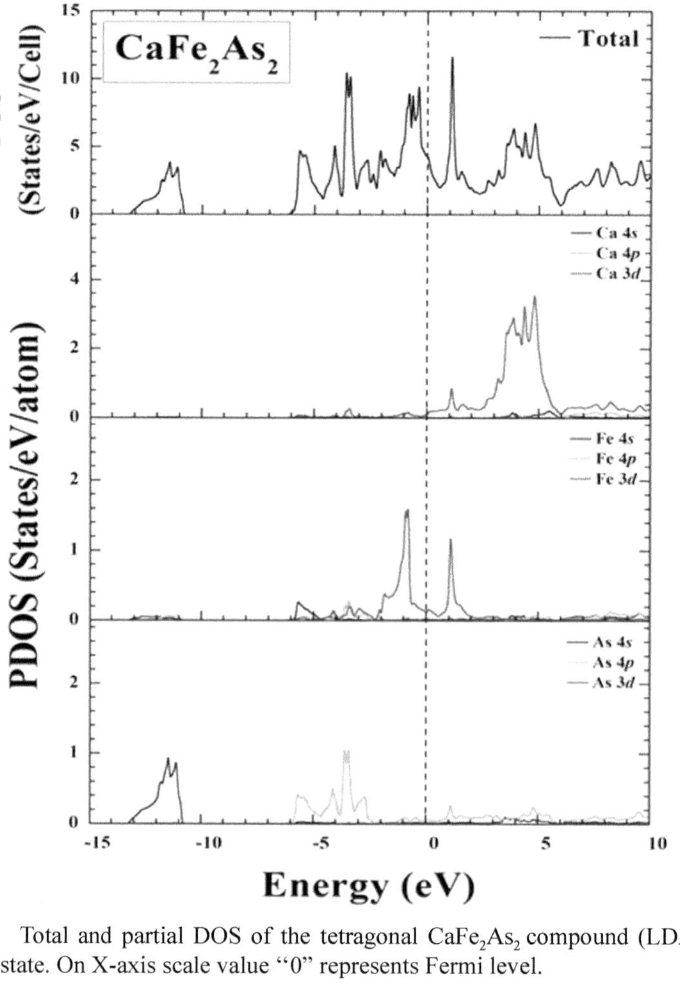

FIGURE 9.9 Total and partial DOS of the tetragonal $CaFe_2As_2$ compound (LDA) in the nonmagnetic state. On X-axis scale value "0" represents Fermi level.

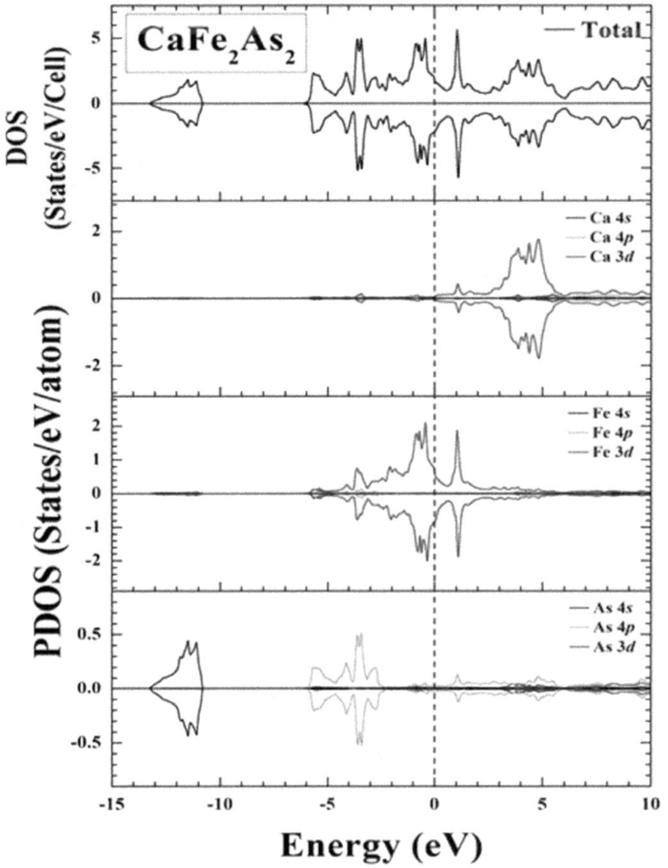

FIGURE 9.10 Spin-polarized total and partial DOS of tetragonal $CaFe_2As_2$ compound (LSDA). On X-axis scale value "0" represents Fermi level.

9.4 CONCLUSION

As a summary, a theoretical study of the electronic structure of iron pnictide $CaFe_2As_2$ compound using the framework LSDA+U with broadening the range of U ($\equiv U_{eff}=U-J$) from the positive to the slightly negative range, has been presented in this investigation. As a consequence, it has been found that the estimated magnetic moment demonstrates good agreements with the experimental results in the negative U_{eff} range. Our analysis covers atomic, electronic, and magnetic properties. Although the orthorhombic Fmmm phase is the focus of the current work, the tetragonal I4/mmm phase is also taken into consideration for completeness. The magnetic ground state of the Fmmm phase was calculated among several spin configurations using a negative Ueff of − 0.18 eV, and the current results demonstrate a respectable agreement with the existing experiment. The magnetic moment of Fe that we estimated closely matches the findings of the experiment. According to the BS and overall DOS, the system behaves metallically for both phases. A sharp peak near the Fermi level is due to Fe-3d states. Hybridized forms of Fe-3d and As-4p are also dominant. Only at lower energy levels does Ca contribution happen. The analysis of the charge density and the knowledge that is currently known regarding charge transfer and bonding have both shown it. Finally, electron transfers from Ca to Fe-As layers and from Fe to As atoms have been seen to occur.

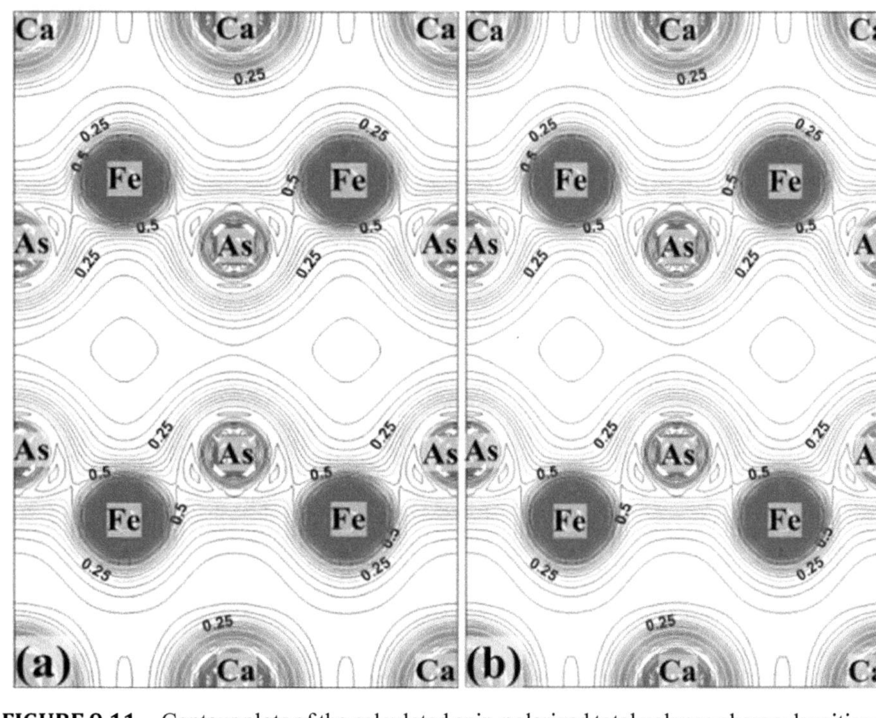

FIGURE 9.11 Contour plots of the calculated spin-polarized total valence charge densities (*a*) spin up and (*b*) spin down, in the $(1\bar{1}0)$ plane of $CaFe_2As_2$ (LSDA+U) in the AFM2 spin state.

ACKNOWLEDGMENTS

One of us (B.B.) acknowledges the Abdus-Salam International Center for Theoretical Physics (ICTP, Trieste, Italy).

KEYWORDS

- **density functional theory**
- **electronic structure**
- **ferromagnetism**
- **ferro-pnictides**
- **LSDA+U**
- **spin-polarization**

REFERENCES

1. Kamihara, Y.; Watanabe, T.; Hirano, M.; Hosono, H. *J. Am. Chem. Soc.***2008,** *130*, 3296–3297.
2. Takahashi, H.; Igawa, K.; Arii, K.; Kamihara, Y.; Hirano, M.; Hosono, H. *Nature* **2008,** *453*, 376–378.
3. Ren, Z.-A.; Lu, W.; Yang, J.; Yi, W.; Shen, X.-L.; Che, G.-C.; Dong, X.-L.; Sun, L.-L.; Zhou, F.; Zhao, Z.-X. *Chin. Phys. Lett.* **2008,** *25*, 2215–2216.
4. Rotter, M.; Tegel, M.; Johrendt, D.; Schellenberg, I.; Hermes, W.; Pöttgen, R. *Phys. Rev. B* **2008,** *78*, 020503.
5. Ni, N.; Nandi, S.; Kreyssig, A.; Goldman, A. I.; Mun, E. D.; Bud'ko, S. L.; Canfield, P. C. *Phys. Rev. B* **2008,** *78*, 014523.
6. Su, Y.; Link, P.; Schneidewind, A.; Wolf, T.; Adelmann, P.; Xiao, Y.; Meven, M.; Mittal, R.; Rotter, M.; Johrendt, D. *Phys. Rev. B* **2009,** *79*, 064504.
7. Goldman, A. I.; Argyriou, D. N.; Ouladdiaf, B.; Chatterji, T.; Kreyssig, A.; Nandi, S.; Ni, N.; Bud'ko, S. L.; Canfield, P. C.; McQueeney, R. J. *Phys. Rev. B* **2008,** *78*, 100506.
8. Wang, Y.; Ding, Y.; Ni, J. *Solid State Commun.* **2009,** *149*, 2125–2129.
9. Kreyssig, A.; Green, M. A.; Lee, Y.; Samolyuk, G. D.; Zajdel, P.; Lynn, J. W.; Bud'ko, S. L.; Torikachvili, M. S.; Ni, N.; Nandi, S.; Leão, J. B.; Poulton, S. J.; Argyriou, D. N.; Harmon, B. N.; McQueeney, R. J.; Canfield, P. C.; Goldman, A. I. *Phys. Rev. B* **2008,** *78*, 184517.
10. Mittal, R.; Pintschovius, L.; Lamago, D.; Heid, R.; Bohnen, K. P.; Reznik, D.; Chaplot, S. L.; Su, Y.; Kumar, N.; Dhar, S. K.; Thamizhavel, A.; Brueckel, T. *Phys. Rev. Lett.* **2009,** *102*, 217001.
11. Rende, M.; Li, Y.; Bai, Z.; Wang, L.; Chen, L. a. *Physica B: Condensed Matter* **2010,** *405*, 4226–4230.
12. Yildirim, T. *Phys. Rev. Lett.* **2009,** *102*, 037003.
13. Kreyssig, A.; Green, M.; Lee, Y.; Samolyuk, G.; Zajdel, P.; Lynn, J.; Bud'Ko, S.; Torikachvili, M.; Ni, N.; Nandi, S. *Phys. Rev. B* **2008,** *78*, 184517.
14. Tomić, M.; Valentí, R.; Jeschke, H. O. *Phys. Rev. B* **2012,** *85*, 094105.
15. Ronning, F.; Klimczuk, T.; Bauer, E. D.; Volz, H.; Thompson, J. D. *J. Phys.: Condensed Matter* **2008,** *20*, 322201.

16. Wu, G.; Chen, H.; Wu, T.; Xie, Y.; Yan, Y.; Liu, R.; Wang, X.; Ying, J.; Chen, X. *J. Phys.: Condensed Matter* **2008,** *20*, 422201.
17. Torikachvili, M. S.; Bud'ko, S. L.; Ni, N.; Canfield, P. C. *Phys. Rev. Lett.* **2008,** 101 (2008) 057006.
18. Rotter, M.; Tegel, M.; Johrendt, D. *Phys. Rev. Lett.* **2008,** *101*, 107006.
19. Sefat, A. S.; Jin, R.; McGuire, M. A.; Sales, B. C.; Singh, D. J.; Mandrus, D. *Phys. Rev. Lett.* **2008,** *101*, 117004.
20. Sasmal, K.; Lv, B.; Lorenz, B.; Guloy, A. M.; Chen, F.; Xue, Y.-Y.; Chu, C.-W. *Phys. Rev. Lett.* **2008,** *101*, 107007.
21. Jeevan, H. S.; Hossain, Z.; Kasinathan, D.; Rosner, H.; Geibel, C.; Gegenwart, P. *Phys. Rev. B* **2008,** *78*, 092406.
22. Li, G.; Hu, W. Z.; Dong, J.; Li, Z.; Zheng, P.; Chen, G. F.; Luo, J. L.; Wang, N. L.; *Phys. Rev. Lett.* **2008,** *101*, 107004.
23. Baek, S. H.; Lee, H.; Brown, S. E.; Curro, N. J.; Bauer, E. D.; Ronning, F.; Park, T.; Thompson, J. D.; *Phys. Rev. Lett.* **2009,** *102*, 227601.
24. Alireza, P. L.; Ko, Y. T. C.; Gillett, J.; Petrone, C. M.; Cole, J. M.; Lonzarich, G. G.; Sebastian, S. E. *J. Phys.: Condensed Matter* **2009,** *21*, 012208.
25. Baek, S. H.; Curro, N. J.; Klimczuk, T.; Bauer, E. D.; Ronning, F.; Thompson, J. D.; *Phys. Rev. B* **2009,** *79*, 052504.
26. Yildirim, T.; *Phys. Rev. Lett.* **2008,** *101*, 057010.
27. Park, T.; Park, E.; Lee, H.; Klimczuk, T.; Bauer, E.; Ronning, F.; Thompson, J. *J. Phys.: Condensed Matter* **2008,** *20*, 322204.
28. Ni, N.; Nandi, S.; Kreyssig, A.; Goldman, A.; Mun, E.; Bud'ko, S.; Canfield, P. *Phys. Rev. B* **2008,** *78*, 014523.
29. Liu, C.; Kondo, T.; Ni, N.; Palczewski, A. D.; Bostwick, A.; Samolyuk, G. D.; Khasanov, R.; Shi, M.; Rotenberg, E.; Bud'ko, S. L.; Canfield, P. C.; Kaminski, A.; *Phys. Rev. Lett.*, 102 (2009) 167004.
30. Poirier, M.; Bilodeau, M.; Lefebvre, S.; Karki, A. B.; Jin, R.; *Phys. Rev. B*, 89 (2014) 155129.
31. Choi, K. Y.; Wulferding, D.; Lemmens, P.; Ni, N.; Bud'ko, S. L.; Canfield, P. C. *Phys. Rev. B* **2008,** *78*, 212503.
32. Adhikary, G.; Biswas, D.; Sahadev, N.; Bindu, R.; Kumar, N.; Dhar, S.; Thamizhavel, A.; Maiti, K. *J. Appl. Phys.* **2014,** *115*, 123901.
33. Dioguardi, A. P.; Crocker, J.; Shockley, A. C.; Roberts-Warren, N.; Lin, C.; Shirer, K. R.; Nisson, D.; Lodhia, A.; Klavins, P.; Curro, N. J. *Superconductor Sci. Technol.* **2013,** *26*, 025012.
34. Ortenzi, L.; Gretarsson, H.; Kasahara, S.; Matsuda, Y.; Shibauchi, T.; Finkelstein, K. D.; Wu, W.; Julian, S. R.; Kim, Y.-J.; Mazin, I. I.; Boeri, L. *Phys. Rev. Lett.* **2015,** *114*, 047001.
35. Mishra, S. N.; Mohanta, S. K.; Davane, S. M.; Kumar, N.; Thamizhavel, A. *J. Magnet. Magnet. Mater.* **2014,** *349*, 240–243.
36. Ji, W.; Yan, X.-W.; Lu, Z.-Y. *Phys. Rev. B* **2011,** *83*, 132504.
37. Colonna, N.; Profeta, G.; Continenza, A.; Massidda, S.; *Phys. Rev. B* **2011,** *83*, 094529.
38. Widom, M.; Quader, K.; *Phys. Rev. B* **2013,** *88*, 045117.
39. Andersen, O. K.; Boeri, L. *Annalen der Physik* **2011,** *523*, 8–50.
40. Goldman, A.; Kreyssig, A.; K. Prokeš, Pratt, D.; Argyriou, D.; Lynn, J.; Nandi, S.; Kimber, S.; Chen, Y.; Lee, Y. *Phys. Rev. B* **2009,** *79*, 024513.
41. Diehl, J.; Backes, S.; Guterding, D.; Jeschke, H. O.; Valentí, R. *Phys. Rev. B* **2014,** *90* (2014) 085110.

42. Zhang, Y.-Z.; Kandpal, H. C.; Opahle, I.; Jeschke, H. O.; Valentí, R. *Phys. Rev. B* **2009,** *80*, 094530.
43. Yao, D.-x. *Front. Phys.* **2011,** *6*, 344–346.
44. Zhang, Y.-Z.; Kandpal, H. C.; Opahle, I.; Jeschke, H. O.; Valentí, R. *Phys. Rev. B* **2009,** *80*, 094530.
45. Wu, X.; Qin, S.; Liang, Y.; Le, C.; Fan, H.; Hu, J. *Phys. Rev. B* **2011,** *91*, 081111.
46. Nakamura, H.; Hayashi, N.; Nakai, N.; Machida, M. *J. Phys. Soc. Jpn* **2008,** *77*, 153–154.
47. Blaha, P.; Schwarz, K.; Madsen, G.; Kvasnicka, D.; Luitz, J. *An Augmented Plane Wave Plus Local Orbitals Program for Calculating Crystal Properties*; Vienna University of Technology: Austria, 2012.
48. Wong, K. M.; Alay-e-Abbas, S. M.; Shaukat, A.; Fang, Y.; Lei, Y. *J. Appl. Phys.* **2013,** *113*, 014304.
49. Wong, K. M.; Alay-e-Abbas, S. M.; Fang, Y.; Shaukat, A.; Lei, Y. *J. Appl. Phys.* **2013,** *114*, 034901.
50. Sjöstedt, E.; Nordström, L.; Singh, D. J. *Solid State Commun.* **2000,** *114*; 15–20.
51. Hohenberg, P.; Kohn, W. *Phys. Rev.* **1964,** 136 B864–B871.
52. Kohn, W.; Sham, L. J. *Phys. Rev.* **1965,** *140*, A1133–A1138.
53. Perdew, J. P.; Wang, Y. *Phys. Rev. B* **1992,** *45*, 13244–13249.
54. Anisimov, V. I.; Solovyev, I. V.; Korotin, M. A.; M.T. Czyżyk, Sawatzky, G. A. *Phys. Rev. B* **1993,** *48*, 16929–16934.
55. Murnaghan, F. D. *Proc. Natl. Acad. Sci.* **1944,** *30*, 244–247.
56. Blöchl, P. E.; Jepsen, O.; Andersen, O. K. *Phys. Rev. B* **1994,** *49*, 16223.
57. Zywietz, A.; Furthmüller, J.; Bechstedt, F. *Phys. Rev. B* **1999,** *59*, 15166–15180.
58. Sarma, D. D.; Mahadevan, P.; T. Saha-Dasgupta, Ray, S.; Kumar, A. *Phys. Rev. Lett.* **2000,** *85*, 2549–2552.
59. Mittal, R.; Rols, S.; Zbiri, M.; Su, Y.; Schober, H.; Chaplot, S. L.; Johnson, M.; Tegel, M.; Chatterji, T.; Matsuishi, S.; Hosono, H.; Johrendt, D.; Brueckel, T. *Phys. Rev. B* **2009,** *79*, 144516.
60. Kurmaev, E. Z.; McLeod, J. A.; Buling, A.; Skorikov, N. A.; Moewes, A.; Neumann, M.; Korotin, M. A.; Izyumov, Y. A.; Ni, N.; Canfield, P. C. *Phys. Rev. B* **2009,** *80*, 054508.
61. Canfield, P.; S. Bud'ko, Ni, N.; Kreyssig, A.; Goldman, A.; McQueeney, R.; Torikachvili, M.; Argyriou, D.; Luke, G.; Yu, W. *Physica C: Superconductivity* **2009,** *469*, 404–412.
62. Phillips, J. C. *Phys. Rev.* **1958,** *112*, 685.
63. Momma, K.; Izumi, F. *J. Appl. Crystallogr.* **2008,** *41*, 653–658.

CHAPTER 10

A Comparative Study of Optoelectronic Behavior of Some Cs-Based Halide Double Perovskites for Solar Cell Applications

ARCHANA and VIPUL SRIVASTAVA

Department of Physics, School of Chemical Engineering and Physical Sciences, Lovely Professional University, Phagwara, Punjab, India

ABSTRACT

Recently, metal halides perovskites have gained interest as semiconductor with favorable properties for optoelectronic devices. Perovskites solar cells have shown improvement in PCE and it is obtained by lead-based perovskites, which are considered to be toxic and instable against temperature and moisture. This has opened new doors for lead-free metal halides perovskites which are currently showing remarkable properties as investigated experimentally and theoretically. The significant attempts have been made mostly using theoretical view. They have suggested formation of novel materials by calculating formation energy. Research is being conducted on a range of halide perovskites without lead to evaluate their optoelectronic, electronic, magnetic, structural, and thermoelectric properties. The focus of this paper is on comparing and analyzing various Cs-based halide double perovskites with the aim of evaluating their electronic and optical characteristics. The considered materials have a space group 225 (Fm-3m) cubic structure under ambient conditions. In their electronic band profile most materials have semiconducting nature except that $Cs_2NaMnCl_6$, $Cs_2NaCoCl_6$, and

Advanced Materials for Next-Generation Technologies: Challenges and New Prospects. Shrikaant Kulkarni, Vipul Srivastava, Rabah Khenata, & Yarub Al-Douri (Eds.)

$Cs_2NaNiCl_6$ have half-metallic nature. This comparative study shows good optical properties and materials use in various optoelectronic applications.

10.1 INTRODUCTION

In recent years, human energy consumption has risen significantly; however, the available sources of energy are unable to meet the growing demands. Hydrocarbons like gas and oil are used for the generation of energy which causes the climate change. Perovskites can be used as alternative energy sources. Numerous perovskite compounds have proven to offer great potential for use in solar cell applications[1,2] and photovoltaic materials,[3] as well as possessing properties like high power conversion efficiency[4] and high carrier mobility. Perovskites are recognized for their half-metallicity, ferromagnetism, optical, and electrical properties, which make them useful for designing mechanical, magnetic, spintronic, memory, photovoltaic, light-emitting, thermoelectric, and other devices.[5–10]

Perovskites have an intriguing structure that can be found in metals, insulators, semiconductors, and superconductors.[11–20] $CaTiO_3$ was the first reported perovskite in 1893 by Gustov Rose.[21] The general formula is ABX_3, where "A" is generally alkaline or rare earth metal presented at the corner of lattice, "B" is transition or nontransition metal presented at the center of the unit lattice, and "X" is either oxides or halides, presented at six faces. The $CH_3NH_3PbX_3$ is a promising candidate for high-efficiency solar cell applications. But toxic Pb and stability problems like temperature sensitivity and moisture limit their applications. To compensate this, the lead-free halide perovskites were designed. In double perovskites, Pb2+ can be substituted with both monovalent B^+ and trivalent B^{3+} ions. The generic formula of double perovskite is $A_2B^+B^{3+}X_6$, where "A" is alkaline metals and "B" site is occupied by two different elements and "X" can be oxygen or halogens. In 2016, the initial successful synthesis of lead-free halide double perovskites was reported by McClure et al., as a possible photovoltaic absorber but similar compounds had been known for a long time as "elpasolites," named after mineral K_2NaAlF_6. Since 2016, numerous halide double perovskites (HDP) have been investigated such as Cs_2AgBiX_6,[22] Cs_2AgInX_6,[23,24] $Cs_2AgSbCl_6$,[25–27] Cs_2AgCrX_6,[28,29] Cs_2NaMCl_6 (M = Mn, Co, Ni, Tl, In),[30,31] $(MA)_2KBiCl_6$,[32] $(MA)_2TlBiBr_6$,[33] and others. Among these compounds $Cs_2AgBiBr_6$,[22] was studied most; it was found that it had small band gap and was stable against heat and moisture. Also its large size single crystal, high quality films, and nanocrystal had been fabricated.[34–42]

When compared to conventional solar cell absorbers, halide perovskites have higher optical absorption coefficients. The elpasolite perovskites such as Cs_2NaMCl_6 (M = In, Tl, Sb, Bi)[43] is an attractive material for optoelectronic applications due to its unique optical properties, as it has discrete energy bands in its band gaps. Direct band gap semiconductors are preferred over indirect band gap semiconductors for optoelectronic applications due to their ability to efficiently absorb and emit light. This is because the minimum energy required for an electron to move from the valence band to the conduction band is equivalent to a specific photon energy, reducing the need for additional phonons and enhancing the overall light emission properties. In contrast, indirect band gap semiconductors tend to be poor emitters of light due to the involvement of phonons. In recent years, there have been numerous theoretical efforts to understand the underlying mechanisms responsible for the exceptional optical properties of double halide perovskites, resulting in significant advancements in this field.

In this review, we have reviewed few of lead-free HDP for their optoelectronic performances; these include Cs_2AgMX_6 (M = Tl, Bi, Cr, In, Sb)[28,29] and Cs_2NaMX_6 (M = Mn, Co, Ni, Tl, Sb, Bi, In, V).[43,44] The review on optoelectronic performances was completed by considering their dielectric function, absorption coefficient, band refractive index, energy band gap, and energy loss function. The motivation behind the present review is to get consolidated information about their ground state to optoelectric performances of some of the double halide perovskites at one place. The provided information may insist one to fabricate optoelectronic devices considering the optical parameters.

10.2 MAIN TEXT

10.2.1 STRUCTURE OF PEROVSKITE MATERIALS

The chemical composition of perovskite materials is usually represented by the formula ABX3, where "A" and "B" are cations and "X" is an anion. Perovskites are characterized by a crystal structure in which BX6 octahedra are corner-shared, and the voids in the structure are occupied by 12-fold coordinated "A" cations. The stability of perovskites depends on stability of BX_6 octahedra,[45] Charge neutrality ($q_A + q_B = -3q_X$),[46] and Goldschmidt tolerance factor. The stability of BX6 octahedra in perovskite materials can be evaluated using the octahedral factor (μ), which is calculated by comparing the radii of the cation occupying the "B"-site and the anion "X."

The μ value is obtained by taking the ratio of the two radii and is a parameter that determines the stability of BX6 octahedra in perovskites.[45]

$$\mu = \frac{r_B}{r_X}$$

For stable halides perovskite materials, the value of μ should be between 0.442 and 0.895.[47,48] Goldschmidt tolerance factor (*t*)[48] is determined by dividing the distance between the "A" cation and the anion "X" by the distance between the "B" cation and the anion "X," assuming an idealized solid sphere model.

$$t = \frac{r_A + r_X}{\sqrt{2}\left(r_B + r_X\right)}$$

The symbols r_A, r_B, and r_X refer to ionic radii of cation "A," cation "B," and anion "X."[49] Its value varies from 0.8 to 1.0 for stable and pure perovskites. The materials with $t = 1.0$ form perfect cubic crystal structure. For lower values of the tolerance factor (0.80–0.89), tetragonal, orthorhombic, and rhombohedral structures are to be formed. But if *t* is greater than 1 then "A" cation is very large whereas for *t* less than 0.8 "A" cation is very small for the formation of structure of perovskites.[50,51]

In 1982, more than two hundred halide elpasolites were known with either cubic or tetrahedra structure.[52] Giustino et al.[53] presented a catalogue of elements from the periodic table that can be employed to synthesize halide double perovskites with a generic formula of $A_2BB'X_6$, as shown in Figure 10.1. A-site cations can be occupied by 7 elements, 8 elements are found B(I), 34 elements can occupy B(III)sites and up to 5 different elements that can occupy the X sites. The majority of them are out of question as their band gaps and carrier transport are suitable for optoelectronics applications. So for finding suitable band gap halide double perovskites, research has been done and some progress has been observed.[54]

In the crystal structure of HDP ($A_2BB'X_6$), there is a resemblance with structure of ABX_3. However, in $A_2BB'X_6$, each "B" cation can be occupied by a pair of cations with different oxidation states, namely B^+ and B^{3+} as shown in Figure 10.2 and form a structure containing alternate B^IX_6 and $B^{III}X_6$ octahedrons with "A" cation at the center of cuboctahedral cavity.[55,56]

The charge neutrality should be obtained for double perovskites as well, so that

$$q_A + q_B + q_{B'} = -6$$

Babel et al.[57] modified the tolerance factor to apply to double perovskite elpasolites, as

$$t = \frac{r_A + r_X}{\sqrt{2}\left(\frac{r_B + r_{B'}}{2} + r_X\right)}$$

The symbols r_A, r_B, $r_{B'}$ and r_X indicate the ionic radius of ions A, B, B', and X.

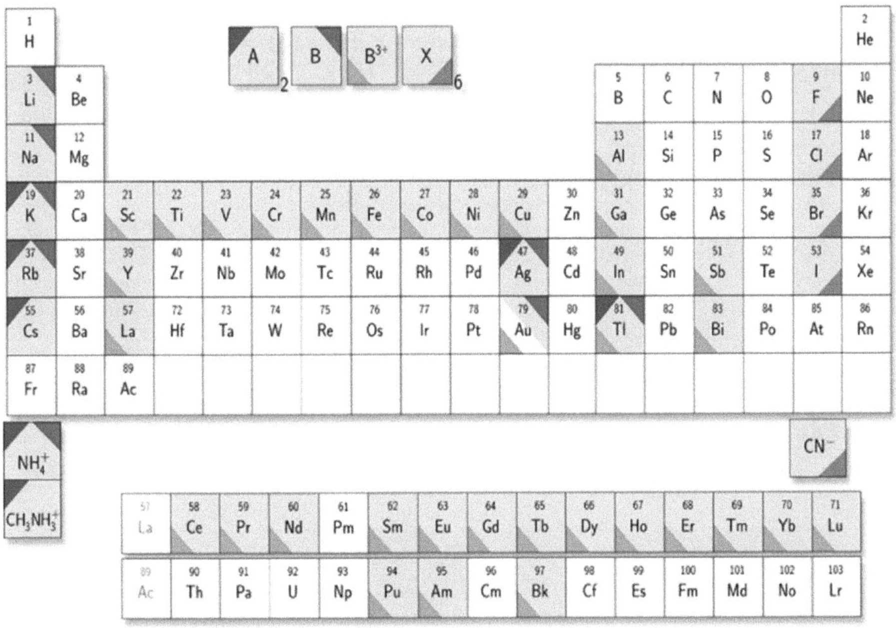

FIGURE 10.1 Elements forming halide double perovskites

Source: Reprinted with permission from Ref. [53].

The structure of materials that belong to the category of HDP is illustrated in Figure 10.2.

10.2.2 CS-AG BASED HDP

McClure et al.[22] reported $Cs_2AgBiBr_6$, first lead-free HDP for photovoltaics application, but it has an indirect band gap of 2.06 eV (1.91 eV) and favors long carrier recombination life time (660 ns).[4] Ning et al.[59] demonstrated $Cs_2AgBiBr_6$ solar cells by using Planar structure and found that its films have large photoexcited carrier diffusion length (110 nm) and have power conversion efficiency over 1%. Cs_2AgBiI_6 has a smaller bandgap but it cannot be synthesized as it has "negative decomposition energy"[60] but recently it has been reported[37] that using anion exchange protocol, Cs_2AgBiI_6 can be synthesized in nanocrystal form from $Cs_2AgBiBr_6$ nanocrystal. Cs_2AgBiI_6 has strong absorption in visible region. Rajeev Kumar et al.[24] investigated Cs_2AgBiX_6 for their electronic, optical, and mechanical properties using Quantum Espresso suite[61] and they used different correlation functionals like GGA,[62] LDA, PBEsol. They found that as "X" changes from Cl to I, their ground state energy and lattice parameter increase whereas bulk modulus decreases; also absorption peaks become sharper and with large magnitude due to smaller width of valence band. These compounds are observed to have ionic bonding and ductile nature.

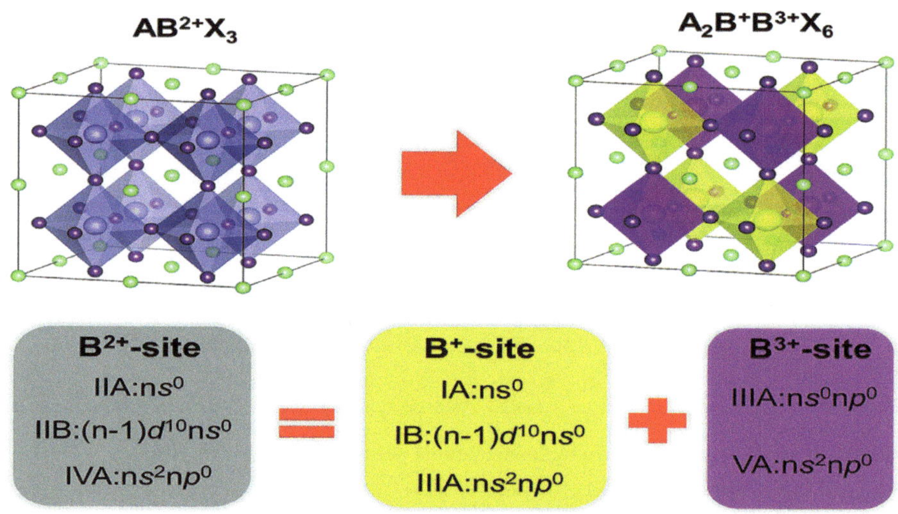

FIGURE 10.2 Perovskites ABX_3 to halides double perovskites $A_2BB'X_6$

Source: Reprinted with permission from Ref. [58] Copyright Royal Society of Chemistry.

These materials have potential for utilization in optoelectronics applications that operate within the visible and ultraviolet spectra. Cs_2AgBiX_6 has large and indirect bandgap which is not efficient for solar cell applications so more research has been conducted for converting to direct band gap and reduces its value. Cs_2AgInX_6[23,63,64] compounds have been studied theoretically as they have direct band gap.

Mathew et al.[23] investigated $Cs_2AgInCl_6$ using Quantum-Espresso Suite[61] and found that the structure is optimized in space group 225, face-centered cubic (Fm-3m) having lattice parameter of 10.565 Å and band gap energy of 1.1 eV. Lua et al.[64] observed from the experiment that there are parity-forbidden transitions and it has large fundamental band gap as PL spectra have emission peak at 2.0 eV. Menedjhi et al.[65] calculated direct band gap 1.427 eV for $Cs_2AgInBr_6$ using B3LYP functional. They observed that $Cs_2AgInBr_6$ shows a band of absorption (2.5–35 eV) which is useful for technological applications. The value of absorption coefficient for $Cs_2AgInBr_6$ has been measured to be greater than 10^4 cm^{-1}. F. Aslam et al. reported Cs_2AgInI_6 to have direct band gap of 0.2 eV and the absorption is observed in visible region for $Cs_2AgInCl_6$, $Cs_2AgInBr_6$ whereas for Cs_2AgInI_6 in infrared region.[66]

G. Murtaza and colleagues[67] utilized the Wien2k code,[68] which is based on the full potential linearized augmented plane wave method,[69] to calculate the electronic and optical properties of $Cs_2AgTlCl_6$ and thermoelectric properties in BoltzTraP code. As the energy of photons increases, the value of $\varepsilon_1(\omega)$ increases and reaches its maximum value within the visible range, after which it rapidly decreases and reaches a minimum in the UV range. The value then rises again for the next resonance. In imaginary part of dielectric constant, single peak behavior is observed. $Cs_2AgTlCl_6$ have high value of figure of merit (0.68) which makes it preferable for thermoelectric applications.

Pradeep R. Varadwaj[28] used SCAN+ rVV10[75] for determining properties of Cs_2AgCrX_6. Due to its tolerance factor (τ = 4.31), Cs_2AgCrI_6 was determined to be unstable. The lattice contraction and expansion is observed as we replace Br atoms by Cl and I. $Cs_2AgCrBr_6$ ($Cs_2AgCrCl_6$) has indirect band gap of 1.27 eV (1.84 eV) and semiconducting nature; however, for Cs_2AgCrI_6, the band gap shifts to a lower energy value of 0.43 eV and becomes direct band gap at the Γ-point.

In Figure 10.3(a), it is evident that Cs_2AgCrI_6 has more screen charged defect as its ε_∞ value is 8.9 while for $Cs_2AgCrCl_6$ and $Cs_2AgCrBr_6$, ε_∞ is 4.5 and 6.1.

The start of optical absorption in ε_2 is around 1.55, 1.15, and 0.45 eV for $Cs_2AgCrBr_6$, Cs_2AgCrI_6, and $Cs_2AgCrCl_6$ respectively. They observed that this has shifted toward low energy infrared region from near infrared region; this shift was observed upon substituting the halogen from Cl to I in Cs_2AgCrX_6. For $Cs_2AgCrCl_6$ the major peaks (Fig. 10.3b) are found at 2.0 and 2.3 eV which is in agreement with the experimental observation (1.6 and 2.2 eV) observed for hexagonal $Cs_2AgCrCl_6$.

Table 10.1 concludes the properties including lattice parameters, formation energy, bulk modulus, equilibrium energy, pressure derivate, and band gap of Cs-Ag based double perovskites.

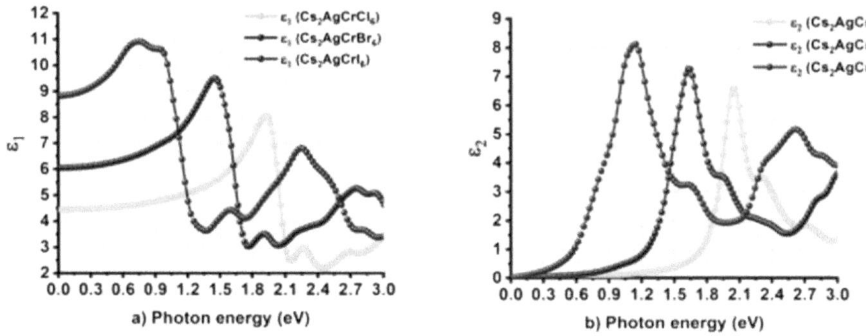

FIGURE 10.3 (a) The real part and (b) imaginary part of the dielectric function for Cs_2AgCrX_6 with respect to photon energy.

Source: Reprinted from Ref. [28] Copyright © 2020 by the author. Licensee MDPI, Basel, Switzerland. http://creativecommons.org/licenses/by/4.0/.

TABLE 10.1 Ground-State Properties of Cs-Ag Based Double Halides Perovskites.

Perovskite Materials	Lattice parameter a_0 (Å)	Formation energy (eV/atom)	Bulk modulus B_0 (GPa)	Pressure derivate B_0'	Equilibrium energy E_0 (Ryd.)	Band gap E_g(eV)	References
$Cs_2AgBiCl_6$	10.7774 10.99	−1.712	23.43	3.79	−396.9754	2.77 1.91	[22, 24]
$Cs_2AgBiBr_6$	11.2711 11.53	−1.515	20.49	5.82	−378.2703	2.19 1.42	[22, 24]
Cs_2AgBiI_6	12.28	−1.126	16.74	14.98	−354.5866	0.89	[24]
$Cs_2AgInCl_6$	10.565 10.43	−1.743				1.1 2.49	[23] [66]
$Cs_2AgInBr_6$	7.94 10.94	−1.528				1.427 1.38	[65] [66]
Cs_2AgInI_6	11.70	−1.119				0.2	[66]
$Cs_2AgCrBr_6$	10.68					1.27 1.46	[28]
$Cs_2AgCrCl_6$	10.13					1.84	[28]
Cs_2AgCrI_6	11.48					0.43	[28]
$Cs_2AgTlCl_6$	10.52	−1.583				0.74	[67]

10.2.3 CS-NA BASED DOUBLE HALIDE PEROVSKITES

Mukhtar et al.[71] computed properties of Cs_2NaInI_6 in Wien2K code, the lattice parameter(a) for Cs_2NaInI_6 was determined to be 11.8 Å in cubic phase and bulk modulus(B) 18.7 GPa. The Poisson parameter ($\upsilon = 3.3$) and Pugh ratio

(B/G = 3.3) indicate the ductility of the material. Cs_2NaInI_6 have direct band gap of 1.11 eV. Dielectric constant components were determined using the Kramer–Kronig relation. The prominent peak in real dielectric is at 4.03 eV and in $\varepsilon_2(\omega)$, it rises from 1.4 eV and reaches maximum to 7.1 eV. In the UV region, the absorption coefficient was determined to be 5948 cm^{-1}, whereas in the visible region it was found to be 368 cm^{-1}. The absorption value in the visible region suggests that the material can be suitable for solar cells whereas absorption in the UV region is important for optoelectronic applications. At room temperature, Cs_2NaInI_6 exhibits a figure of merit of 0.79, indicating its potential for use in thermoelectric applications.

Yasir Saeed et al.[72] found $Cs_2NaGaBr_6$ band gap to be direct (0.451 eV), with spin orbit coupling, band gap reduces to 0.38 eV. The effective masses m_h* and m_e* are measured to be 0.186 and 0.176 due to which it has greater transport of charge carriers. $Cs_2NaGaBr_6$ are brittle in nature because it has Pugh ratio (1.65) less than the critical value of 1.75 and negative value of Cauchy's pressure (−0.36). The value of static dielectric constant is observed at 1.6 eV whereas for static refractive index values observed at 1.8. It has high optical conductivity in visible light and absorption coefficient starts from 1.76 eV and reaches to 3.72 eV which makes it suitable for solar cell applications.

Shabir Ahmad Mir et al.[30] investigated Cs_2NaBCl_6(B = Mn, Co, Ni) for their magnetic and optoelectronic properties. It was observed that these materials exhibit stability in a cubic structure with $Fm\bar{3}m$ symmetry as tolerance factor is 0.99 and of brittle nature.

From the band structure they observed that in the spin down direction, $Cs_2NaNiCl_6$ and $Cs_2NaMnCl_6$ show semiconducting nature, while in the spin up direction they show metallic character whereas $Cs_2NaCoCl_6$ shows opposite behavior and these double perovskites have integral magnetic moment as 4, 1, 4μ_B for Co-, Ni-, and Co-based double perovskites; this results in half metallicity. The behavior of optical parameters such as the optical conductivity, absorption coefficient, and dielectric constant displays a trend of absorption in the UV-Visible region with respect to energy.

Xiangrong Cao et al.[44] prepared Cs_2NaVCl_6 by hydrothermal and solid reaction method. Cs_2NaVCl_6 exhibits a cubic structure having lattice parameter a(Å) = 10.3497, also they investigated for antiferromagnetic phase using DFT and this calculation indicates that Cs_2NaVCl_6 belongs to direct band gap (2.64 eV) semiconductor. Two prominent absorption peaks were detected at 558 and 900 nm and it is stable up to 650°C. It is observed that Cs_2NaVCl_6 can be used in solar energy application.

Imad Khan et al.[43] used different exchanges. Correlation approximation like LDA, GGA, GGA+SOC, mBJ, mBJ with SOC was employed to determine the electronic properties; it was observed that spin–orbit coupling reduces the band gaps. Cs_2NaMCl_6(M = In, Tl) have direct band gaps and Cs_2NaMCl_6(M = Sb, Bi) are indirect band gap semiconductors. These materials are ductile, anisotropic in nature. The application of spin–orbit coupling resulted in the generation of distinct energy bands in the conduction and valence bands because distances between cations are large and large electronegative difference. High light yield and effective impact ionization are caused by discrete energy bands.

TABLE 10.2 Ground State Properties of Cs-Na Based Double Halides Perovskites.

Perovskite Materials	Lattice parameter a_0 (Å)	Formation energy (eV/atom)	Bulk modulus B_0(GPa)	Pressure derivate B_0'	Band gap E_g(eV)	References
$Cs_2NaMnCl_6$	10.36		28.48		2.06 GGA 5.03 GGA-mBJ	[30]
$Cs_2NaCoCl_6$	10.33		24.08		4.25 GGA 6.07 mBJ	[30]
$Cs_2NaNiCl_6$	10.21		22.25		0.96 GGA 2.9 GGA-mBJ	[30]
$Cs_2NaTlCl_6$	10.58	−1.840			3.52 GGA 2.03 GGA+SOC 4.00 mBJ 4.09 mBJ+SOC	[43]
$Cs_2NaSbCl_6$	10.73 10.652	−1.918			3.90 GGA 3.02 GGA+SOC 4.13 mBJ 3.75 mBJ+SOC 3.13	[43] [73]
$Cs_2NaSbBr_6$	11.223				2.54	[73]
Cs_2NaSbI_6	12.056				1.90	[73]
$Cs_2NaBiCl_6$	10.80 10.837 21.72	−1.980	24.52	4.46	3.61 GGA 2.83 GGA+SOC 4.73 mBJ 3.85 mBJ+SOC 3.73 3.70	[43] [73] [31]
$Cs_2NaBiBr_6$	11.364 21.75		19.31	4.36	3.07 2.6	[31] [56]
Cs_2NaBiI_6	12.199 21.75		15.63	5.17	2.23 2.0	[31] [56]

TABLE 10.2 *(Continued)*

Perovskite Materials	Lattice parameter a_0 (Å)	Formation energy (eV/atom)	Bulk modulus B_0(GPa)	Pressure derivate B_0'	Band gap E_g(eV)	References
$Cs_2NaInCl_6$	10.47 10.37	−2.019	32.21		3.58 GGA 3.10 GGA+SOC 4.67 mBJ 4.09 mBJ+SOC 2.491	[43] [31]
$Cs_2NaAlCl_6$	9.99	−2.166	36.24		3.039	[72]
$Cs_2NaAlBr_6$	10.66		25.25		1.728	[72]
$Cs_2NaInBr_6$	10.97	−1.772	20.64		1.401	[72]
$Cs_2NaGaBr_6$	10.96	−1.723	22.80		0.451 1.762 mbj+SOC	[72]
$Cs_2NaGaCl_6$	10.09	−1.984	33.66		1.691	[72]
Cs_2NaInI_6	11.8	−1.321	18.7		1.11	[73]

They observed that at energy of 16 eV, $\varepsilon_1(\omega)$ goes below zero due to reflection of incident photons from medium; this indicates metallic nature. The multiple peaks were observed in $\varepsilon_2(\omega)$ because of discrete electronic bands. No energy loss is obtained for energy lower than band gap whereas maximum loss is obtained for range 20–23 eV.

Shuai Zhao et al.[73] observed that values obtained suggest that Cs_2NaSbI_6, $Cs_2NaSbBr_6$, and Cs_2NaBiI_6 show high absorption efficiency in the visible region due to its narrow band gaps and calculated formation enthalpy of halides which show only chloride compounds have stable double perovskites phase.

Absorption coefficient of double halide perovskites is illustrated in Figure 10.4. The compounds having small band gaps have broad absorption in the visible region as absorption is inversely proportional to the band gap. In the visible region, Cs_2NaSbI_6, Cs_2NaBiI_6, and $Cs_2NaSbBr_6$ demonstrate effective absorption. $Cs_2NaBiCl_6$ has potential to be used in the development of UV sensors or detectors. Therefore, the present study reports the involvement of different optical parameters for making them applicable for optoelectronic device applications.

10.3 CONCLUSIONS

We have reviewed some Cs-based HDP for their optoelectronic applications. The PCE of perovskites solar cells has improved up to 20%, but these perovskites are lead based which is considered to be toxic and instable. There has been considerable advancement in the development of halide perovskites that are free of lead. $Cs_2AgBiBr_6$ have been studied experimentally as well as theoretically. The material exhibits an indirect band gap, and its magnitude precludes effective solar cell performance whereas $Cs_2AgInCl_6$ has direct band gap but it has parity forbidden transitions and large fundamental band gap; as a result it is unsuitable for photovoltaic applications. Sodium-based double halide perovskites like $Cs_2NaGaBr_6$ and Cs_2NaInI_6 have direct band gap of 0.451 and 1.11 eV. Various optical parameters like dielectric constant, absorption coefficient have been plotted with respect to energy which suggests absorption in UV-visible region.

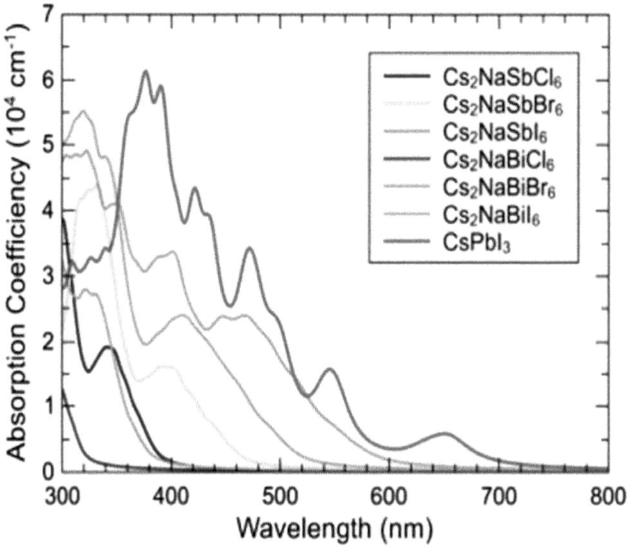

FIGURE 10.4 Absorption coefficient of Cs_2NaMCl_6(M = Sb, Bi) and $CsPbI_3$

Source: Reprinted from Ref. [73].

For Cs_2NaInI_6, in the UV region, the absorption coefficient of the material was determined to be 5948 cm^{-1} while in the visible region, it is 368 cm^{-1}. Absorption in the visible region is acceptable for solar cells whereas absorption in the UV region is important for optoelectronic applications. These perovskites have applications like solar energy applications (Cs_2NaVCl_6, $Cs_2NaGaBr_6$), UV absorber ($Cs_2NaBiCl_6$), and optoelectronics (Cs_2NaInI_6).

KEYWORDS

- **double halide perovskites**
- **optoelectronics**
- **density functional theory**
- **lead-free halide perovskites**

REFERENCES

1. Park, N. G. Perovskite Solar Cells: An Emerging Photovoltaic Technology. *Mater. Today* **2015,** *18* (2), 65–72.
2. Chen, Y.; Zhang, L.; Zhang, Y.; Gao, H.; Yan, H. Large-Area Perovskite Solar Cells—A Review of Recent Progress and Issues. *RSC Adv.* **2018,** *8* (19), 10489–10508.
3. Burschka, J.; Pellet, N.; Moon, S. J.; Humphry-Baker, R.; Gao, P.; Nazeeruddin, M. K.; Grätzel, M. Sequential Deposition as a Route to High-Performance Perovskite-Sensitized Solar Cells. *Nature* **2013** *499* (7458), 316–319.
4. Slavney, A. H.; Hu, T.; Lindenberg, A. M.; Karunadasa, H. I. A Bismuth-Halide Double Perovskite with Long Carrier Recombination Lifetime for Photovoltaic Applications. *J. Am. Chem. Soc.* **2016,** *138* (7), 2138–2141.
5. Kulkarni, A.; Ciacchi, F. T.; Giddey, S.; Munnings, C.; Badwal, S. P. S.; Kimpton, J. A.; Fini, D. Mixed Ionic Electronic Conducting Perovskite Anode for Direct Carbon Fuel Cells. *Int. J. Hydrogen Energy* **2012,** *37* (24), 19092–19102.
6. Wang, J.; Zhang, C.; Liu, H.; McLaughlin, R.; Zhai, Y.; Vardeny, S. R.; et al. Spin-Optoelectronic Devices Based on Hybrid Organic-Inorganic Trihalide Perovskites. *Nat. Commun.* **2019,** *10* (1), 129.
7. Chen, Y.; Zhang, L.; Zhang, Y.; Gao, H.; Yan, H. Large-Area Perovskite Solar Cells—A Review of Recent Progress and Issues. *RSC Adv.* **2018,** *8* (19), 10489–10508.
8. Park, N. G. Perovskite Solar Cells: An Emerging Photovoltaic Technology. *Mater. Today* **2015,** *18* (2), 65–72.
9. Zabek, D.; Morini, F. Solid State Generators and Energy Harvesters for Waste Heat Recovery and Thermal Energy Harvesting. *Thermal Sci. Eng. Progress* **2019,** *9*, 235–247.
10. Jaziri, N.; Boughamoura, A.; Müller, J.; Mezghani, B.; Tounsi, F.; Ismail, M. A Comprehensive Review of Thermoelectric Generators: Technologies and Common Applications. *Energy Rep.* **2020,** *6*, 264–287.
11. Stoumpos, C. C.; Malliakas, C. D.; Peters, J. A.; Liu, Z.; Sebastian, M.; Im, J.; et al. Crystal Growth of the Perovskite Semiconductor CsPbBr3: A New Material for High-Energy Radiation Detection. *Crystal Growth Design* **2013,** *13* (7), 2722–2727.
12. Li, L.; Liu, X.; He, C.; Wang, S.; Ji, C.; Zhang, X.; et al. A Potential Sn-Based Hybrid Perovskite Ferroelectric Semiconductor. *J. Am. Chem. Soc.* **2020,** *142* (3), 1159–1163.
13. Dar, S. A.; Srivastava, V.; Sakalle, U. K. Ab Initio High Pressure and Temperature Investigation on Cubic $PbMoO_3$ Perovskite. *J. Electron. Mater.* **2017,** *46*, 6870–6877.

14. Dar, S. A.; Srivastava, V.; Sakalle, U. K.; Parey, V.; Pagare, G. DFT Investigation on Electronic, Magnetic, Mechanical and Thermodynamic Properties Under Pressure of Some $EuMO_3$(M = Ga, In) Perovskites. *Mater. Res. Express* **2017,** *4* (10), 106104.
15. He, T.; Huang, Q.; Ramirez, A. P.; Wang, Y.; Regan, K. A.; Rogado, N.; et al. Superconductivity in the Non-Oxide Perovskite $MgCNi_3$. *Nature* **2001,** *411* (6833), 54–56.
16. Dar, S. A.; Sharma, R.; Srivastava, V.; Sakalle, U. K. Investigation on the Electronic Structure, Optical, Elastic, Mechanical, Thermodynamic and Thermoelectric Properties of Wide Band Gap Semiconductor Double Perovskite Ba_2InTaO_6. *RSC Adv.* **2019,** *9* (17), 9522–9532.
17. Kumari, P.; Sharma, R.; Lilhore, U.; Khenata, R.; Srivastava, V. First-Principles Study on Structural, Electronic, Elastic, Mechanical, Thermodynamic, and Thermoelectric Properties of $RbSnX_3$(X = F, Cl, and Br) Perovskites. *Int. J. Energy Res.* **2022,** *46* (15), 23893–23907.
18. Dar, S. A.; Srivastava, V.; Sakalle, U. K. A First-Principles Calculation on Structural, Electronic, Magnetic, Mechanical, and Thermodynamic Properties of $SrAmO_3$. *J. Superconductivity Novel Magnet.* **2017,** *30* (11), 3055–3063.
19. Kumari, P.; Srivastava, V.; Khenata, R.; Dar, S. A.; Naqib, S. H. A First-Principles Prediction of thermophysical and Thermoelectric Performances of $SrCeO_3$ Perovskite. *Int. J. Energy Res.* **2022,** *46* (3), 2934–2946.
20. Kumari, P.; Dar, S. A.; Sharma, R.; Srivastava, V. Quantum Mechanical Calculations on Mechanical and Elastic Properties of Cubic $SrCeO_3$ Perovskite. *Eur. Phys. J. Plus* **2022,** *137* (3), 313.
21. Bhalla, A. S.; Guo, R.; Roy, R. The Perovskite Structure—A Review of Its Role in Ceramic Science and Technology. *Mater. Res. Innov.* **2000,** *4* (1), 3–26.
22. McClure, E. T.; Ball, M. R.; Windl, W.; Woodward, P. M. Cs2AgBiX6 (X = Br, Cl): New Visible Light Absorbing, Lead-Free Halide Perovskite Semiconductors. *Chem. Mater.* **2016,** *28* (5), 1348–1354.
23. Mathew, N. P.; Kumar, N. R.; Radhakrishnan, R. First Principle Study of the Structural and Optoelectronic Properties of Direct Bandgap Double Perovskite Cs2AgInCl6. *Mater. Today: Proc.* **2020,** *33*, 1252–1256.
24. Kumar, N. R.; Radhakrishnan, R. Electronic, Optical and Mechanical Properties of Lead-Free Halide Double Perovskites Using First-Principles Density Functional Theory. *Mater. Lett.* **2018,** *227*, 289–291.
25. Tran, T. T.; Panella, J. R.; Chamorro, J. R.; Morey, J. R.; McQueen, T. M. Designing Indirect–Direct Bandgap Transitions in Double Perovskites. *Mater. Horiz.* **2017,** *4* (4), 688–693.
26. Deng, W.; Deng, Z. Y.; He, J.; Wang, M.; Chen, Z. X.; Wei, S. H.; Feng, H. J. Synthesis of $Cs_2AgSbCl_6$ and Improved Optoelectronic Properties of $Cs_2AgSbCl_6/TiO_2$ Heterostructure Driven by the Interface Effect for Lead-Free Double Perovskites Solar Cells. *Appl. Phys. Lett.* **2017,** *111* (15), 151602.
27. Zhou, J.; Rong, X.; Molokeev, M. S.; Zhang, X.; Xia, Z. Exploring the Transposition Effects on the Electronic and Optical Properties of $Cs_2AgSbCl_6$ via a Combined Computational-Experimental Approach. *J. Mater. Chem. A* **2018,** *6* (5), 2346–2352.
28. Varadwaj, P. R. $A_2AgCrBr_6$ (A = K, Rb, Cs) and Cs_2AgCrX_6(X = Cl, I) Double Perovskites: A Transition-Metal-Based Semiconducting Material Series with Remarkable Optics. *Nanomaterials* **2020,** *10* (5).
29. Saeed, M.; Haq, I. U.; Saleemi, A. S.; Rehman, S. U.; Haq, B. U.; Chaudhry, A. R.; Khan, I. First-Principles Prediction of the Ground-State Crystal Structure of Double-Perovskite Halides Cs_2AgCrX_6(X = Cl, Br, and I). *J. Phys. Chem. Solids* **2022,** *160*, 110302.

30. Mir, S. A.; Gupta, D. C. Systematic Investigation of the Magneto-Electronic Structure and Optical Properties of New Halide Double Perovskites Cs_2NaMCl_6(M = Mn, Co and Ni) by Spin Polarized Calculations. *RSC Adv.* **2020,** *10* (44), 26277–26287.
31. Choudhary, S.; Tomar, S.; Kumar, D.; Kumar, S.; Verma, A. S. Investigations of Lead Free Halides in Sodium Based Double Perovskites Cs_2NaBiX_6(X = Cl, Br, I): an Ab Intio Study. *East Eur. J. Phys.* **2021,** (3), 74–80.
32. Wei, F.; Deng, Z.; Sun, S.; Xie, F.; Kieslich, G.; Evans, D. M. et al. The Synthesis, Structure and Electronic Properties of a Lead-Free Hybrid Inorganic–Organic Double Perovskite $(MA)_2KBiCl_6$(MA = methylammonium). *Mater. Horiz.* **2016,** *3* (4), 328–332.
33. Deng, Z.; Wei, F.; Sun, S.; Kieslich, G.; Cheetham, A. K.; Bristowe, P. D. Exploring the Properties of Lead-Free Hybrid Double Perovskites Using a Combined Computational-Experimental Approach. *J. Mater. Chem. A* **2016,** *4* (31), 12025–12029.
34. Pan, W.; Wu, H.; Luo, J.; Deng, Z.; Ge, C.; Chen, C. et al. $Cs_2AgBiBr_6$ Single-Crystal X-ray Detectors with a Low Detection Limit. *Nat. Photon.* **2017,** *11* (11), 726–732.
35. Lozhkina, O. A.; Murashkina, A. A.; Elizarov, M. S.; Shilovskikh, V. V.; Zolotarev, A. A.; Kapitonov, Y. V. et al. Microstructural Analysis and Optical Properties of the Halide Double Perovskite $Cs_2BiAgBr_6$ Single Crystals. *Chem. Phys. Lett.* **2018,** *694*, 18–22.
36. Bartesaghi, D.; Slavney, A. H.; Gélvez-Rueda, M. C.; Connor, B. A.; Grozema, F. C.; Karunadasa, H. I.; Savenije, T. J. Charge Carrier Dynamics in $Cs_2AgBiBr_6$ Double Perovskite. *J. Phys. Chem. C* **2018,** *122* (9), 4809–4816.
37. Creutz, S. E.; Crites, E. N.; De Siena, M. C.; Gamelin, D. R. Colloidal Nanocrystals of Lead-Free Double-Perovskite (Elpasolite) Semiconductors: Synthesis and Anion Exchange to Access New Materials. *Nano Lett.* **2018,** *18* (2), 1118–1123.
38. Zhou, L.; Xu, Y. F.; Chen, B. X.; Kuang, D. B.; Su, C. Y. Synthesis and Photocatalytic Application of Stable Lead-Free $Cs_2AgBiBr_6$ Perovskite Nanocrystals. *Small* **2018,** *14* (11), 1703762.
39. Yang, B.; Chen, J.; Yang, S.; Hong, F.; Sun, L.; Han, P. et al. Lead-Free Silver-Bismuth Halide Double Perovskite Nanocrystals. *Angew. Chem.* **2018,** *130* (19), 5457–5461.
40. Bekenstein, Y.; Dahl, J. C.; Huang, J.; Osowiecki, W. T.; Swabeck, J. K.; Chan, E. M. et al. The Making and Breaking of Lead-Free Double Perovskite Nanocrystals of Cesium Silver–Bismuth Halide Compositions. *Nano Lett.* **2018,** *18* (6), 3502–3508.
41. Ning, W.; Wang, F.; Wu, B.; Lu, J.; Yan, Z.; Liu, X. et al. Long Electron–Hole Diffusion Length in High-Quality Lead-Free Double Perovskite Films. *Adv. Mater.* **2018,** *30* (20), 1706246.
42. Wu, C.; Zhang, Q.; Liu, Y.; Luo, W.; Guo, X.; Huang, Z. et al. The Dawn of Lead-Free Perovskite Solar Cell: Highly Stable Double Perovskite $Cs_2AgBiBr_6$ Film. *Adv. Sci.* **2018,** *5* (3), 1700759.
43. Khan, I.; Haq, I. U.; Ali, A.; Ali, Z.; Ahmad, I. Elastic and Optoelectronic Properties of Cs_2 $NaMCl_6$ (M = In, Tl, Sb, Bi). *J. Electron. Materi.* **2021,** *50*, 456–466.
44. Cao, X.; Kang, L.; Guo, S.; Zhang, M.; Lin, Z.; Gao, J. Cs_2NaVCl_6: A Pb-Free Halide Double Perovskite with Strong Visible and Near-Infrared Light Absorption. *ACS Appl. Mater. Interfaces* **2019,** *11* (42), 38648–38653.
45. Travis, W.; Glover, E. N. K.; Bronstein, H.; Scanlon, D. O.; Palgrave, R. G. On the Application of the Tolerance Factor to Inorganic and Hybrid Halide Perovskites: A Revised System. *Chem. Sci.* **2016,** *7* (7), 4548–4556.

46. Qian, J.; Xu, B.; Tian, W. A Comprehensive Theoretical Study of Halide Perovskites ABX_3. *Organ. Electron.* **2016,** *37*, 61–73.
47. Huang, T. J.; Thiang, Z. X.; Yin, X.; Tang, C.; Qi, G.; Gong, H. (CH_3NH_3) 2PdCl4: A Compound with Two-Dimensional Organic–Inorganic Layered Perovskite Structure. *Chem.—Eur. J.* **2016,** *22* (6), 2146–2152.
48. Li, C.; Lu, X.; Ding, W.; Feng, L.; Gao, Y.; Guo, Z. Formability of ABX_3 (X = F, Cl, Br, I) Halide Perovskites. *Acta Crystallogr. Sect. B: Struct. Sci.* **2008,** *64* (6), 702–707.
49. Nim, G. K.; Bansal, P.; Kar, P. Nitrate Ion-Incorporated Stable Perovskite Nanocrystals by a Solvent-Free Mechanochemical Reaction. *ACS Omega* **2019,** *4* (13), 15678–15683.
50. Uribe, J. I.; Ramirez, D.; Osorio-Guillén, J. M.; Osorio, J.; Jaramillo, F. $CH_3NH_3CaI_3$ Perovskite: Synthesis, Characterization, and First-Principles Studies. *J. Phys. Chem. C* **2016,** *120* (30), 16393–16398.
51. Kieslich, G.; Sun, S.; Cheetham, A. K. Solid-State Principles Applied to Organic–Inorganic Perovskites: New Tricks for an Old Dog. *Chem. Sci.* **2014,** *5* (12), 4712–4715.
52. Meyer, G. The Synthesis and Structures of Complex Rare-Earth Halides. *Progress Solid State Chem.* **1982,** *14* (3), 141–219.
53. Giustino, F.; Snaith, H. J. Toward Lead-Free Perovskite Solar Cells. *ACS Energy Lett.* **2016,** *1* (6), 1233–1240. https://pubs.acs.org/doi/10.1021/acsenergylett.6b00499.
54. Zhao, X. G.; Yang, D.; Ren, J. C.; Sun, Y.; Xiao, Z.; Zhang, L. Rational Design of Halide Double Perovskites for Optoelectronic Applications. *Joule* **2018,** *2* (9), 1662–1673.
55. Kaltzoglou, A.; Antoniadou, M.; Perganti, D.; Siranidi, E.; Raptis, V.; Trohidou, K. et al. Mixed-Halide $Cs_2SnI_3Br_3$ Perovskite as Low Resistance Hole-Transporting Material in Dye-Sensitized Solar Cells. *Electrochim. Acta* **2015,** *184*, 466–474.
56. Xiao, Z.; Song, Z.; Yan, Y. From Lead Halide Perovskites to Lead-Free Metal Halide Perovskites and Perovskite Derivatives. *Adv. Mater.* **2019,** *31* (47), 1803792.
57. Babel, D.; Haegele, R.; Pausewang, G.; Wall, F. Ueber kubische und hexagonale elpasolithe $AI_2BIMIIIF_6$. *Mater. Res. Bull.* **1973,** *8* (12), 1371–1382.
58. Zhang, P.; Yang, J.; Wei, S. H. Manipulation of Cation Combinations and Configurations of Halide Double Perovskites for Solar Cell Absorbers. *J. Mater. Chem. A* **2018,** *6* (4), 1809–1815.
59. Ning, W.; Wang, F.; Wu, B.; Lu, J.; Yan, Z.; Liu, X. et al. Long Electron–Hole Diffusion Length in High-Quality Lead-Free Double Perovskite Films. *Adv. Mater.* **2018,** *30* (20), 1706246.
60. Xiao, Z.; Meng, W.; Wang, J.; Yan, Y. Thermodynamic Stability and Defect Chemistry of Bismuth-Based Lead-Free Double Perovskites. *ChemSusChem* **2016,** *9* (18), 2628–2633.
61. Giannozzi, P.; Baroni, S.; Bonini, N.; Calandra, M.; Car, R.; Cavazzoni, C. et al. Quantum Espresso: A Modular and Open-Source Software Project for Quantum Simulations of Materials. *J. Phys.: Condensed Matter* **2009,** *21* (39), 395502.
62. Perdew, J. P.; Burke, K.; Ernzerhof, M. Generalized Gradient Approximation Made Simple. *Phys. Rev. Lett.* **1996,** *77* (18), 3865.
63. Volonakis, G.; Filip, M. R.; Haghighirad, A. A.; Sakai, N.; Wenger, B.; Snaith, H. J.; Giustino, F. Lead-Free Halide Double Perovskites via Heterovalent Substitution of Noble Metals. *J. Phys. Chem. Lett.* **2016,** *7* (7), 1254–1259.
64. Luo, J.; Li, S.; Wu, H.; Zhou, Y.; Li, Y.; Liu, J. et al. $Cs_2AgInCl_6$ Double Perovskite Single Crystals: Parity Forbidden Transitions and Their Application for Sensitive and Fast UV Photodetectors. *ACS Photon.* **2018,** *5* (2), 398–405.

65. Menedjhi, A.; Bouarissa, N.; Saib, S.; Bouamama, K. Halide Double Perovskite $Cs_2AgInBr_6$ for Photovoltaic's Applications: Optical Properties and Stability. *Optik* **2021,** *243*, 167198.
66. Aslam, F.; Sabir, B.; Hassan, M. Structural, Electronic, Optical, Thermoelectric, and Transport Properties of Indium-Based Double Perovskite Halides Cs_2InAgX_6 (X = Cl, Br, I) for Energy Applications. *Appl. Phys. A* **2021,** *127*, 1–12.
67. Murtaza, G.; Alshahrani, T.; Khalil, R. A.; Mahmood, Q.; Flemban, T. H.; Althib, H.; Laref, A. Lead Free Double Perovsites Halides $X_2AgTlCl_6$ (X = Rb, Cs) for Solar Cells and Renewable Energy Applications. *J. Solid State Chem.* **2021,** *297*, 121988.
68. Blaha, P.; Schwarz, K.; Tran, F.; Laskowski, R.; Madsen, G. K.; Marks, L. D. WIEN2k: An APW+ lo Program for Calculating the Properties of Solids. *J. Chem. Phys.* **2020,** *152* (7), 074101.
69. Blaha, P.; Schwarz, K.; Sorantin, P.; Trickey, S. B. Full-Potential, Linearized Augmented Plane Wave Programs for Crystalline Systems. *Comput. Phys. Commun.* **1990,** *59* (2), 399–415.
70. Peng, H.; Yang, Z. H.; Perdew, J. P.; Sun, J. Versatile van der Waals Density Functional Based on a Meta-Generalized Gradient Approximation. *Phys. Rev. X* **2016,** *6* (4), 041005.
71. Mukhtar, M. W.; Ramzan, M.; Rashid, M.; Naz, G.; Imran, M.; Fahim, F. et al. New Lead-Free Double Perovskites A_2NaInI_6 (A = Cs, Rb) for Solar Cells and Renewable Energy; First Principles Analysis. *Mater. Sci. Eng.: B* **2021,** *273*, 115420.
72. Saeed, Y.; Amin, B.; Khalil, H.; Rehman, F.; Ali, H.; Khan, M. I. et al. $Cs_2NaGaBr_6$: A New Lead-Free and Direct Band Gap Halide Double Perovskite. *RSC Adv.* **2020,** *10* (30) 17444–17451.
73. Zhao, S.; Yamamoto, K.; Iikubo, S.; Hayase, S.; Ma, T. First-Principles Study of Electronic and Optical Properties of Lead-Free Double Perovskites Cs_2NaBX_6 (B = Sb, Bi; X= Cl, Br, I). *J. Phys. Chem. Solids* **2018,** *117*, 117–121.

CHAPTER 11

First Principle Study to Determine the Structural and Elastic Properties of Metal Nitride (XN; X=Ta, W)

AGNIBHA DAS MAJUMDAR[1,2], OM MISHRA[1], and NEHA MUNJAL[1]

[1]*Department of Physics, School of Chemical Engineering and Physical Sciences, Lovely Professional University, Phagwara, Punjab, India*

[2]*Department of Physics, Madanapalle Institute of Technology and Science, Madanapalle, Andhra Pradesh, India*

ABSTRACT

The current study investigates the structural and elastic characteristics of two metal nitrides, TaN and WN, using the linear combination atomic orbital (LCAO) method. The lattice constant (a_0) and bulk modulus (B) for the B1, B2, and B3 structures were calculated, and the values for different elastic constants (C11, C12, and C13) were determined. The study employs the Brich–Murnaghan equation to find the approximate values of the bulk modulus. The study used the density functional theory with the support of CRYSTAL coding. The results of the study provided a comprehensive understanding of the structural and elastic properties of TaN and WN, which have potential applications in various fields such as electronics, aerospace, and energy.

11.1 INTRODUCTION

11.1.1 TRANSITION METALS

A chemical substance present with the W (atomic number = 74) is tungsten, often popular as wolfram. Rare metal tungsten is only naturally present on

Advanced Materials for Next-Generation Technologies: Challenges and New Prospects. Shrikaant Kulkarni, Vipul Srivastava, Rabah Khenata, & Yarub Al-Douri (Eds.)

Earth, and it is virtually always found in combinations with other elements. It was first separated as a metal in 1783 after being discovered as a new element in 1781.[1] Scheelite and wolframite, which give the element its other name, are two of its significant ores. Due to the fact that it has the highest known melting point, that is—3,422 °C (3,695 K)—the free element is notable for its resilience. The boiling point of this substance is the greatest at 5,930 °C (6,203 K). Its density, which is 19.30 g/cc, is higher (approximately 1.7 times) than that of lead and comparable to that of uranium and gold. Carbon has no melting point because it sublimes at atmospheric pressure rather than melting at greater temperatures than tungsten. However, the most reliable crystal phase for tungsten exhibits no structure generated by high-pressure changes for pressures up to at least 364 gigapascals. Of all pure metals, tungsten has the least thermal expansion coefficient.[2,3] The strong metallic connections that the 5d electrons create between the tungsten atoms are what give tungsten its high melting point, moderate thermal expansion, and high tensile strength. Steel becomes significantly more robust when tungsten is added in modest amounts as an alloy. It is challenging to convert polycrystalline tungsten into metal since it is an inherently brittle and hard substance (under normal circumstances, while uncombined). Pure single-crystalline tungsten may, however, be cut with a hard-steel hacksaw and is more ductile.

Radiation shielding, super-alloys, electrodes in gas tungsten arc welding, incandescent light bulb filaments, and X-ray tubes are having the application of the several alloys that include tungsten and have numerous uses. Due to its extreme hardness and high density, tungsten (W) is well suited for use in military piercing projectiles. Industrial catalysts frequently use tungsten compounds. The only third transition series metal that is known to be found in biomolecules is tungsten, which is found in a few types of archaea and bacteria.[4–6]

Chemical compounds made of tungsten can have oxidation states ranging from -II to VI. Middle-level oxidation states are frequently linked to metal clusters, whereas extremely low oxidation levels are frequently linked to CO complexes. Its terrestrial occurrence and biological functions are connected to higher oxidation states, which are always oxides. The chemical compositions of tungsten are heavier than their lighter congener, chromium and molybdenum exhibit significant differences. For instance, the relative prevalence of the chromium (III) compounds contrasts with the relative rarity of tungsten(III). In tungsten(VI) oxide, the maximum oxidation state is visible (WO_3).[7]

The manufacturing of hard materials, such as tungsten carbide, uses around half of the tungsten, with alloys and steels being its principal uses for the remainder. In other chemical compounds, less than 10% is utilized. Due to tungsten's high ductile-brittle transition temperature, chemical vapor deposition powder metallurgy, hot isostatic pressing, spark plasma sintering and thermoplastic processes are frequently used to create the metal's products. The production of tungsten carbide (WC), one of the toughest carbides, is the main application of tungsten. Compared to WC, W2C is a less efficient electrical conductor.[4] WC is used in the manufacturing of wear-resistant abrasives and "carbide" cutting tools such circular saws, knives, dies, and many more in various sectors. WC particles are held in place by metallic cobalt, which serves as a binding (matrix) material in carbide tooling, a ceramic/metal composite. Around 60% of the world's current tungsten consumption is due to this type of industrial application.

Tungsten's density, which is comparable to that of gold, enables its use as a substitute for gold or platinum in jewellery. Metallic tungsten, though not as hard as tungsten carbide, is harder than gold alloys and is hypoallergenic. This makes it an ideal material for scratch-resistant rings, particularly those with brushed-finish designs.

Furthermore, due to its affordability and similar density to gold (with only a 0.36% difference), tungsten has been employed in the production of counterfeit gold bars. This involves techniques such as plating tungsten bars with gold, a practice documented since the 1980s, or replacing portions of genuine gold bars with tungsten rods by drilling holes in them. Although gold and tungsten have slightly different densities and properties, gold-plated tungsten can pass basic authenticity tests.[8,9]

Ta represents tantalum, an element in the periodic table with an atomic number of 73. It was previously referred to as tantalium and derived its name from Tantalus. Tantalum is classified as a transition metal, known for its exceptional hardness, ductility, glossy appearance, and distinctive blue-gray color. Additionally, tantalum exhibits remarkable resistance to corrosion. It belongs to the family of metals known as refractory metals, which are frequently employed to make strong alloys with high melting points.[10] It is a group 5 element that also occurs in the minerals tantalite, columbite, and coltan. It constantly appears in geologic sources along with the related element niobium, which is having almost same chemical characteristics. Since Jean Charles Galissard de Marignac's discovery of the technique in 1866, commercial tantalum from niobium separation has been accomplished with fractional crystallization of potassium oxypentafluoroniobate monohydrate and potassium heptafluorotantalate. Solvent extraction from tantalum solutions that contain fluoride has replaced this approach. Anders Ekeberg made the discovery of tantalum in Sweden in 1802 while examining two mineral samples, one from Finland and the other from Sweden.[11,12]

Tantalum finds its primary application as a metal powder in the manufacturing of electronic components, particularly high-power resistors and capacitors. Tantalum electrolytic capacitors utilize the unique property of tantalum to form a protective surface layer through electrolytic solution or conductive solid, creating a dielectric oxide layer. In these capacitors, tantalum powder is compressed into pellet form to serve as one of the capacitor plates. Numerous alloys with high melting temperatures, strength, and ductility are created using tantalum as well. It is also used to produce super-alloys for jet engine parts, chemical process machinery, containers, heat exchangers, and vessels when alloyed with other metals to create carbide tools for metal-working equipment. NASA used tantalum to protect spaceship parts from radiation, including Voyager 1 and Voyager 2. The metal is used to make hoover boiler components because of its high melting point and resistance to oxidation. Due to its high level of inertness, tantalum is used to create a variety of corrosion-resistant components, including tantalum bolts, thermowells, and valve bodies.[13,14]

11.1.2 TRANSITION METAL NITRIDE

Titanium nitride (TiN), vanadium nitride (VN), chromium nitride (CrN), and other nitrides of transition metals are a group of compounds where nitrogen atoms acquire spaces within the lattice of the parent metal. The common transition metal nitrides (TMN) are known to exhibit three different crystal structures: hexagonally closed packed (HCP), face-centered cubic (FCC), and simple hexagonal (HEX). Transition metal nitrides, regardless of their structure, often possess more metallic and covalent attributes compared to similar oxides and sulfides, providing them with finest properties in various aspects. These common traits include metallic electrical conductivity, good thermal stability, and chemical bonding. Consequently, metal nitrides have a lot of potential in the energy industry, like photo-catalytically induced pollution elimination, and energy conversion. Previously, the synthesis of transition metal nitrides involved reacting transition metals with flowing nitrogen at high temperatures. The unbreakable nonpolar bond and the substantial activation barrier of the nitrogen triple bond (bond energy: 940.95 kJ mol^{-1}) prevented this approach from being used on all nitrides.

Science and technology advancements have produced new, less complicated synthetic processes, such as ammonolysis. The development of metal nitrides in electrocatalysis, energy storage and conversion, fuel cells, and other domains has been the subject of recent reviews. It has also been published on the most recent developments in nanostructured transition metal nitrides for energy storage. When utilized as active materials in energy-related applications, metal nitrides have benefits over metal oxides, metal sulfides, and metal carbides, including low electrical resistance and great chemical stability.[15]

It is essential to conduct additional experimental and theoretical calculations, such as scanning transmission microscope, in order to determine the accurate metal nitride reaction mechanisms for energy storage and catalysis, which are still unknown. To comprehend the reaction mechanism taking place while applying in case of metal nitrides, new advanced characteristic techniques must be developed, such as in-situ microscopes and spectra. Metal nitrides function at all pH extremes when used in supercapacitor or other electrolyte-required settings. To increase metal nitrides' environmental endurance, it is therefore crucial to combine them with other stable materials. Another trend is the search for appropriate substrates for metal nitrides in order to simultaneously achieve excellent performance and good endurance. The crucial charge and discharge processes for metal air batteries are OER (oxygen evaluation reaction) and ORR (oxygen reduction reaction). OER and ORR's separate electrodes result in sluggish kinetics. Bimetallic nitrides or multi-metal nitrides are therefore highly desirable to operate as both OER electrodes and ORR electrodes in order to boost the reaction activity. Interface engineering should be carefully thought out because strong coupling of the metal nitride/substrate/other composite parts' interface can lead to more active sites, quick electron and ion transport channels, and hetero-junctions, all of which are beneficial for enhancing energy storage and catalysis performance.[16–20]

At visible and longer wavelengths, metal nitrides have metallic characteristics. These interstitial, non-stoichiometric chemicals have high free-carrier concentrations. These substances are also refractory, stable, and rigid, and their optical qualities can be adjusted by modifying their composition. They are essential in terms of technology since several of them are currently used in silicon CMOS technology as barrier layers in copper-Damascene processes as well as gate metals for n-type and p-type transistors. Typically, in the dual Damascene process' barrier layer is made of titanium nitride. In order to integrate plasmonics and nano-electronics, metal nitrides have benefits in manufacture and integration. Tough materials are those with a strong resistance to failure that combine hardness with ductility.[21] Toughness can be easily determined for bulk materials by calculating the area under the stress–strain curve during uniaxial compression or tensile tests, but it is less clear for coatings. The design and development of such materials have always been challenging since an increase in hardness is often followed by an increase in brittleness rather than ductility. One of the key goals of material scientists and the hard coating industry has been the development of hard materials with great temperature stability. Several of these nitrides are epitaxially growable on a variety of substrates, including Si (100). Moreover, TiN, HfN, ZrN, and TaN use cubic substrates like MgO; they might all be epitaxially grown and have cubic lattices. Numerous nitrides, including TiN, can also develop as pseudomorphic epitaxial layers on c-sapphire substrates.[22] Several of the cubic rock-salt nitrides generate domain-matched epitaxial layers on silicon substrates. The production of exceptionally smooth and thin films is made possible via epitaxial growth. Noble metals perform terribly and thin, high-quality coatings with little losses are desirable. Although a lot of hardness is required, most applications also call for ductility to prevent the brittle failure brought on by splitting in coatings subjected to high pressures. Brittle refractory ceramics made of hard materials, including transition metal nitrides, are the norm. In polycrystalline materials, grain borders have weaker structural stability, which makes them fracture propagation hotspots.[23] Because the crack size is often correlated with the grain size, it is also possible to lessen the tendency for fracture propagation by shrinking grains or flaws to the nanoscale. Such nano-structuring can be created by synthesizing nano-scale composites using a variety of methods, such as hardening and toughening ceramic coatings.[22] The origins of toughness are not well understood at the fundamental level, and advancement in this field can only be made by studying the origins of brittleness vs. ductility at the atomic and electronic levels. Yet, there are not many of these investigations. Ab initio theoretical studies, employing DFT calculations, are particularly suited for such design since they reveal the connection between several properties. This research intends to take advantage of recent DFT toughness improvement studies to forecast tough TM nitrides that might be later deposited as coatings. Transition metal nitride (TMN)-based hard and wear-resistant coatings have primarily been used as decorative or protective overlayers, such as in the machining, engine component, and cutting and forming tool industries. By chemical composition, fine-tuning the nano-, micro-, and numerous scientific breakthroughs have resulted in a notable enhancement of their hardness or thermal stability.[24,25]

Transition metal nitride (TMN)-based hard and wear-resistant coatings have primarily been used as decorative or protective overlayers, such as in the machining, engine component, and cutting and forming tool industries. By fine-tuning the micro-, nano-, and chemical composition, numerous scientific breakthroughs have resulted in a notable enhancement of their hardness or thermal stability. Phase stability adjustment, interface design (superlattices, nanocomposites), and multicomponent alloying are a few effective strategies. The development of tough and hard coatings is a significant task in materials science since it helps minimize brittle failure when these materials are subjected to external forces. Recent research has improved our knowledge of thin-film toughening in ternary alloys based on TiN and VN with cubic NaCl structures. They discovered that raising the VEC promotes dislocation gliding. Metal nitrides have great corrosion resistance and are very chemically stable. They are frequently utilized in the industry, for example, as cutting tools, as a result of these conditions. Moreover, they may find use in high power energy industries, optoelectronics, and information storage technologies for coating magnetic sheets. They also adopt a variety of crystal forms in addition to a number of fascinating physical characteristics. Metal nitrides can crystallize in various crystal lattices and create various phases.[26–28]

11.2 LITERATURE REVIEW

Tantalum nitride (TAN) stands out among transition metal nitrides due to its unique CoSn-type structure at room temperature (1:1 ratio). Our research focuses on creating densely packed tantalum nitride phases through both static high-pressure compression and shock investigations. Shock compression techniques have enabled the fabrication of single-phase bulk bodies of tantalum nitride in the Bl-type structure, maintaining proper stoichiometry.

Furthermore, through static compression, we have successfully generated single-phase bulk bodies of tantalum nitrides in the B1 and WC types. The WC-type phase is a high-pressure phase that forms at low temperatures, while the Bl-type phase is a high-temperature phase, as observed in our static-compression quenching studies. These B1- and WC-type bulk bodies exhibit significantly higher hardness compared to the CoSn-type phase and exhibit excellent stability at high temperatures.

An interesting observation from our shock compression experiments is the emergence of superconductivity in the Bl-type tantalum nitride bulk bodies, with a relatively high critical temperature.[29]

Using the density-functional theory, the electrical and mechanical characteristics of LaN to AuN 5d transition metal mononitrides are thoroughly investigated. Six structures—zinc blende, rocksalt, NiAs, CsCl, WC, and wurtzite structures—are taken into account for each nitride. The most stable of the structures under consideration is the rocksalt structure for HfN, AuN, and LaN followed by WC for NiAs, TaN, WN, and wurtzite for OsN, ReN, and PtN, IrN. Each nitride has a mechanically stable structure that is the most stable. From LaN to AuN, the formation enthalpy increases. While the formation enthalpy is positive from WN to AuN, with the exception of the wurtzite structure in ReN, it is negative for every molecule that is taken into consideration. They are entirely metallic, as evidenced by the DOS calculations. The bulk modulus of ReN in the NiAs structure is the highest at 418 GPa. The TaN in the WC structure has the highest shear modulus, 261 GPa.[30]

The studies focused on estimating the mixing enthalpy (Cij) using first-principles pseudopotential calculations for ternary metal nitride alloys, specifically TaxMe1-xN, where Me represents either Mo or W. The calculations were performed within the framework of the cubic B1-rocksalt structure, and we investigated the lattice constants and single-crystal elastic constants. To model disordered ternary alloys, we utilized the VCA and constructed alloy pseudo-potentials based on a first-principles VCA method. For ordered structures, we employed the supercell method (SC) to analyze the effects of clustering. The results demonstrated that the mixing enthalpy remained negative throughout the entire composition range, indicating the stability of the cubic TaxMo1-xN and TaxW1-xN ordered solid solutions. We also examined the evolution of TaN's mechanical properties over time and how the alloying of Mo and W influenced them. Specifically, we investigated the mechanical stability of the ternary structure and its brittle/ductile behavior in relation to the TaN and MeN alloys. This analysis involved studying parameters like the shear-to-bulk modulus ratio (G/B), the Cauchy pressure (C12–C44), and the effective shear elastic constant (C44). To identify the transition from an unstable to a stable B1-rocksalt structure, we applied the elastic stability criterion while alloying MeN with Ta. Furthermore, we estimated the averaged constants (Cij) for polycrystalline TaxMe1-xN alloys using a homogenization method within the framework of anisotropic elasticity. We specifically considered the case of an isotropic medium without any crystallographic texture, which presented a unique scenario in our analysis.[31]

Body-centered bc transition metals, as well as B1-type mononitrides and monocarbides of IIIb-VIb transition metals, are studied in terms of their lattice dynamics using generalized gradient approximation to the exchange-correlation functional, ultra soft pseudopotentials, and first-principles perturbation theory. With the created ultrasoft pseudopotentials, the ground-state characteristics of transition metals and their compounds are accurately replicated. The inelastic neutron scattering experiment results and the estimated phonon spectra of the bcc metals agree remarkably well. We demonstrate how the phonon spectra of transition metal nitrides (TMN) and transition metal carbides (TMC) are related to anomalies, and how the amount of valence electrons in crystals is related to the anomalies of the spectra. For TM, TMC, and TMN, the estimated electron–phonon interaction constants are in very good agreement with the values found by experimentation. The cubic NaCl- and hexagonal WC-type structures' phonon spectra for a number of transition metal mononitrides and monocarbides are anticipated. The dynamic stability and superconducting properties of ScC, YC, and VC ideal stoichiometric B1 crystals are anticipated. YN being a semiconductor, we also deduce.[26]

An efficient design approach to promote toughening in hard nitride-based coatings is multicomponent alloying. We give experimental evaluation of the elastic characteristics of a number of ternary transition metal nitride coatings as well as calculations using the density functional theory. Reactive magnetron co-sputtering was used to create $M1_{1}\text{-}_{x}M2_{x}N$ films. X-ray reflectivity, electron probe microanalysis, X-ray diffraction, and SEM were used to evaluate the structural and growth morphological evolution as a function of film thickness and chemical composition. To determine trends in bonding character, brittleness vs. ductility, and elastic anisotropy, we use the Blackman diagram along with the Pugh and Pettifor criterion. It is described how microstructure and preferred orientation affect polycrystalline films' elastic response. The creation of a nanocomposite system made up of a blend of cubic and hexagonal phases contributes favorably to toughening for Ta-rich TiTaN and TaZrN systems.[27]

First-principles calculations have been used to examine the elastic properties and electronic structures of 4d-transition metal mononitrides and 5d-transition metal mononitrides with both B1 and B3 structures. It is possible to obtain cohesive energies, band structures, bulk moduli, and lattice constants. As a function of the number of d electrons, trends in lattice constants, cohesive energies, and electrical characteristics are shown.[28]

The chemical, mechanical, and thermal properties of tungsten nitrides are distinct from or superior to those of other transition metal nitrides. The integration of nitrogen into the tungsten lattice is thermodynamically unfavorable at atmospheric pressure, which makes it difficult to manufacture these nitrides. The majority of W–N system materials now exist as thin films made by nonequilibrium processes, which significantly restricts their utilization in various technological applications. Here, we report the synthesis of tungsten nitrides using novel methods, including solid-state ion exchange and under-pressure nitrogen degassing. We introduce several new nitrides, such as hexagonal and rhombohedral W_2N_3. Phase-pure and well-crystallized in bulk, the finished products are.[32]

So, from the introduction and literature it can be easily predicted that very few research has been done on the transition metal nitrides. The structural and elastic attribute determination has been done in many conventional ways, but the computational studies have its own importance. Many approaches of DFT first-principles studies have been taken into consideration for different attribute determinations of transition metal nitrides, although the LCAO method has been applied least. Present research work has been applied LCAO method along with the hybrid scheme to study the structural and elastic properties of concerned compounds.

11.3 COMPUTATIONAL METHOD

For the purpose of electronic structural investigation of any atom density functional theory (DFT) is useful in solid-state physics for many times. In the time period of 1990s it became very useful and famous in quantum chemistry. This theory allowed most complex systems to be handled then by traditional first-principles methods, while retaining much of their accuracy. But, DFT is not just an alternative way to solve the Schrödinger wave equation or it is not just a method for parameterizing the empirical results. Density functional theory is a differently unique, formally rigorous, way for solving any interacting problem. The methodology of this theorem is applied in a large multiple of fields to many different problems. In this case the ground-state electronic structure problem is the most common problem.[33]

Density functional theory is basically a quantum computing mechanical modeling process generally used in the field of Physics, Chemistry, and Material science. This is basically a process for the investigation of the electronic or nuclear structure of any N-body system. According to this theory, through the functional many-electron system's properties can be characterized. Here, this term functional intentionally refers to the density of an electron. Hence, the name of this theory is known and popular as density functional theory (DFT). DFT is the mostly used, known, also versatile method available in computation physics and computational chemistry. As this theory became popular in the 1970s, but still till the 1990s it was not considerable as a useful method until it was not enough to calculate the results in quantum chemistry. The costing in case of computational methods is comparatively less, as compared with the generally used experimental methods.[34]

For any given atomic system the quantum mechanical wave-function represents about all the details. For any simple two-dimensional systems it can be defined by any square potential well or it can also be a hydrogen atom. After that the determination of the allowed energy states of that particular system can be done. Regrettably, it is impossible for solving any Schrödinger wave equation for any N-body system. Now, some approximations to solve the complex problem can be introduced to solve that problem easily.

Density functional theory basically is advancement to the Hartree–Fock theory. So, after doing some modifications to HF theory DFT come. DFT describes Fermions in terms of density. It is based on the electron density functional, i.e., function of space and time is a fundamental property utilized in this theory in contrary to the many body wave function used in Hartree–Fock theory. Hence, the name density functional theory is given to it. It obeys the Pauli principle and there is exchange and correlation interaction between the electrons with parallel and anti-parallel spins due to their columbic repulsion. Electron density is only the function of three spatial coordinates x, y, and z whereas the many body electronic wave function used in the HF theory is the function of 3N variables.

The structural and elastic characteristics of metal nitrides (TaN and WN) compounds have been ascertained in this work using a first-principles analysis of the linear combination of atomic orbitals. The LCAO method was derived from quantum mechanics. Linear combination of atomic orbitals, as the name implies, is the superposition of orbitals. The electrical configuration of an atom is characterized as a wave function if the quantum case is taken into account. Yet, this specific wave function in mathematics provides information on an atom's electrons. The fundamental concept of the LCAO approach is that the total number of atomic orbitals and the number of molecular orbitals must be equal. In a different sense, atomic orbitals are combined to create molecular orbitals.[35,36]

The density functional theory used in the CRYSTAL06 programme will solve the Kohn–Sham Hamiltonian for the calculations. Researchers from the Computational Materials Sciences group at CLRC (Daresbury, UK) and the Theoretical Chemistry Department in Torino (Italy) created a very helpful and effective code with significant scientific value in 1978. In the context of HF, DFT, and other hybrid estimates, the CRYSTAL programme calculates the electrical as well as the structural properties of periodic systems. "*CRYSTAL is a program for exploring the crystalline solids.*" The Becke exchange programme and the PBE correlation technique to construct the Kohn–Sham Hamiltonian. Moreover, to determine the values of the bulk modulus, use the Brich–Murnaghan equation's standard form. The results will be checked against earlier experimental and computational findings after calculation.[37]

In the current study, the structural and elastic properties of transition metal nitride compounds, XN; X=Ta, W, are determined using the first-principles (LCAO) approach. Numerous instances have shown that this technique is effective.[38–40] For the purpose of determining the energy values in the CRYSTAL code, the Kohn–Sham Hamiltonian equation under the density functional theory was solved. The tolerance has been employed with a sufficient tolerance of 84k points, and 55% mixing has been done to ensure self-consistency throughout the course of 10 cycles.

11.4 RESULTS AND DISCUSSIONS

11.4.1 STRUCTURAL PROPERTIES

The total energy is expressed as a function of cell volume to obtain the Birch Murnaghan equation and is given as

$$E = E_0 + \frac{9V_0B_0}{16}\left[\left\{\left(\frac{V_0}{V}\right)^{\frac{2}{3}} - 1\right\}^3 B' + \left\{\left(\frac{V_0}{V}\right)^{\frac{2}{3}} - 1\right\}^2 \left\{6 - 4\left(\frac{V_0}{V}\right)^{\frac{2}{3}}\right\}\right]$$

where E(V) represents the energy corresponding to particular volume, E_o is the equilibrium energy, B_o is the bulk modulus, and V_o is the equilibrium volume. We may determine the ground-state parameters, such as the bulk modulus and the lattice constant, using this equation. The graph is presented in the figures, and Table 11.1 summarizes the final results for the measured bulk modulus and lattice constant.

Figures 11.1 and 11.2 represent the variation of the volume of unit cell of lattice structure with change in energy. Two cases for two different binary compounds TaN and WN have been taken into consideration for the structural analysis. A compound's stability is attained with the least amount of energy of its unit cell. By reducing energy in relation to unit cell volume, structural parameters of metallic nitride (TaN and WN) binary compounds have been computed. The findings of the software-generated values are compared to the values from the Birch–Murnaghan equation. The results acquired and the previously completed experimental results were in perfect agreement. The study has been done for three different crystal structures, i.e., B1, B2, and B3. B1 stands for the crystals with NaCl structure, whereas B2 and B3 stand for CsCl and Zinc blende structure respectively.

TABLE 11.1 Calculated Values of Lattice Constant and Bulk Modulus of XM (X= Ta and W) Compound.

	WN (B1)	WN (B2)	WN (B3)	TaN (B1)	TaN (B2)	TaN (B3)
Lattice constant						
Present	4.27	2.71	4.54	4.36	2.67	4.64
Other calculations	4.31 [31], 4.35[24]	–	–	4.31[31], 4.38[25]	–	–
Bulk modulus						
Present	364.57	314.7	295.56	377.8	375.76	298
Other calculations	368.9 [24], 348.8 [41]	–	–	389 [25], 328.3 [42]	–	–

A study has been done to relate the relationship between two structural factors that is bulk modulus and lattice constant and presented in Figures 11.4 and 11.5 The data make it very evident that the structural parameters vary with the B2, B1, and B3 structure, respectively. With this phase order the bulk modulus changes linearly, but with negative slope and lattice constant it changes with linear positive slope. DFT was used to study changes in the bulk modulus and lattice constant.

11.4.2 ELASTIC PROPERTIES

From the literature it is very obvious that B1 or Nacl (Racksalt) structure is the most stable structure for this compound. So, for that structure every elastic property of these two transition metal nitride compounds has been calculated. In the elastic properties Young's modulus E (GPa), stiffness constants Cij (GPa), Poisson's ratio v, shear modulus G (GPa), and bulk modulus B (GPa) have been taken into consideration. When examining the stability and stiffness of materials, the elastic constants of solids are crucial. There are just three independent elastic constants in a cubic crystal: C11, C12, and C44. They are computed from the total energy as a function of strain.

The process of calculating the elastic constants Cij requires knowledge of the type of strain provided by the tensor,

$$\varepsilon = \begin{matrix} \delta & 0 & 0 \\ 0 & -\delta & 0 \\ 0 & 0 & \dfrac{\delta^2}{(1-\delta^2)} \end{matrix}$$

The values of stiffness constants have been calculated with DFT calculations in software and it has been established that the cubic crystal's elastic stability criteria ($C_{11} - C_{12} > 0$, $C_{11} + 2C_{12} > 0$, $C_{44} > 0$, $C_{12} < B < C_{11}$, $C_{11} > 0$) are true. The other elastic constants, i.e., Poisson's ratio ν, shear modulus G (GPa), Young's modulus E (GPa), and bulk modulus B (GPa) have been determined with some mathematic formulas, using the Voigt–Reuss–Hill approximations,

$$B = \frac{C_{11} + 2C_{12}}{3}$$

$$G_R = \frac{5C_{44}\,(C_{11} - C_{12})}{4C_{44} + 3(C_{11} - C_{12})}$$

$$E = \frac{9BG}{G + 3B}$$

$$\nu = \frac{3B - 2G}{2(3B + G)}$$

The values of all calculated elastic properties have been shown in Table 11.2.

TABLE 11.2 Calculated Values of Elastic Parameters of XM (X= Ta and W) Compounds.

Compound		C11	C12	C44	B	G	E	ν
TaN	Present	854	137	74	376	108.41	296.71	0.37
	Other literature	898.4 [25], 817 [43]	130.7 [25], 132 [44]	63.8 [25], 65 [44]	389 [25], 347 [43]	143.8 [25], 144 [43]	383.8 [25], 380 [43]	0.33 [25], 0.32 [44]
WN	Present	234	399	−69	344	−73.83	−238.56	0.62
	Other literature	215.5 [24], 259.2 [41]	245.6 [24], 243.6 [41]	−88.5 [24], −87 [41]	368.9 [24], 348.8 [41]	–	–	–

Pettifor suggested using Cauchy pressure, which is represented by the equation $p = C_{12} – C_{44}$, as a tool for determining whether a material is brittle or ductile. The atomic bonding's characteristics and changes in p are related. For more brittle materials with directed bonding, $p < 0$, while for more metallic bonding, $p > 0$. In the present study, for TaN and WN the value of p is 63 and 468 respectively. So the results have indicated that both the compounds are hard metal in nature.

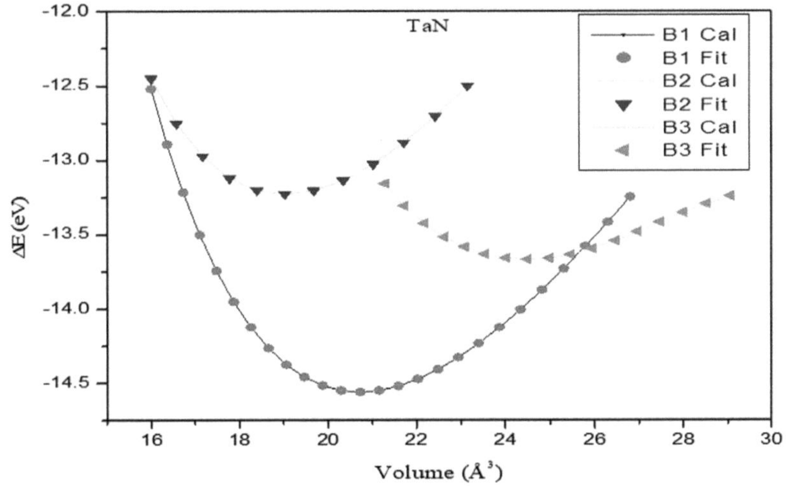

FIGURE 11.1 Variation of unit cell volume with energy for TaN binary compound.

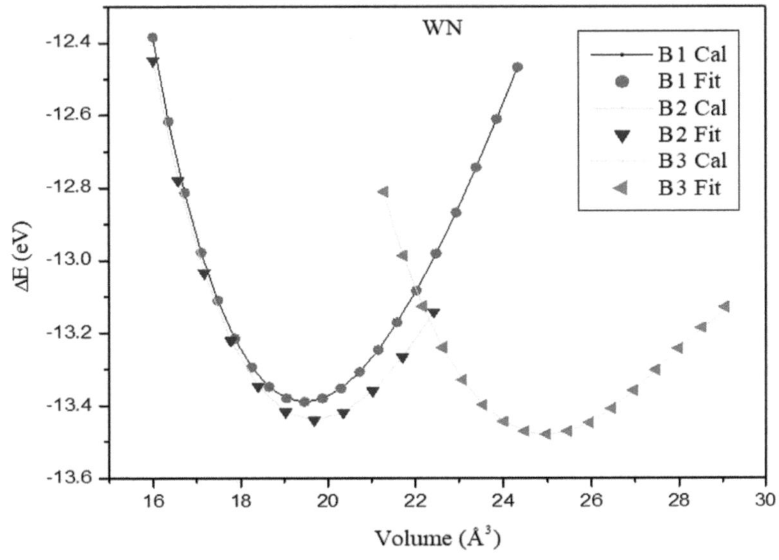

FIGURE 11.2 Variation of unit cell volume with energy for WN binary compound.

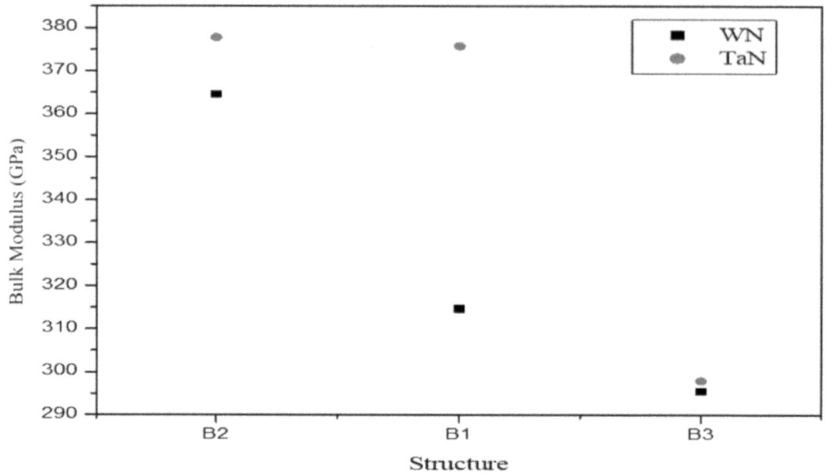

FIGURE 11.3 Variation of the bulk modulus with different crystal structures of unit cell of transition metal nitride compounds.

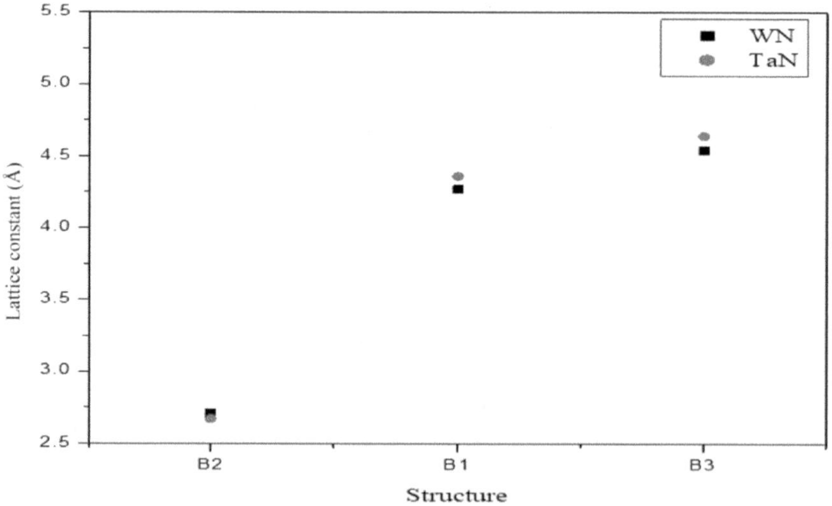

FIGURE 11.4 Variation of the lattice constant with different crystal structures of unit cell of transition metal nitride compounds.

11.5 CONCLUSION

In conclusion, the various structural properties of transition metallic nitride compounds are calculated using the ab-initio LCAO approach in conjunction with the B3LYP hybrid model. Figures 11.1 and 11.2 explain the change in total energy as a function of volume and Figures 11.3 and 11.4 also display the variation in the bulk modulus and lattice constants of XN compounds (where X= Ta and W) as a function of structural phase. The lattice constant rises and bulk modulus decreases with the phases B2, B1, and B3 of the transition metal nitride compounds. Also, all the compounds have shown elastic stability. Additionally, other elastic parameters were predicted and might be used as a guide in other investigations.

KEYWORDS

- **density functional theory**
- **LCAO**
- **metal nitrides**
- **elastic properties**
- **structural properties**

REFERENCES

1. Brown, T.; Pitfield, P. Tungsten. *Critical Metals Handbook*; 2014; pp 385–413.
2. Strigul, N.; Koutsospyros, A.; Arienti, P.; Christodoulatos, C.; Dermatas, D.; Braida, W. Effects of Tungsten on Environmental Systems. *Chemosphere* **2005,** *61* (2), 248–258.
3. Roundy, D.; Krenn, C. R.; Cohen, M. L.; Morris Jr, J. W. The Ideal Strength of Tungsten. *Phil. Magaz. A*, **2001,** *81* (7), 1725–1747.
4. Lide, D. R. (Ed.). *CRC Handbook of Chemistry and Physics*, Vol. 85; CRC Press, 2004.
5. Tolias, P.; EURO fusion MST1 Team. Analytical Expressions for Thermo Physical Properties of Solid and Liquid Tungsten Relevant for Fusion Applications. *Nucl. Mater. Energy*, **2017,** *13*, 42–57.
6. Lassner, E.; Schubert, W. D. *Tungsten: Properties, Chemistry, Technology of the Elements, Alloys, and Chemical Compounds*; Springer Science & Business Media, 1999.
7. Bevers, L. E.; Hagedoorn, P. L.; Hagen, W. R. The Bioinorganic Chemistry of Tungsten. *Coord. Chem. Rev.* **2009,** *253* (3–4), 269–290.
8. Ding, J.; Feng, A.; Li, X.; Ding, S.; Liu, L.; Ren, W. Properties, Preparation, and Application of Tungsten Disulfide: A Review. *J. Phys. D: Appl. Phys.* **2021,** *54* (17), 173002.
9. Cong, S.; Geng, F.; Zhao, Z. Tungsten Oxide Materials for Optoelectronic Applications. *Adv. Mater.* **2016,** 28 (47), 10518–10528.
10. Moseley, P. T.; Seabrook, C. J. The Crystal Structure of β-Tantalum. *Acta Crystallogr. Sect. B: Struct. Crystallogr. Cryst. Chem.* **1973,** *29* (5), 1170–1171.

11. Lambert, J. B. Tantalum and Tantalum Compounds. *Kirk-Othmer Encyclopedia of Chemical Technology*; 2000.
12. Bleiwas, D. I.; Papp, J. F.; Yager, T. R. *Shift in Global Tantalum Mine Production, 2000–2014* (No. 2015–3079); US Geological Survey, 2015.
13. Buckman Jr, R. W. New Applications for Tantalum and Tantalum Alloys. *Jom* **2000,** *52* (3), 40–41.
14. Albrecht, W. W. Production, Properties and Application of Tantalum, Niobium and Their Compounds. In: *Lanthanides, Tantalum and Niobium: Mineralogy, Geochemistry, Characteristics of Primary Ore Deposits, Prospecting, Processing and Applications Proceedings of a workshop in Berlin, November 1986*. Springer: Berlin Heidelberg, 1989; pp 345–358.
15. Luo, Q.; Lu, C.; Liu, L.; Zhu, M. A Review on the Synthesis of Transition Metal Nitride Nanostructures and Their Energy Related Applications. *Green Energy Environ.* 2022.
16. Xue, Z.; Lv, L.; Tian, Y.; Tan, S.; Ma, Q.; Tao, K.; Han, L. Co_3S_4 Nano-Plate Arrays Decorated with Oxygen-Deficient CeO_2 Nanoparticles for Supercapacitor Applications. *ACS Appl. Nano Mater.* **2021,** *4* (3), 3033–3043.
17. Zhou, P.; Lv, X.; Xing, D.; Ma, F.; Liu, Y.; Wang, Z.; Huang, B. High-Efficient Electrocatalytic Overall Water Splitting Over Vanadium Doped Hexagonal $Ni_{0.2}Mo_{0.8}N$. *Appl. Catal. B: Environ.* **2020,** *263*, 118330.
18. Zhang, F. F.; Cheng, C. Q.; Wang, J. Q.; Shang, L.; Feng, Y.; Zhang, Y.; Du, X. W. Iridium Oxide Modified with Silver Single Atom for Boosting Oxygen Evolution Reaction in Acidic Media. *ACS Energy Lett.* **2021,** *6* (4), 1588–1595.
19. Chen, P.; Xu, K.; Fang, Z.; Tong, Y.; Wu, J.; Lu, X.; Xie, Y. Metallic Co_4N Porous Nanowire Arrays Activated by Surface Oxidation as Electrocatalysts for the Oxygen Evolution Reaction. *Angew. Chem.* **2015,** *127* (49), 14923–14927.
20. Dong, Y.; Wu, Y.; Liu, M.; Li, J. Electrocatalysis on Shape-Controlled Titanium Nitride Nanocrystals for the Oxygen Reduction Reaction. *ChemSusChem* **2013,** *6* (10), 2016–2021.
21. Chaudhuri, K.; Shaltout, A.; Shah, D.; Guler, U.; Dutta, A.; Shalaev, V. M.; Boltasseva, A. Photonic Spin Hall Effect in Robust Phase Gradient Metasurfaces Utilizing Transition Metal Nitrides. *ACS Photon.* **2018,** *6* (1), 99–106.
22. Abadias, G. Stress and Preferred Orientation in Nitride-Based PVD Coatings. *Surface Coat. Technol.* **2008,** *202* (11), 2223–2235.
23. Joelsson, T.; Hultman, L.; Hugosson, H. W.; Molina-Aldareguia, J. M. Phase Stability Tuning in the $Nb_xZr_{1-x}N$ Thin-Film System for Large Stacking Fault Density and Enhanced Mechanical Strength. *Appl. Phys. Lett.* **2005,** *86* (13), 131922.
24. Qin, J.; Zhang, X.; Xue, Y.; Li, X.; Ma, M.; Liu, R. Structure and Mechanical Properties of Tungsten Mononitride Under High Pressure from First-Principles Calculations. *Computational Mater. Sci.* **2013,** *79*, 456–462.
25. Chang, J.; Zhao, G. P.; Zhou, X. L.; Liu, K.; Lu, L. Y. Structure and Mechanical Properties of Tantalum Mononitride Under High Pressure: A First-Principles Study. *J. Appl. Phys.* **2012,** *112* (8), 083519.
26. Isaev, E. I.; Simak, S. I, Abrikosov, I. A.; Ahuja, R.; Vekilov, Y. K.; Katsnelson, M. I.; Lichtenstein, A. I.; Borjë J. Phonon Related Properties of Transition Metals, Their Carbides, and Nitrides: A First-Principles Study. *J. Appl. Phys.* **2007,** *101*, 123519.
27. Abadias, G.; Djemia, P.; Belliard, L. Alloying Effects on the Structure and Elastic Properties of Hard Coatings Based on Ternary Transition Metal (M= Ti, Zr or Ta) Nitrides. *Surface Coat. Technol.* **2014,** *257*, 129–137.

28. Chen, W.; Jiang, J. Z. Elastic Properties and Electronic Structures of 4d-and 5d-Transition Metal Mononitrides. *J. Alloys Compd.* **2010,** *499* (2), 243–254.
29. Mashimo, T.; Tashiro, S.; Nishida, M.; Miyahara, K.; Eto, E. B1-Type and WC-Type Phase Bulk Bodies of Tantalum Nitride Prepared by Shock and Static Compressions. *Phys. B: Condensed Matter* **1997,** *239* (1–2), 13–15.
30. Zhao, E.; Wu, Z. Electronic and Mechanical Properties of 5d Transition Metal Mononitrides via First Principles. *J. Solid State Chem.* **2008,** *181* (10), 2814–2827.
31. Bouamama, K.; Djemia, P.; Benhamida, M.; First-Principles Calculation of the Structural and Elastic Properties of Ternary Metal Nitrides $Ta_xMo_{1-x}N$ and $Ta_xW_{1-x}N$. In: *J. Phys.: Conf. Ser.* **2015,** *640* (1), 012022.
32. Wang, S.; Yu, X.; Lin, Z.; Zhang, R.; He, D.; Qin, J.; Zhu, J.; Han, J.; Wang, L.; Mao, H. K.; Zhang, J. Synthesis, Crystal Structure, and Elastic Properties of Novel Tungsten Nitrides. *Chem. Mater.* **2012,** *24* (15), 3023–3028.
33. Orio, M.; Pantazis, D. A.; Neese, F. Density Functional Theory. *Photosynth. Res.* **2009,** *102*, 443–453.
34. Burke, K. Perspective on Density Functional Theory. *J. Chem. Phys.* **2012,** *136* (15), 150901.
35. Engel, E.; Dreizler, R. M. Density Functional Theory. *Theoret. Math. Phys.* **2011,** 11–56.
36. Makkar, P.; Ghosh, N. N. A Review on the Use of DFT for the Prediction of the Properties of Nanomaterials. *RSC Adv.* **2021,** *11* (45), 27897–27924.
37. Dovesi, R.; Saunders, V. R.; Roetti, C.; Orlando, R.; Zicovich-Wilson, C. M.; Pascale, F.; Civalleri, B.; Doll, K.; Harrison, N. M.; Bush, I. J.; D'Arco, Ph.; Llunell, M. *CRYSTAL06 User's Manual*; University of Torino: Torino, 2006.
38. Majumdar, A. D.; Munjal, N. High-Pressure Structural Phase Transition of Alkali Hydride Compounds: An Ab Initio Study. In: *Advanced Functional Materials and Devices: Select Proceedings of AFMD 2021*; Springer: Singapore, 2022; pp 139–148.
39. Majumdar, A. D.; Munjal, N.; Kamboj, U.; Dogra, K. Investigation of Structural and Electronic Properties of BaX (X= S, Se, and Te): A DFT study. In: *Advances in Materials Science and Engineering: Select Proceedings of ICFMMP 2019*; Springer: Singapore, 2020; pp 281–289.
40. Majumdar, A. D.; Kaur, P.; Munjal, N. A First Principle Study of Structural Properties of KX_xI_{1-x} (X= Cl, Br and x= 0.00, 0.25, 0.50, 0.75, and 1.00). In: *Advances in Functional and Smart Materials: Select Proceedings of ICFMMP 2021*; Springer Nature: Singapore, 2022; pp 175–185.
41. Suetin, D. V.; Shein, I. R.; Ivanovskii, A. L. Elastic and Electronic Properties of Hexagonal and Cubic Polymorphs of Tungsten Monocarbide WC and Mononitride WN from First-Principles Calculations. *Physica Status Solidi (B)*, **2008,** *245* (8), 1590–1597.
42. Cao, C. L.; Hou, Z. F.; Yuan, G.; First-Principles Study of the Structural Stability and Electronic Structures of TaN. *Physica Status Solidi (B)* **2008,** *245* (8), 1580–1585.
43. Zhao, E.; Hong, B. O.; Meng, J.; Wu, Z. First Principles Investigation on the Ultra-Incompressible and Hard TaN. *J. Computational Chem.* **2009,** *30* (14), 2358–2363.
44. Ren, F.; Wang, Y.; Pressure-Induced Phase Transition of Tantalum Mononitride. *Thin Solid Films* **2011,** *519* (11), 3954–3958.

CHAPTER 12

Electronic Energy Loss and Ranges of Different Ions in Germanium: A Comparative Study

KAWALJEET SINGH SAMRA and MUZAFAR SALAM

Department of Physics, Lovely Professional University, Phagwara, Punjab, India

ABSTRACT

Electronic energy loss and ranges of different projectiles: Ti, Li, Pb, and Xe in a semiconductor target, that is, Germanium, have been measured by using SRIM and data tables given by Northcliffe and Schilling. A mutual comparison of those values has been done and, in a few cases, the experimental values are also used for the validation. In this investigation we tried to consider the maximum range available in SRIM and Northcliffe and Schilling tables for comparison. A discussion on the agreements and differences is reported in this manuscript.

12.1 INTRODUCTION TO ION MATTER INTERACTION

The study of ion matter interaction was started by Bohr, Bethe, and Bloch by using the classical approach, which suggests that, when ions pass through matter, inelastic interactions between the projectile and target atoms are the main cause of energy drop over a broad range of velocity.[1] The term inelastic depicts that the energy is strayed during the interactions which may result in agitations among bounded electrons of the medium, and in the electron-cloud of Refs. [2, 3]. As we know the study of the collision of

Advanced Materials for Next-Generation Technologies: Challenges and New Prospects. Shrikaant Kulkarni, Vipul Srivastava, Rabah Khenata, & Yarub Al-Douri (Eds.)

charged particles with matter is itself a rudimentary science problem.[4,5] The study of energy losses for various ion–target permutations is needed for a wide range of investigations in the atomic and nuclear physics. Its utilization is prime in areas like radiation-protection, ion-implantation, and nuclear-medicine.[6,7] It has been seen that charged-particle sensors are accountable for almost all experimental data in entire range of high-energy-physics and in understanding of the stopping-power and range connection is one of the most indispensable needs for the use and calibration of any charged-particle sensor.[8]

Till now a considerable number of experimental and theoretical works have been reported and then compiled in various data collections for a wide range of ions and target materials encompassing a large energy range with the objective of getting appropriate stopping power and projected range values for enhancing trial and semi-pragmatic data base and ameliorating the ion-beam research and irradiation methods.[7] In 1980, Bimbot et al. calculated the stopping powers of Kr, Xe, Pb, and U ions in solid media, including some plastic films. This study was done by recoil proton technique at Unilac accelerator, GSI Darmstadt.[7] The experimental data was compared with the values reported in the tables of Northcliffe and Schilling. Variation of 30–60% was reported for light detectors. Andero et al. in 1986 calculated the values of stopping power for high-energy photon-beams using the Monte Carlo method. Results were mapped with stopping power reported in various dosimetry protocols and with the reported experimental data. The accordance with most dosimetry protocols was within about 1% and with experimental results it was better than 0.5%.[9]. Sharma et al. (1993–1994) did a stopping-power comparison (between the experimental and theoretical values) of heavy ions like U, Pb, Au, Xe, Mo. Kr, Cu, Ca, Ar, P, O, N, Li etc., in different targets (Be, C, AI, Si, Ti, Ni, Cu, Zr, Ag, Ta, Au, Bi, U, Mylar, Hostaphan, etc.) at different energies. It was observed that the Hubert's formulation was best mapped with the experimental data for all target and projectile permutations in the studied energy range.[10] Later in the year 2000, the same group had measured the stopping power of Mylar for heavy ions, whose atomic number lied between 11 and 29, having energy between 0.3 and 2.3 MeV/u by elastic-recoil-detection (ERD) method.[11] The stopping-power values measured experimentally were compared with the semi-empirically calculated values from Northcliffe and Schilling tables, Ziegler's model, and LSS theory. The pros and cons of these semiempirical models were spotlighted. Diwan et al. in 2002 studied the deceleration of MeV heavy ions with atomic numbers between 6–29 in polyethylene naphthalate by using the reflection mode of the ERD technique.[12] These measurements were carried out by applying the 140 MeV beam of Ag^{13+} ions from the Pelletron at Inter-University Accelerator Centre (IUAC), New Delhi, India. The stopping-power data based on Lindhard, Scharff and Schiott (LSS) theory, Northcliffe and Schilling tabulations, Ziegler's SRIM code and extended Hubert's formulations were compared with the experimental results. It was revealed that the calculated values from the formulations of Hubert and SRIM-2000.38 mapped well with the experimental data. Fertman et al.[13] measured the stopping power of Cu ion in hydrogen and nitrogen. They determined the stopping powers and compared them with those of SRIM-2000/2003, and with the STIP code.[13] In the year 2006, Diwan et al. studied the slowing down and straggling of Li, C, and O ions in Polyethylene terephthalate and Polycarbonate films by varying their thicknesses in the energy range between 3.0 and 6.0 MeV/u.[14] These experimental values were compared with values calculated from different formulations like Bethe–Livingston, Bohr, Lindhard and Scharff. Devendrappa et al. (2016) investigated the slowing down of alpha particles of 5.48 MeV, in polystyrene film.[16,15] It was reported that with the increase of transmitted energy, the stopping power of alpha particles in polystyrene film decreases, but the projected range of alpha-particles increased considerably. Linaries et al. experimentally measured the stopping powers of Ti, V, and Cr ion in Ge and Au, in the energy span between 150 to 500 keV/u and validated them with the previously done measurements.[16] The experimental results of the Ge film deviated from the theoretical values. Yoneda et al.[17] studied about the interaction of C ions (4.8–6.8 MeV) along the [1 1 0] and [1 1 1] directions in Si by using the backscattering spectra. Their experimental results were almost mapped with the calculated values of stopping power of the CasP code.[17] Recently, in 2019, Mammeri et al. investigated the straggling and energy loss of He+ ions (0.9–3.0 MeV) in Bismuth foils. Rutherford back-scattering spectroscopy was utilized to find the relation between the straggling and stopping power.[18] The measured stopping powers were in good accord with those calculated numerically by SRIM-2013 and CasP 5.2 codes. In the same year, Iqbal et al. explored about the energy drop of protons (1–12 MeV) in lead and beryllium and compared them with percent mass-stopping power calculated from the models proposed by Bethe-Bloch and Bichsel-Sternheimer. They reported the deviation of experimental data with theoretical values at lower energy range.[19]

In the present study, the comparison of electronic energy loss and projected range measurements over the incident ion energy range for Li, Ti, Xe, and Pb ions incident on the Germanium target is done. In our study we took help of SRIM and Northcliffe and Schilling (NS) tabulations for extracting stopping power and range values and compared a few of those values with the already reported experimental values.

12.2 DIFFERENT MODELS

Stopping and range of ions in matter (SRIM) was introduced in 1985 by Ziegler, Biersack and Littmark and has been continuously upgraded.[1] The textbook "SRIM—The Stopping and Range of Ions in Matter" elaborates the basic physics behind this semiempirical model.[2] Since 1985 updations have been done by considering newly reported experimental results.[3] Essential changes imbibed in SRIM after regular intervals of time, there were so many updations, which were incorporated in this model like, in 1995 a complete revamp of the slowing down of relativistically moving light ions at 1 MeV/u or above. In 2010, notable changes were done to rectify the energy loss of ions in compounds, which is regularly applied for quantitative analysis in applied fields of science and technology. Still, the model contains significant variation between calculated and experimentally determined values for heavier projectiles due to extrapolation limitations and scarcity of experimental results available for heavy ions. In our work, the range of ions we took using SRIM-2013 is from 10 keV to 10^6 keV.

The NS tables were developed in the 1970s and are regularly utilized to extract the quantitative values of stopping powers and ranges of various ions in different targets. It is developed by applying interpolation/extrapolation method, on the experimental data available at that time, to generate multiple tables for the stopping power of various ions in 24 elements.[4] The foundation of the NS tables is the assumption that the mapping for the stopping power of two different targets depends on the velocity but not on the atomic number of the projectile. In the present study, we took the complete range available in this tabulation.

12.3 COMPARISON OF STOPPING AND RANGE OF IONS IN MATTER (SRIM) AND THE NORTHCLIFFE AND SCHILLING TABLES

The electronic energy loss of Li, Ti, Xe, and Pb ions in the Ge target, calculated by using SRIM, in the energy range 10 keV to 10^6 keV is displayed in Figure 1. It is revealed from Figure 1 that the electronic energy loss of Pb ions is maximum and of Li ions is minimum in the investigated ion energy range, which may be attributed to the size of the ions. It is observed that for all ions electronic energy loss initially increases with ion energy, reaches maximum, and then starts decreasing with further increase of ion energies. This behavior may be attributed to the dominance of inelastic collisions at intermediate ion energies, but elastic scattering contributes more at lower and higher ion energies. First impression of the behavior of electronic energy loss calculated from the NS table, of all four ions in Ge in the studied ion energy range, as revealed from Figure 2, is almost similar. A close look at Figure 3 reveals various dissimilarities among the stopping power behaviors of four ions in the Germanium target. It is observed that for all ions at the lower ion energies, stopping power behaviors of both the codes coincide, but a noticeable difference is observed at the higher ion energies. The magnitude of the values of electronic energy loss of Ti and Xe ions measured by using SRIM is comparatively larger as compared to those of the NS table, but an opposite trend is observed for Pb ion in the same target. This may be attributed to the different assumptions taken by the codes for higher energy regime. Figure 4 displays the comparison of electronic stopping power of Ge toward Ti ions having varying energy, measured experimentally[16] and from SRIM and NS table. The experimental data agrees with the values fetched from both codes, that is, SRIM and NS table.

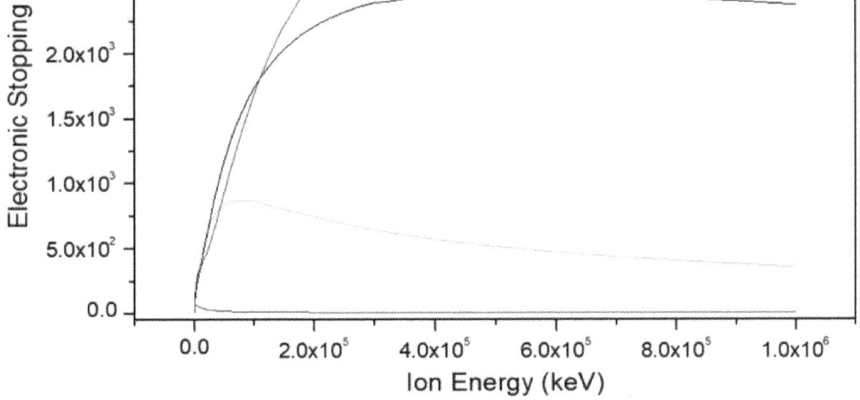

FIGURE 12.1 Electronic stopping power of Ge calculated from SRIM.

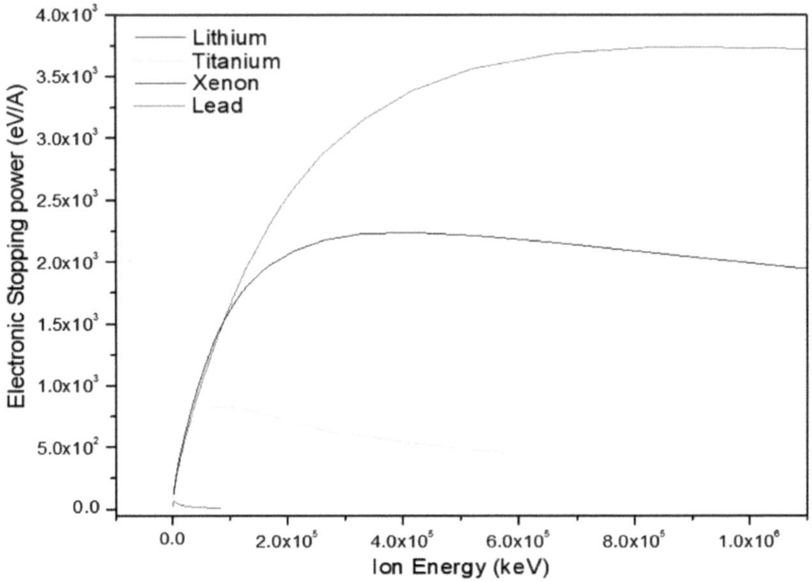

FIGURE 12.2 Electronic stopping power of Ge calculated from the NS table.

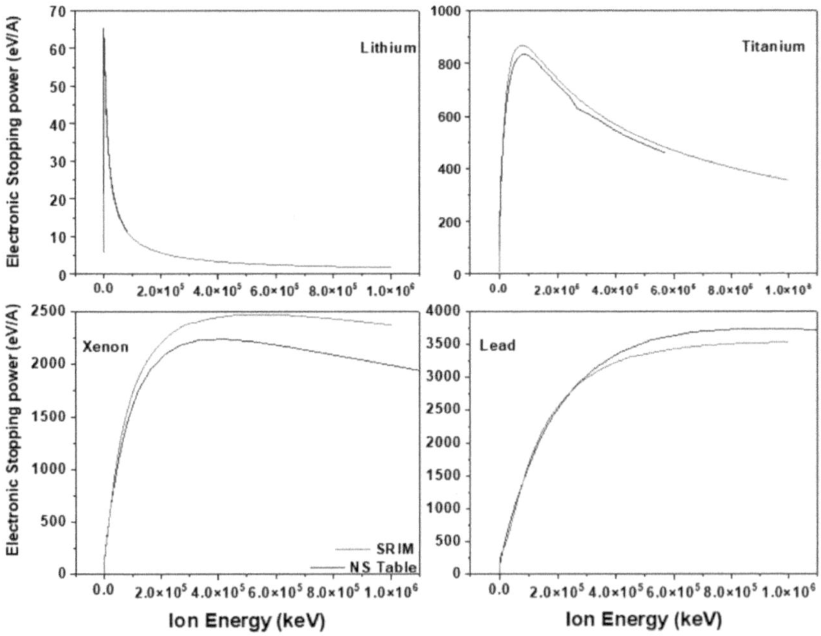

FIGURE 12.3 Comparison of electronic energy of various ions in Ge calculated from SRIM IM and NS table.

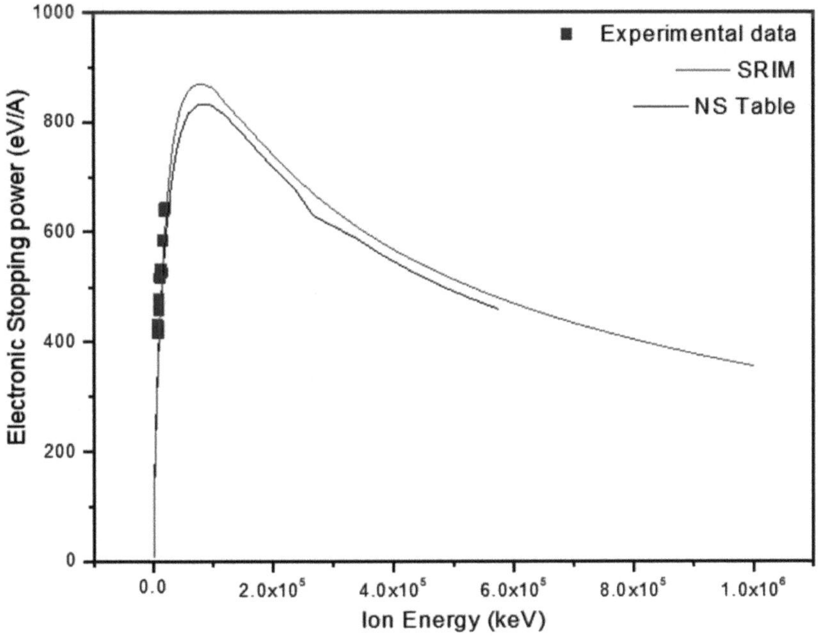

FIGURE 12.4 Comparison of electronic stopping power of Ge measured experimentally[16] and by using SRIM and NS table.

A comparison of ranges of various ions in Ge, calculated by using SRIM and NS table, is plotted in Figure 5. It is revealed from Figure 4 that the range of Li ions is maximum and that of Pb ions is minimum in the investigated ion energy range, which may be owed to the size of the ions. It is further observed that for all ions at the lower ion energies, magnitudes of values of ranges fetched from both codes coincide, but a noticeable difference is observed at the higher ion energies. The magnitude of the range of Ti and Xe ions, measured by using the NS table, is comparatively larger as compared to those of SRIM, but an opposite trend is observed for Pb ion in the same target.

12.4 CONCLUSION

In this article the electronic energy loss and ranges of Li, Ti, Xe, and Pb ions in the Germanium target at varying ion energies have been calculated from SRIM and NS table. A mutual comparison of these values has been done. It is observed that for all ions at the lower ion energies, magnitudes of values of electronic energy loss and ranges fetched from both codes coincide, but a noticeable difference is observed at higher ion energies.

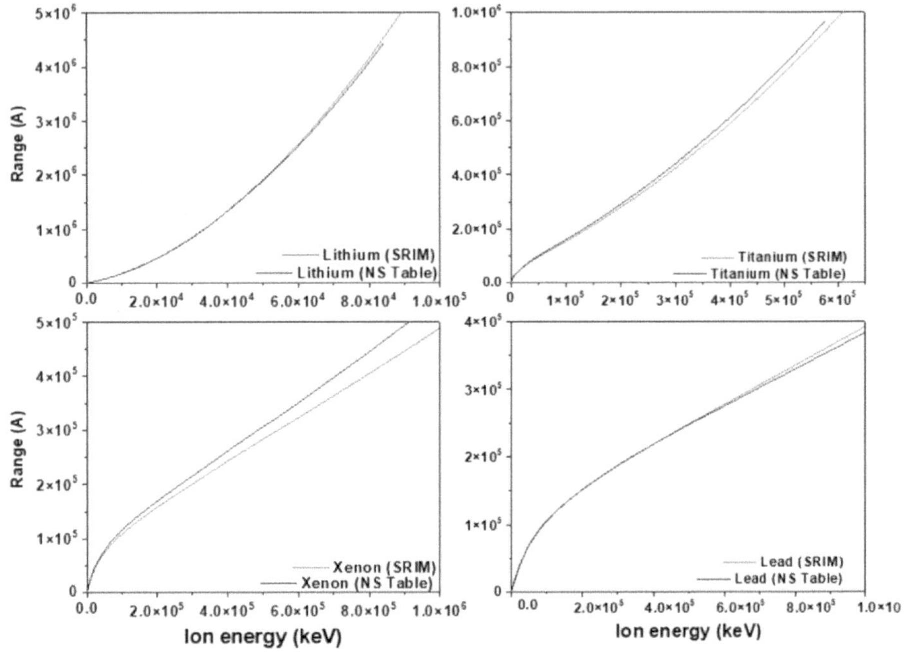

FIGURE 12.5 Comparison of ranges of various ions in Ge calculated by using SRIM and NS table.

ACKNOWLEDGMENT

Lovely Professional University, Punjab, is highly acknowledged for providing the adequate environment and opportunity to pursue this work.

KEYWORDS

- **electronic stopping power**
- **range**
- **SRIM & NS table**

REFERENCES

1. Ziegler, J. F.; Biersack, J. *The Stopping and Range of Ions in Matter*; Pergamon Press: New York, 1985.
2. Paul, H.; Schinner, A. Judging the Reliability of Stopping Power Tables and Programs for Heavy Ions. *Nucl. Instruments Methods Phys. Res. Sect. B* **2003,** *209*, 252–258.
3. Northcliffe, L. C.; Schilling, R. F. Range and Stopping-Power Tables for Heavy Ions. *Atom. Nucl. Data Tables* **1970,** *7* (3–4), 233–463.
4. Brandolini, F.; Medina, N. H.; Poli, M. D.; Pavan, P.; Wilhelm, M.; Dewald, A.; Pascovici, G. Stopping Power of Nd ions in Pb Determined from γ-Ray Line Shape Analysis in Coulomb excitation. *Nucl. Instruments Methods Phys. Res. Sect. B* **1997,** *132* (1), 11–17.
5. Ziegler, J. F. SRIM—The Stopping and Range of Ions in Matter. *Nucl. Instruments Methods Phys. Res. Sect. B* **2010,** *268*, 1818–1823.
6. Ziegler, J. F. SRIM—2003. *Nucl. Instruments Methods Phys. Res. Sect. B* **2004,** *219–220*, 1027–1036.
7. Bimbot, R.; Cabot, C.; Gardes, D.; Gauvin, H.; Orliange, I.; de Reilhac, L.; Hubert, F. Stopping Power of Gases for Heavy Ions: Gas-Solid Effect: II. 2–6 MeV/u Cu, Kr and Ag Projectile. *Nucl. Instruments Methods Phys. Res. Sect. B* **1989,** *44-1*, 19–34.
8. Brandolini, F.; Ribas, R. Doppler Shift Attenuation Method Analysis with the Narrow Gate on Transitions Below Procedure. *Nucl. Instruments Methods Phys. Res. Sect. A* **1998,** *417*, 150–157.
9. Andreo, P.; Brahme, A. Stopping Power Data for High-Energy Photon Beams. *Phys. Med. Biol.* **1986,** *31* (8), 839–858.
10. Sharma, S. K.; Kumar, S.; Yadav, J. S.; Sharma, A. P. Stopping Power of Heavy Ions in Solids: A Comparative Study. *Appl. Radiat. Isotopes* **1995,** *46* (1), 39–52.
11. Sharma, A.; Kumar, S.; Sharma, S.; Diwan, P.; Nath, N.; Mittal, V.; Avasthi, D. K. Stopping Power of Mylar for Heavy Ions up to Copper. *Nucl. Instruments Methods Phys. Res. Sect. B* **2000,** *170* (3–4), 323–328.
12. Diwan, P.; Kumar, S.; Sharma, V.; Sharma, S. K.; Mittal, V. K.; Sannakki, B.; Mathad, R. D.; Kumar, K. U.; Khan, S. A.; Avasthi, D. K. Slowing Down of MeV Heavy Ions with Z=6–29 in PEN ($C_7H_5O_2$). *Nucl. Instruments Methods Phys. Res. Sect. B* **2003,** *201* (2), 389–395.
13. Fertman, A. D.; Mutin, T. Y.; Basko, M. M.; Golubev, A. A.; Kulevoy, T. V.; Kuybeda, R. P.; Sharkov, B. Y. Stopping Power Measurements for 100-keV/u Cu Ions in Hydrogen and Nitrogen. *Nucl. Instruments Methods Phys. Res. Sect. B* **2006,** *247* (2), 199–204.
14. Diwan, P. K.; Sharma, V.; Aggarwal, S.; Kumar, S.; Sharma, S. K.; Mittal, V. K.; Sannaki, B.; Mathad, R. D.; Khan, S. A.; Avasthi, D. K. Energy Loss Straggling of Li, C and O Ions in Mylar and Polycarbonate Absorber Foils. *Nucl. Instruments Methods Phys. Res. Sect. B* **2006,** *244* (1), 289–293.
15. Devendrappa, M.; Nagaraja, S.; Mathad, R. D.; Sannakki, B. Energy Loss, Range and Straggling of Light Ions in Polystyrene Film. *Mater. Today: Proc.* **2016,** *3* (10), 3966–3969.
16. Linares, R.; Ribas, R. V.; Oliveira, J. R. B.; Medina, N. H.; Santos, H. C.; Seabra, C. C.; Sigaud, L.; Cybulska, E. W.; Seale, W. A.; Allegro, P. R. P.; Touffen, D. L.; Silveira, M. A. G. Electronic Stopping Power of Ti, V and Cr Ions in Ge and Au at 150–500 keV/u Energies. *Nucl. Instruments Methods Phys. Res. Sect. B* **2017,** *413*, 1–5.
17. Yoneda, T.; Horikawa, J.; Saijo, S.; Arakawa, M.; Yamamoto, Y.; Yamamoto, Y. Stopping Power for 4.8–6.8 MeV C Ions Along [1 1 0] and [1 1 1] Directions in Si. *Nucl. Instruments Methods Phys. Res. Sect. B* **2018,** *425*, 1–3.

18. Mammeri, S.; Dib, A.; Ammi, H. Stopping Power and Energy Loss Straggling Data of Bismuth Thin Film for (0.9–3.0) MeV $^{4}He^{+}$ Swift Ions. *Nucl. Instruments Methods Phys. Res. Sect. B* **2019,** *459*, 15–21.
19. Iqbal, A.; Ullah, N.; Ur Rahman, A. Density-Dependent Energy Loss of Protons in Pb and Be Targets and Percent Mass-Stopping Power from Bethe-Bloch Formula and Bichsel-Sternheimer Data Within 1–12 MeV Energy Range. *J. Med. Imag. Radiat. Sci.* **2018,** *50* (1), 149–156.

Index

X

Z